A GRAND COMPLICATION

A GRAND COMPLICATION

~

*The Race to Build the World's
Most Legendary Watch*

STACY PERMAN

ATRIA BOOKS

New York   London   Toronto   Sydney   New Delhi

 BOOKS

A Division of Simon & Schuster, Inc.
1230 Avenue of the Americas
New York, NY 10020

First Atria Books hardcover edition February 2013

ATRIA BOOKS and colophon are trademarks of Simon & Schuster, Inc.

For information about special discounts for bulk purchases,
please contact Simon & Schuster Special Sales at 1-866-506-1949
or business@simonandschuster.com.

The Simon & Schuster Speakers Bureau can bring authors to your live event.
For more information or to book an event contact the Simon & Schuster Speakers Bureau
at 1-866-248-3049 or visit our website at www.simonspeakers.com.

Manufactured in the United States of America

10   9   8   7   6   5   4   3   2   1

Library of Congress Cataloging-in-Publication Data

Perman, Stacy.
A grand complication : the race to build the world's most legendary watch / Stacy Perman.
p. cm
Includes bibliographical references and index.
1. Pocket watches—Design and construction—History—20th century. 2. Clocks and watches—
History. 3. Packard, James Ward, 1863–1928. 4. Graves, Henry, 1868–1953. 5. Industrialists—
United States—Biography. 6. United States—Social life and customs—20th century. I. Title.
TS543.U6P47 2013   338.092—dc23   [B]   2012034555

ISBN 978-1-4391-9008-1
ISBN 978-1-4391-9010-4 (ebook)

For E. and M.

CONTENTS

CONTENTS

❧

CHAPTER ONE

Lot 7

Obsession is a demanding mistress. On December 2, 1999, Philippe Stern, president of the venerable Swiss watchmaker Patek Philippe, found himself once again in her grasp. Keeping his steely Genevese reserve, he glanced at his watch, an impeccably crafted Patek Philippe, purportedly the perpetual calendar made in 1943, the one given to him by his father, Henri, who had preceded him as president. At sixty-one, Stern remained the very picture of understated elegance. A gentle man with thinning white hair, he had quietly slipped away from his office at the company's headquarters in Plan-les-Ouates, a sleepy village on the outskirts of Geneva, well before the appointed hour for the auction of the legendary Supercomplication, the most exquisitely complex mechanical watch ever created. Given the time, it was probably best to take the telephone call from his family's villa in Anières on the shores of Lake Geneva. In any event he would be more comfortable there.

The time quickly approached 8:30 p.m., nearly 2:30 p.m. in New York City. There was still some time before the auction began. It had been thirty years since the Supercomplication had last surfaced. Having already eluded the family once, the moments until bidding opened were relatively short.

Outside, the wind scythed across the lake in the spectral dark of winter. Here on Lac Léman, Stern had won a record seven regattas as a young sailor. A fierce competitor, he had also been a member of the Swiss national ski team. Along with his wife, Gerdi, herself a European mushing champion, Stern had competed in international dog sled races. The roster of pastimes enjoyed by the head of one of the oldest Swiss watchmakers, as it had been noted, all revolved around beating the clock. Stern had once described his hobbies as good business training, saying that they had taught him "to go fast and take risks sometimes."

With the anticipation of victory, Stern put the telephone to his ear. Four thousand miles away, his aide-de-camp Alan Banbery was waiting for him.

•   •   •

Everyone who was anyone in the insular world of watch collecting filed into Sotheby's ten-story glass-and-steel tower at 1334 York Avenue on Manhattan's Upper East Side on that unseasonably warm afternoon. Just three weeks before Christmas, thermometers registered 60 degrees. At the Botanical Gardens, fragrant white blooms had broken out five months prematurely, and a ring of yellow forsythia blossoms graced the city with a touch of spring. As the mercury soared, so did the stock market. The Dow pushed past 11,000, marking the end of the millennium with record gains. The entire city appeared to levitate on its great good fortune. Watching the buoyant crowd descend upon Sotheby's, a betting man could wager with ease that, before the day was over, more than one record would be broken.

As in previous years, the major auction houses dangled before an excitable audience the spectacle of record-smashing greatness for paintings, jewels, and *objets d'art,* all with impeccable provenance or rarity. Each season brought an increasing fervor. A daisy chain of superlatives wrapped around each sale. The items featured were not just *rare* but *extremely rare.* They were not merely *important* but the *most important.* More than offering desirables of a certain vintage, the auction houses were peddling history bound in shiny catalogues and generated in regularly scheduled cycles. And the press and public responded like hounds to blood.

For weeks, expectations followed the Dow over Sotheby's "Masterpieces of the Time Museum." Eighty-one watches and clocks were set to go on the block. Culled from the 3,500-piece collection of Seth G. Atwood, a wealthy entrepreneur and something of an eccentric, these were some of the most extraordinary and important timepieces created since man began calculating the minutes and hours and fashioning devices by which to measure them. A quick rifle through the handsomely appointed catalogue yielded numerous affirmations that the auction's billing offered more than hyperbole. "Sometimes sales catalogues are mutton dressed as lamb," one Sotheby's veteran dryly noted at the sale. "This was not the case."

By any measure the sale was unprecedented. Fascinated with the art and the science of time, Atwood had become one of the world's most significant collectors and horological scholars. He hailed from an old-line Illinois family that made its fortune in manufacturing, banking, and real estate, which afforded him the leisure to pursue his interests wherever they took him. He had spent the past three decades circling the globe, cherry-picking the most important clocks and watches in history, mounting a private collection that came to rival that of the British Museum—in importance if not scope. To house his collection, Atwood had constructed the Time Museum, located in his hometown, far from all known capitals (horological or otherwise), in the basement of his Clock Tower Inn in Rockford, Illinois, a green little pocket of nowhere straddling the Rock River, off Interstate 90.

While universally acknowledged as an endearingly kind and charming fellow, Atwood had rightfully earned a reputation as a bull in the world's china shop of clocks. Rivals collapsed when he was in the game. And once a timepiece was in Atwood's possession the game was over. Knowledgeable horophiles accepted that it was as likely that the sun would change course and rise in the West as that Atwood's collection could ever be broken up and sold in lots. Yet at age eighty-two, Atwood had resolved to do just that. He was shutting down his museum (a process that would be filled with a number of unforeseen twists and turns and take several years) and, somewhat unsentimentally, deaccessioning its priceless contents.

Few could recall an auction of this caliber and range. Atwood's assemblage represented an astonishing array of ancient astrolabes, astronomical chiming clocks, musical automata, and other devices of wonder from every period, important watchmaker, and country from the beginning of timekeeping. "It was mind-blowing," described one British dealer who arrived with instructions to bid on a number of the lots on behalf of his deep-pocketed clients. "Just looking at it set the heart aflutter."

Lot 22 contained the famous Ormolu-mounted red boulle Sympathique no. 128, made in 1835 by the House of Breguet for the Duc d'Orléans. Watchmaker to Louis XVI, Napoleon, Marie Antoinette, and the sultans of the Ottoman Empire, the Swiss-born Parisian Abraham-Louis Breguet remains the greatest watchmaker in history. It was Breguet who invented the tourbillon, circa 1795, an ingeniously delicate and tiny gyroscope-like device that turns about every minute to compensate for gravity's pull on a movement, dramatically improving accuracy and revolutionizing timekeeping. The wittily inventive Sympathique was designed with a removable pocket watch that rested in a cradle atop the main clock. A series of pins locked the pocket watch into position, while overnight a mechanism inside the clock movement rewound, readjusted, and reset its companion with jumping hours, a quarter repeater that chimed precisely on the quarter hour, and a power reserve indicator. The clock was one of only eleven ever produced.

Even in the seventeenth century, when the manufacture of any timepiece was far from routine, the final lot, the Tompion no. 381, was quite something. A gilt-brass, mounted red tortoiseshell table clock featuring both *grande sonnerie* and quarter repeating chimes, it was one of only three such pieces manufactured by the brilliant seventeenth-century British clockmaker Thomas Tompion. The Englishman first earned a name in 1676 when King Charles II commissioned him to create two faultlessly accurate clocks for the Royal Observatory in Greenwich, which notably became the center of world time in 1884. By the time of Tompion's death in 1713, the eminent clockmaker had become a national treasure and was buried with royalty in the nave of Westminster Abbey.

In an auction with many rare and ingenious instruments, the sale's crown jewel belonged to Lot 7. Known as the Graves Supercomplication, the double-dial pocket watch was a miniature masterpiece of mechanical engineering. First commissioned in 1925 by the enigmatic New York financier Henry Graves, Jr., from Patek Philippe & Co., the pocket watch, measuring three and a half inches wide and scarcely one inch deep, was the storied Geneva watchmaker's magnum opus, requiring three years to contemplate its design and nearly another five to produce.

Elaborately constructed in eighteen-carat gold and jeweled pivots, the Supercomplication boasted the skill and expense expended by Patek Philippe. The watch's enamel dials were protected by crystal with the hardness of sapphire, penetrable only by a diamond. Crafted at a time when watchmakers hand-tooled their own parts, the Supercomplication consisted of over 900 individual pieces—most no bigger than a grain of rice or thicker than a strand of hair. An orchestra of 430 screws, 110 wheels, more than 120 various movable parts, and 70 jewels produced a symphony of 24 complications. This dazzling array included a different chronological function for each hour of the day, a perpetual calendar (requiring resetting only in the year 2100), and a split-second chronograph. Smaller sub-dials indicated the exact time of sunrise and sunset calibrated for New York City, while another measured the phases of the moon. A minute repeater struck the hour, half-hour, and quarter-hour, playing the same tintinnabulary melody as heard ringing out from Big Ben, the great bell clock rising over London's Palace of Westminster. These were but a fraction of the watch's virtues. A midnight-blue celestial chart also mapped the nighttime sky over Manhattan, complete with the exact magnitudes of the stars and the Milky Way gliding across the sky in tandem with the actual heavens.

For this horological Willy Wonka factory in miniature, Graves paid 60,000 Swiss francs, the equivalent of $15,000 at the time, a staggering amount. Measured in 1999 dollars, the price equaled some $265,000, but in effect such an undertaking would cost closer to $2 million.

The Supercomplication was more than just a golden toy for the Gilded

Age. In the history of timekeeping, less than a handful of *grandes compli-cations*—including Breguet's fabled Marie-Antoinette, the crystal pocket watch made for the ill-fated French queen—have reached the highest levels of *haute horlogerie,* earning the title of a "supercomplication." Possessing twenty-four complications, the Graves eclipsed them all. Even with the march of years and the advent of modern technology, the watch is a technological achievement of the highest order. Entirely calculated and crafted by hand, the Supercomplication was a powerful computer of time before there were computers.

Produced on a dare, Patek Philippe's tour de force inspired marvel and desire. It was as if Merlin captured Galileo, Shakespeare, Newton, and Beethoven, encased their essence in gold, and then slipped the mechanism inside a man's waistcoat, where their collective genius might tick through eternity.

In the annals of timekeeping the Supercomplication loomed large in any important collector's understanding of the field. Its technical triumph, however, was only part of its luster. The pocket watch had been the victor in a strange contest that took place long ago between Graves, a fiercely private society tycoon, and his rival, James Ward Packard, the automobile magnate, during the early years of the twentieth century, when enterprising Americans amassed huge fortunes in banking, auto making, coal, oil, steel, and the railways. Their newly minted prosperity soon eclipsed the lucre accumulated over a millennium by European royalty. Keen to demonstrate the cultured face of capitalism's excess, America's moneyed moguls scooped up European patrimonies, paying enormous sums for Old Master paintings and sculptures, antique tapestries, ancient manuscripts, and ingenious mechanical pocket watches.

Throughout history, these beautiful instruments represented the most innovative, extravagant gadgets that money could buy, available to an elite few. King Farouk I of Egypt, intrigued by their beauty and exotic novelty, possessed a collection of erotic automata pieces. On one of his most famous, the king calculated the time by counting the number of sexual thrusts executed by the figures inside the watch's movement, visible with the press of a button on the lid. In the early seventeenth century, Swiss, French, and English watchmakers flocked to Constantinople, vying for the patronage of the sultans who acquired thousands of highly ornate timepieces as astrological amusements.

Once the prerogative of the aristocracy, watches had become the ultimate status symbol of wealth, and America's financial princes found themselves not merely bewitched by their beauty but transfixed by their illustrious pedigrees. The country's most powerful banker, the cigar-chomping John Pierpont Morgan, spent enormous sums to purchase 240 European timepieces dating to the sixteenth and seventeenth centuries. In love with his own legacy, Morgan eventually bequeathed his collection to the Metropolitan Museum of Art.

Henry Graves, Jr., a private man of fine tastes and even grander possessions, and James Ward Packard, the brilliant engineer, both possessed the passion and limitless purse to acquire the finest instruments ever produced. Timepieces from antiquity scarcely roused their attention, but over the course of three decades the pair engaged in a surreptitious collecting duel to own the most complicated watch of their era. Their patronage spurred Swiss watchmakers, chief among them Patek Philippe, to craft ever more innovative and ingenious timepieces, pushing the boundaries of mathematics, astronomy, craftsmanship, technology, and physics. Their contest produced a series of transcendently unique *grandes complications*. Almost a century later, very few of their timepieces had ever come to market. The rather gauzy tale of how "the Graves" ended up in the hands of Seth Atwood added another layer to the watch's already considerable mystique.

Among horophiles, the Supercomplication was considered the *Mona Lisa* of timepieces, and most believed they were likely to see Leonardo da Vinci's masterpiece appear on the market before the Graves ever went under the hammer. Should the legendary piece leave the Graves family and come onto the open market, it was widely accepted that it would land in Patek Philippe's private collection. Yet in a stunning reversal, Seth Atwood acquired the Supercomplication on March 26, 1969. Thirty years later, even the Stern family had to admit that the pocket watch's path from Henry Graves, Jr.'s Fifth Avenue vault to Sotheby's wood-paneled seventh-floor salesroom was as unexpected and random as the watch was obsessively crafted and meticulously built.

Surveying the room, Daryn Schnipper, Sotheby's director of worldwide watches and clocks, found it difficult to contain her composure. Securing the Time Museum sale and with it the Graves had been a coup for the auction house. For nearly twenty years, the petite, bright-eyed brunette had traveled between Europe and New York and, more recently, Asia, building up Sotheby's slice of the growing global watch market. When she first joined the auction house in 1980, timepieces remained the exclusive domain of a small group of collectors whose obsessive tastes ran to historical clocks and pocket watches. Mechanical wristwatches were regarded mostly as an afterthought. In recent years, however, a vortex of money and rumors had catapulted vintage wristwatches into serious collectibles, fetching prices in the six figures, nudging them above their previous station in the art world as expensive trinkets for eccentrics and ephemera for curiosity seekers. The mounting interest drew in new buyers while helping to pull all horology from the periphery to the center, where the big money resided.

The commission to sell the Supercomplication was an auctioneer's dream. Schnipper had never sold anything like it. She was about to witness the biggest sale of her life, one that had the potential to drive Sotheby's ahead of its great rival, Christie's, in dominating the watch market. This was a once-in-a-lifetime sale that could change the field. In her mind, there were two possible outcomes: spectacular triumph or humiliating failure.

Schnipper first laid eyes on the Supercomplication during a visit to the Time Museum in 1986. She had traveled to Illinois to assist with Atwood's single catalogue dispersal of 370 pocket watches for Sotheby's. At the time, she had little notion that she was witnessing the beginning of the end of Atwood's entire museum. There, Atwood produced a fitted tulipwood box and handed it to Schnipper so that she could inspect its contents. Inlaid with ebony and centered by a mother-of-pearl panel engraved with the Graves family coat of arms, it held a gold pocket watch. In awe, Schnipper removed it and turned it over to see the inscription: *Made for Henry Graves, Jr., New York 1932 by Patek Philippe & Co., Geneva, Switzerland.* "It was," she later recalled, "like holding a beating heart in the palm of my hand."

Sotheby's had no clear picture of who exactly might pursue the pocket watch. In most instances, an auction house can predict with a fair degree of confidence which collectors and dealers might push a sale, particularly at this level, for which the pool of potential buyers is notably small and included a handful of Americans, a few well-placed Germans, petro-sheiks, and the Stern family. The Russian oligarchs and the Chinese had yet to enter the art world with their centrifugal force. Yet the Supercomplication introduced a high degree of uncertainty, for, although only a handful of people could afford to spend the kind of money that Sotheby's was hoping the pocket watch would bring at auction, to date none of them had ever done so.

During the weeks before the curtain rose, Sotheby's aggressively promoted the sale. The auction house re-created a replica of the Time Museum on its tenth floor, high above York Avenue, where interested parties viewed the sale's lots in an environment primed to stoke buyers' interest. The Supercomplication sat alone in a bulletproof case, separate from the other masterpieces. Parties interested in a closer inspection were accompanied by one of Sotheby's watch experts, given a pair of latex gloves, and taken to a small room, where they examined the piece under the intense gaze of security.

As part of the presale activities, Sotheby's had toured the prized Super-complication to select clients in a series of private viewings. John Reardon, a young vice president in Sotheby's Watch Department, who had given up a stint in the Peace Corps after becoming a watch enthusiast, was charged with showing the watch around the country. Reardon had developed an enormous appreciation for Henry Graves, Jr., as a collector and an even deeper fascination for the *chef d'oeuvre* of his collection. Formal security remained discreet. Reardon later recounted that he had created an impregnable bag out of a bulletproof and fireproof camera case; he stuffed it with foam and cut a hole in the middle in order to hold the delicate timepiece in place. "If the plane went down," he said, "the Super-complication would survive."

During a stop at O'Hare Airport in Chicago, Reardon nearly missed his flight when security agents at the x-ray machine thought the double-dial watch looked like a bomb. In a moment of quick thinking Reardon convinced the agents not to pry it open and instead offered a private screening. In a small room he gave them an impromptu lesson on the most complicated watch ever created. In Reardon's telling, the security agents sat rapt for nearly an hour. "It captured their imaginations," he recalled, describing his brief confinement at O'Hare as "one of those magical horological moments."

In New York, despite palpable anticipation, the exhibition did not draw the kind of presale crowds that Sotheby's had expected. Schnipper remembered that in 1986, when the house had handled Atwood's initial sale of fine pocket watches, they had to turn away private viewers. The dearth of spectators was troubling, given the overall hype surrounding the sale. Even more puzzling, it mirrored another odd phenomenon: few people were talking up the price of the Supercomplication. Usually speculation runs high, with any number of individuals offering their opinion on where a price might strike, and with luck—along with a discreet and subtle push from the auction house—the excitement turns to fevered bidding. There was a disquieting silence, however, about the Supercomplication.

In truth, the Supercomplication posed a conundrum for Sotheby's. In the run-up to the Time Museum sale, the art market's unrestrained hubris had led to excessive estimates and even wilder sales. The Australian billionaire entrepreneur and yachtsman Alan Bond shattered all previous records in 1987 when he snapped up Van Gogh's *Irises* for a cool $54.9 million.* While astonishing, this record was destroyed three years later when the Japanese multimillionaire Ryoei Saito dropped $82.5 million on Van Gogh's *Portrait du Dr. Gachet* during a Sotheby's auction in New York. Within days, the head of Daishowa Paper Manufacturing bid $78.1 million for Renoir's *Au Moulin de la Galette*. After spending a few hours with his masterpieces, the highflying Saito put them in foam-padded crates and shipped them to a climate-controlled vault in a secret Tokyo warehouse.

These were crazy days in a manic decade. In 1997 an anonymous bidder, widely rumored to be the Austrian-born investment fund manager Wolfgang Flötti, bought Picasso's 1932 *Le Rêve* for $48.4 million at Christie's in New York. The following year, Microsoft's founder Bill Gates purchased Winslow Homer's *Lost on the Grand Banks* for $30 million.

Wine, precious objects, and jewels were also creating monetary spectacles. In 1985 the Forbes publishing dynasty paid $156,450 for a 1787 Château Lafite reportedly bottled for Thomas Jefferson but which was later claimed to be a fraud. Just four years before the Supercomplication was to go on the block, Sotheby's smashed all former jewelry auctions with the sale of the Star of the Season, a 100.1-carat pear-shaped colorless diamond for $16.5 million, sold to an anonymous buyer.

The auction market for vintage watches had also begun to heat up in recent years, but no timepiece had entered the stratosphere of a Monet or a bottle of Bordeaux. Only three years earlier, in 1996, a single watch had passed the million-dollar mark at auction. The piece was a one-of-a-kind Patek Philippe Calatrava astronomic minute-repeating gentleman's wristwatch with perpetual calendar and moon-phase indication, with a platinum bracelet and case. Manufactured in 1939, the watch fetched $1,715,000 in a New York auction held by the Geneva-based house Antiquorum.

Placing a price on the Graves pocket watch posed something of a dilemma for Schnipper and her team at Sotheby's. Prior to any sale, auction houses put estimates, one low and one high, on the objects to be sold, based on comparable sales and the current market. But nothing of similar caliber, rarity, or provenance had ever come on the market. In more ways than one, the Supercomplication was without rival. "Everyone was nervous as fish," recalled William Andrewes, the Time Museum's former curator.

---

*Soon after the sale, Bond was unable to come up with the funds to pay for the painting and it had to be resold. The Getty Museum in Los Angeles purchased the Van Gogh for an undisclosed price in 1990.

Schnipper knew that a brilliant result was in no way a foregone conclusion. She had hoped the watch might go for at least $2 or $3 million. In Geneva, one month before the sale, a prominent client casually suggested to her that the Supercomplication could easily fetch $8 million. The figure rocked her composure, but she refused to allow herself such flamboyant imaginings. Anxiously, she asked herself, "Who's going to buy it? The world is going crazy." In the end, Sotheby's put its low estimate on the watch at $3 million, with $5 million at the high end. Still, doubt nagged at Schnipper. Did the price bracket reflect an unreasonable aspiration? Not infrequently, auctions placed a too-high estimate on works that did not slot easily. It was a gamble. There was the danger that a wild estimate might kill the sale altogether.

Osvaldo Patrizzi, the flamboyant six-foot-four head of the auction house Antiquorum, had little doubt the Supercomplication would not only sell but would sell big. Suave and charming, with an encyclopedic knowledge of watches and the instincts of a born salesman, Patrizzi joined the ranks at thirteen, restoring clocks and watches in Milan. By the time he was twenty-nine, he had founded his first auction house, focusing on clocks, watches, and *objets d'art*. In a relatively short time, he carved out his own role as one of the watch world's rainmakers. Under Patrizzi, vintage mechanical wristwatches began their ascent as collectibles at auction as early as 1980; he helped push open the Asian market, holding one of the first watch auctions in Hong Kong, in 1979. He was famous for unfurling thematic watch sales accompanied by glossy catalogues and glittering cocktail parties, and his influence reached far and wide. "The watch, for me, is not just one timepiece," he was quoted as saying. "It is the accumulation of experience. It is about savoir faire."

Quietly Patrizzi cast about to get a sense of interest in the Supercomplication. Another specialist had told him the watch would go for $4 million. Patrizzi knew better. If a pair of motivated collectors started bidding against each other, anything was possible. Under the right circumstances the sky was the limit. Patrizzi joined up with three investors who collectively raised $5 million for their bid. Still, he thought the odds in his favor were slim to none. "We knew we had no chance," he lamented. "But then again, in an auction one never knows."

Alan Banbery strode into Sotheby's supremely confident. In the space of one week the curator of Patek Philippe's private collection had flown from Geneva to New York, where he was given a private viewing of the Supercomplication, and then back again. In Switzerland days earlier, he had discussed the pocket watch with Philippe Stern, and the two men agreed that he should acquire the piece that had long been in the family's sights.

Banbery first joined Patek Philippe in 1965 as its director of sales in English-speaking markets. A tall gentleman with an aristocratic bearing, he would have been comfortable mingling with the characters in an E. M. Forster novel. Born in London, Banbery earned a degree in horology at the famed Watchmaking School of Geneva and worked on the bench at Universal Geneva, where he specialized in the complicated Compax and Tri-Compax movements. When President Gamal Abdel Nasser of Egypt nationalized the Suez Canal in 1956, Banbery was called to Her Majesty's service. Following the military campaign he returned to England and opened a shop in the town of Windsor, specializing in watches and jewelry. Eventually he took a position with the British house Garrard, a prestigious appointment. Established in 1735, Garrard was one of the world's oldest jeweler's. In 1843 Queen Victoria appointed the shop with the warrant of Crown Jeweler. While working for Garrard Banbery garnered the attention of another esteemed house, Patek Philippe. It was no less than Henri Stern who quietly approached him and offered him the position that took him back to Geneva. In 1970 Banbery was named curator of the company's private collection.

For more than thirty years, almost from the moment he began his association with Patek Philippe, Banbery assisted the Sterns in amassing a colossally important collection of timepieces that had begun a generation earlier with Henri's father, Charles. Initially Banbery traipsed the planet buying back both at auction and through private deals key pieces that Patek Philippe had produced going back centuries. On behalf of the company, Banbery purchased the one-of-a-kind Calatrava for more than $1.7 million at the Antiquorum sale in 1996, breaching the seven-figure mark for a wristwatch at auction. The *maison*'s ambitions soon grew beyond Patek Philippe pieces to include miniature portraits on enamel, automata, watches of historical significance dating back to the sixteenth century, examples from the Geneva tradition, and a broader selection of the world's horological masterpieces.

After Henri retired in 1990, his son Philippe came to see himself as the steward not only of Patek Philippe but of the Swiss tradition, and as something of a caretaker for the world's timepieces. In 1989, as the firm celebrated its 150th anniversary, the idea to memorialize their private collection in a museum quickly gained favor within the family. Work was already under way to transform one of Patek Philippe's workshops, Les Ateliers Réunis in the Plainpalais district, into what the Sterns envisioned as "an epic poem to time." Undoubtedly the Supercomplication would prove to be the *pièce de résistance* for the new Patek Philippe Museum, scheduled to open in the fall of 2001, two years after the Sotheby's sale. It was a matter of pride as well as patrimony.

After all, the firm had only recently and rather spectacularly reclaimed the masterpiece that its watchmakers had crafted seventy-one years earlier for Graves's great rival, James Ward Packard. Without the watch known as "the Packard," there would likely be no "Graves."

In New York expressly for the auction, Banbery was ushered through a side door and up to one of the auction house's skyboxes, suspended above the sales gallery, where he could bid by telephone in absolute anonymity. Sotheby's offers these private salons to their most important clients for whom discretion is paramount.

Pressing back into his seat, Banbery held the satellite phone that connected him to Philippe Stern in Switzerland in one hand and in the other a phone that connected him to the telephone banks on the sales floor below. He was not the only eager presence weighing his chances behind closed curtains. A sober-minded German occupied another of the skyboxes where he too could comfortably and anonymously participate in the proceedings. A collector in his own right, he had registered for two paddles and was said to be making bids on behalf of a pair of prominent museums. Infectiously exuberant, Seth Atwood held forth in yet another skybox, surrounded by his family. Considering that his life's passion was about to go on the block, Atwood remained quite unsentimental, treating the afternoon like an excursion to the theater where he happened to run into old friends. Below them all, the salesroom was packed. Not one of the plastic folding chairs lining the floor stood empty. The earlier silence that embraced the sale had been misleading.

A quick survey of the room revealed a caste system of taste and money among the collectors, dealers, pawnbrokers, hobbyists, and spectators. Catalogue grazers sized each other up, sniffing out who was purely a moneyed, prestige collector and who was the true connoisseur—a distinction that had grown more stark in recent years. Usually an auction such as this was remarkable for who you didn't see because important collectors preferred to hide behind their dealers, who were themselves cloaked behind the anonymity of the telephone banks. As one of Sotheby's European watch experts groaned, "The big shots sit on the phone." The Time Museum sale, however, had drawn some heavy hitters in the flesh. Collectors who normally skipped New York, generally preferring the Geneva and Hong Kong sales, had made an exception, along with esteemed dealers and museum representatives from around the globe. A number of individuals long thought retired from collecting also surfaced at Sotheby's that afternoon.

Schnipper and others noted that a large number of faces were new to the auction house. The uptick in global wealth over the past two decades had introduced a new generation to mechanical watches. These were men who, like Graves's contemporaries, had accumulated colossal wealth in a short amount of time and represented a new breed of horophile, an aggressive rookie class that thought nothing of spending thousands of dollars on a suit and viewed such timepieces as the ultimate symbol of affluence. Historically, possession of such objects had gone from royalty and the aristocracy to wealthy industrialists to serious collectors and now to new-issue moguls. They all believed it was possible to know the full flavor of an individual by the timepiece he strapped on his wrist or pulled out of his pocket. This was a new pack of dogs in the hunt.

The room settled into an electric silence. An Egyptian limestone stelophorous astrolabe dated between 1400 and 1300 BC was the first to go. Hammered down for an impressive $31,625, the funerary stone that ancient Thebans used to determine the sun's position fetched more than twice Sotheby's high-end estimate. Within a few minutes, the statuette was a memory. Lot 6 also doubled the house's high-end estimate, quickly moving the auction into seven-figure territory. An eighteenth-century English brass-and-silvered mechanical clock that measured the celestial equator, once owned by the Earl of Ilay, realized $1,047,500.

Ian Irving, the auctioneer overseeing the sale, stood at the podium, his sharp blue eyes framed in thick black spectacles scanning the room. Nattily dressed in a dark suit, purple shirt, and neon pink-orange tie, Irving had gotten his start in Sotheby's Silver Department. The Englishman was one of the few individuals who had spent time out in front of the public with the gavel and more recently behind the scenes in private sales. As the head of Sotheby Ventures, Irving had been intimately involved in the Time Museum consignment; now he was presiding over its sale. Eleven years earlier the dapper auctioneer had overseen the estate sale of Andy Warhol, a record-breaking ten-day affair that unloaded ten thousand of the pop artist's items, everything from cookie jars to plastic chairs, grossing $25.3 million. Although he was no stranger to headline-making auctions, Irving admitted, "This collection was extraordinary."

The weight of anticipation over the Supercomplication was particularly heavy on Irving, even though he was quite practiced in the auction dance. "I am always nervous until the hammer goes down," he later confessed. "Everyone is looking at you. You can't make any mistakes. You don't want to leave any money on the floor." The Supercomplication added a new layer of stage fright. "Here we were projecting a record price," he explained. "We were telling the market to sit up and take notice."

As bidding on the Supercomplication was about to commence, Irving looked up to the left and then the right, acknowledging "the power players in the little discreet rooms" above. A large screen to his left illuminated a massive image of the pocket watch. Below him were a long, elevated wood-paneled desk and the bank of telephones where anonymous bidders placed their tenders. Situated just behind him was another screen, with a money chart simultaneously displaying bids in ten different currencies.

Irving opened the sale at $2.8 million, and the bidding started at $3 million. After a tentative start, the tenders came in like quick bursts of gunfire. Irving moved the hammer along in $250,000 increments. The blue paddles dipped up and down at $3.25 million, $3.5 million, creating a wave that nearly repositioned the molecules in the room. Most of Sotheby's staff crowded into the back of the hall, far from the front-row seats reserved for the collecting hierarchy and well beyond the reach of the curtained skyboxes. At each new benchmark gasps were heard. The entire room strained to see from where the bid came. Excitement shot across the floor as the bids ricocheted between the chairs and the telephone banks. Schnipper had positioned herself near the gavel, and from her perch she could observe the entire floor. As the price edged higher, her chest tightened and her throat went dry during what she would later call the longest few minutes of her life.

The Graves Supercomplication goes under the hammer at Sotheby's New York on December 2, 1999.
Photograph courtesy of Sotheby's, Inc. © 2012.

Caught up in the moment, Irving tried to jump the bidding to $500,000 increments, but the room gently pushed him back. He nervously laughed out loud, briefly breaking the tension. Most on the floor had no idea of the activity above them. The German who had purchased one of the earlier lots, a sixteenth-century gilt-brass compendium for $299,500, found the Supercomplication too rich for his patrons' blood and dropped out.

Within a few tense minutes, the price had tipped well beyond Schnipper's most hopeful musings. When Irving called $4 million, the bidding ranks thinned considerably. At this point, the small group of American collectors began to waver. By the time the price reached $5 million, just five bidders remained in the fight. All the Americans had withdrawn their paddles. Schnipper went numb. As Irving called $6 million, four of the remaining bidders dropped out. Then Alan Banbery stepped in to make his move.

Above the gallery floor, Philippe Stern's lieutenant patiently sat in his skybox. Banbery had nursed a bit of a grudge against Seth Atwood. In the clubby, cloistered universe of watch collecting, the two men's interests often overlapped and collided, making them either friends or rivals, depending on who sat where during the chase. Over time, their personal contact remained cordial but limited. Several years earlier Banbery had dined with Atwood at Le Richemond, one of Geneva's nineteenth-century palace hotels that sits like a perfumed dowager overlooking the Brunswick Gardens on Lake Geneva. As their evening drew to a close, Banbery turned to the American collector like a courteous suitor leaving his calling card. "If you ever consider selling the Graves watch," he said discreetly, "please call upon me as your first option." The call, of course, never came. As he held a phone in each hand, the memory of that evening burned through Banbery like a lit cigarette in his breast pocket.

In close contact with Stern throughout the auction, Banbery waited until the price hit $6 million. The number of challengers able to bid that high would be mercifully few. Indeed when Banbery stepped in, the contest was reduced to two telephone bidders—a boxing match on the head of a pin. All eyes turned to the phone banks in the front. From this point on, the prices moved along at a caffeinated velocity, from $6 million to $7 million to $8 million. "I have $8.25 million on the telephone," Irving trilled. "I have $8.5 million on the telephone." When the price of the pocket watch reached $9.5 million, Irving stuttered, stumbling over the mouthful of zeros.

Over the satellite phone Stern instructed Banbery to bid $9.75 million. Instantly a $10 million counteroffer came in. Banbery paused. It was the first time since the bidding began that there was a real break in the pro-

ceedings. Irving looked directly at Banbery's telephone proxy. "Are you
sure?" he gently quipped. "You've come so far." Silence.

The confidence that had carried Banbery up into the skybox scarcely
minutes earlier had dissolved. Over the years he had grown deft at recognizing his opponents' thresholds. It had become obvious to him that his rival
for the Supercomplication was more than determined. Rather than bringing
him closer to victory, each bid pulled him further into the gloaming. Regret-
fully, Banbery realized his dogged rival simply had no limit. Rather plain-
tively, he advised Stern of his conclusion: "I reckon we should step down."

Irving cast a look at the telephone bank and puckishly inquired, "Any
more?," and the room broke into laughter. "I have ten million." Banbery
pressed the satellite phone hard to his ear. On the other end Stern uttered
just two words: "We're out."

Irving brought down the final gavel on the Graves Supercomplication.
The buyer's premium brought the total to $11,002,500. Only then did
Schnipper realize that she had stopped breathing. The room erupted in
flashbulbs. Applause thundered across the room. The price achieved was
more than astounding; it was historic. The Graves watch held title as the
most complicated timepiece ever created. Adding to its already considerable pedigree, it was now the most expensive watch ever sold.

The Supercomplication remains
the most expensive timepiece
ever sold at auction.
Photograph courtesy of Patek Philippe.

• • •

For the second time in thirty years the Supercomplication slipped away from Patek Philippe. "I was absolutely floored" was how Banbery recalled the end. "It was a very, very big disappointment for me personally and for Mr. Stern." As the underbidder, the unflappable Brit would come to rue the event as one of his greatest failures. "The watch escaped me. It was an awfully bad evening."

Save for his bright shirt and tie, all of the color had drained from Irving. Following the sale, an elderly gentleman crossed the room and approached the auctioneer. "Congratulations," he beamed, extending his hand. "You've become a member of the ten-million-dollar club. It is a club that in thirty years as an auctioneer, I have never had the fortune of joining." Irving later admitted, "I wanted a drink after that."

Amid the frenzy over the Graves, the fact that the entire sale realized $28,285,050, more than double Sotheby's $13.7 million high estimate, was almost entirely glossed over. None of the eighty-one works offered failed to find a buyer, but the Supercomplication had carried the day, capturing everyone's imagination.

Philippe Stern contemplated his defeat in Anières as the victor savored his win from a safe and unknown distance, insisting his identity remain private. In consolation, the buyer agreed to loan the Graves Supercomplication to the Patek Philippe Museum in Geneva for a period of time. The generosity was bittersweet, as this would in all likelihood be the last time the watch would ever be seen by the public.

Just as only a small circle knows the codes for launching a nuclear strike, only a few at the auction house were allowed to know the name of the buyer. Outside, speculation was rife. In less than five minutes, a new player had changed the game forever. A day before the sale this mystery man, who had scarcely entered his thirties, showed up to view the Supercomplication. A ravenous art collector of varied tastes, he was known to leave entire inventories empty in his wake, paying enormous sums for art and antiquities in a number of fields—except for mechanical watches, until now. Generally understood to work through European dealers, he was, quite simply, a one-man market. Arriving at Sotheby's prior to the sale, this intriguing collector made his way up to the tenth floor and dead straight for the Graves, his security entourage trailing behind him. Darting quickly inside, he picked up the watch, tilted it to the side, put it down, and remarked, "I'll buy it," before flitting away just minutes after he first appeared.

Once the initial excitement settled over the record-shattering bid and the man who made it, other questions among the growing ranks of new collectors and curiosity seekers emerged. Chief among them: Who was Henry Graves, Jr., and how did he come to possess the most complicated watch ever made?

More than a century earlier, long before the Graves Supercomplication made global headlines, there was another man and another watch. Far from the roaring excesses of New York City in the 1920s and farther still from Patek Philippe's ateliers in Switzerland, the story of the Supercomplication began with the American automobile magnate and inventor James Ward Packard at the close of the nineteenth century in the quiet town of Warren, Ohio.

With the players long gone and the evidence—an extraordinary series of mechanical watches—now scattered across the globe in places both known and unknown, the obscure story had been mostly forgotten. Largely ignored by the public, various fragmented tales had been passed down through the years. One horological chronicler described the two men as "adversaries in pursuit of their greatest passion: collecting Patek Philippe masterpieces," another as the "driving force behind the complications that still hold horologists in thrall." Cheryl Graves, one of Henry Graves, Jr.'s great-granddaughters, remembered sifting through watches with her father, Henry Dickson Graves, before he died in 1996. She recalled that he had said his grandfather was "in some kind of a contest with this man Packard to see who could have the better watch. They wanted to see who could come up with the more interesting timepiece. They were out to out-do each other."

In fact their story was slightly more labyrinthine. The two men never met, although their lives intersected. Obsessed with watches, they crossed swords in the dark, each attempting to possess the most ingeniously complicated watch ever produced.

CHAPTER TWO

❧

The First Tick

On the evening of March 30, 1922, the rain dropped a blanket of cold over the Mahoning Valley as James Ward Packard and his wife, Elizabeth, welcomed her father, Judge Thomas I. Gillmer, and a guest, Dr. John S. Kingsley, into their home, a handsome red-brick Colonial Revival on Park Avenue in Warren, Ohio, not far from the Victorian grandeur of the downtown courthouse. Outside, the streetlamps cast refracted halos of light onto the damp air. Inside, Ward, as he was known, dressed as he usually did, meticulously. Barely five-feet-four, he wore one of his custom-made wool suits, a waistcoat, and a bow tie. His slight physical stature notwithstanding, nobody stood taller in this town than the man whose decoratively embossed letterhead read "J. W. Packard."

Eight months shy of fifty-nine, Ward had a full head of hair carefully combed into a three-quarter side part and greased back with tonic. He wore round, wire-frame spectacles on the bridge of his nose, and on his pinky finger an ornately engraved, eighteen-carat gold ring. Measuring about one inch long and half an inch wide, the ring was actually a small rectangular case. It framed a hexagon-shaped crystal that revealed an even smaller white enamel dial with black Arabic numerals and a minuscule winding stem crowning the top. Five years earlier, Patek Philippe had made this tiny mechanical watch, with the movement no. 174659, expressly for Ward, undoubtedly one of the benefits of his long-standing patronage of the Swiss watchmaker.

A handsome ebony walking stick stood against the wall in the foyer, another example of Ward's enduring, unique relationship with Patek Philippe. In good weather, Ward went for brisk walks several times a week, usually following a path along the railroad tracks that fanned out northwest from Warren in the direction of Leavittsburg or southwest toward Newton Falls. During these regular jaunts he carried in his breast pocket a slim, worn blood-red leather Excelsior diary; there he recorded

James Ward Packard.
Courtesy of Special Collections, Lehigh University.

the weather, the distance, and the time it took to walk from one point to another. Without fail, he also carried one of his walking sticks in one hand and a watch in his pocket.

When possible, Elizabeth joined her husband. Sharing long strolls together had been one of the ways that, twenty-one years earlier, Ward had courted the woman who had strenuously vowed never to marry. A Vassar graduate, Elizabeth had studied medicine at Northwestern Women's Medical College in Chicago, read law in her father's chambers, and taught public school for two years in Arizona. She stood at least a head taller than Ward and was every inch his intellectual equal. The times and her prominent family defined marriage as a woman's place in society, but Elizabeth found the idea execrable.

During the couple's courtship, however, Ward slowly won her over. They found comfort in shared silence and the opportunity to discuss a spectrum of ideas. Fittingly, the pair had first met over books at Warren's library, where her father was president of the Warren Library Association. Ward often brought Elizabeth science books; once he gave her a barometer. They were both gracious but not without tempers. Neither was given to trifles or open displays of emotion. Much to her surprise, they developed a deep bond based on intellectual kinship and caring. Following one of their walks, Elizabeth wrote in her diary, "Ward & I were very prosaic and dull. We never let that worry us however, that is one of the charms of our camaraderie, we take each other just as we happen to be and never make any effort to entertain one another. That is real friendship."

The interplay of wit, substance, and banality had its desired effects. Three years after writing in her diary, "If I ever marry—but there's no use of discussing that for I never shall," Elizabeth relented. The pair became husband and wife in a hastily arranged wedding that came together on the morning of August 31, 1904, at her parents' Italianate mansion on Mahoning Avenue. The bride wore an embroidered blue going-away gown and plumed hat. She was thirty-two, and Ward was not quite forty-one, both well past the blush of young love. Indeed the union came as such a shock that their wedding announcement read in part, "It was a complete surprise to many of their friends who are nonetheless quick to extend congratulations." Not given to florid expressions, Ward noted the occasion in his diary as simply "Wedding."

A photograph taken on the occasion shows the newlyweds in a Packard Model L touring car on the way to their honeymoon on the shores of Lake Chautauqua in Lakewood, New York. Stiffly pressed against the car's high-backed tufted leather seats, with Ward at the wheel, the pair stared into their future, captured by the unknown photographer, still wearing their wedding finery and the faintest of smiles. Despite being married for twenty-four years, Elizabeth and Ward would remain childless.

As with nearly all of the objects in his possession, Ward owned multiple canes. During the summer of 1917, months after he received his ring-watch, he commissioned Patek Philippe to produce for him a special walking stick. It was two months after the United States formally entered World War I when he made the request, and the piece took more than

Ward drives his bride, Elizabeth Gillmer Packard, to their honeymoon in his Packard Model L, August 1904.
Courtesy of Betsy Solis.

a year to manufacture. On September 18, 1918, Ward took delivery of the beautiful polished ebony cane topped with a winding sterling silver watch as its handle. The enamel dial had black Roman numerals and was designed to twist off the cane and be replaced with a mushroom-shaped knob of ivory, depending on his mood. On Ward's instructions, the movement, the internal mechanism of the watch, no. 174826, was engraved *Made By Patek Philippe & Co., Geneva Switzerland for J. W. Packard 1918*. Ward put his new, one-of-a-kind walking stick to use at once.

Though he considered himself a mechanical engineer, Ward had spent the previous three decades pioneering some of the country's most important industries of the Industrial Revolution and, along with his older brother, William Doud Packard, created entire businesses around them. His prolific ventures made him both a mogul and a millionaire several times over. His first effort, the Packard Electric Company, had put Warren on the map. Established in 1890 as a producer of high-quality incandescent lightbulbs, the company eventually grew into one of the biggest manufacturers of automotive and electrical systems. Nonetheless it was the Packard Motor Car Company that magnified the Packard name across the globe.

In 1899 Ward designed his first horseless carriage, a single-cylinder, single-seat roadster with a tiller for steering. Twenty-three years later, the Packard reigned supreme as America's premiere luxury car. Favored by royalty, industrialists, and movie stars, the Packard was renowned for its elegant lines and superb engineering. "Ask the man who owns one," exclaimed the company's famous tagline. Indeed, on March 4, 1921, Ohio-born Warren G. Harding rode to his inauguration in a shiny Packard Twin Six with its regal lined hood and brass-rimmed windshields, the first U.S. president to travel by motorcar to his swearing-in ceremony.

On this cold spring evening in 1922, as the Packards and their guests sat down for dinner, the fortunes of both Ward and the country seemed buoyant. Before the year was over Packard Motor would report record sales of $38 million (in today's dollars, an amount equal to more than $518 million), and its directors would reward shareholders with a 100 percent dividend on the company's common stock. America was ebullient. Having survived the deadly influenza epidemic of 1918 and World War I, the country roared into the 1920s. (Packard Motor served the country admirably. In the Allies' air corps, the Liberty aircraft were powered by Packard eight- and twelve-cylinder engines, originally built for the Packard Twin Six.)

Exuberant jazz music wafted from radios and drifted on plumes of smoke out of nightclubs. Prohibition spawned illegal speakeasies. Women began dropping their social constraints while raising their hemlines, largely untroubled by the new libertine spirit or the scruples that preceded them. It was a time of mass production, massive wealth generation, and mass consumption. The era would later be recalled as the Jazz Age, the Age of Intolerance, and the Age of Wonderful Nonsense. Whatever the period was called, America was undeniably at the start of the modern age and Europe began to recede in its shadow. This was a time of new heights and numerous firsts.

For Ward, an inventor of prolific ability, there were still a few more firsts to pull out from under his crisp, cuff-linked sleeve.

The evening at home marked one of the small and occasional intellectual gatherings organized by Ward, who several years earlier had shifted his attention from the day-to-day operations of his companies to focus on a range of interests and new inventions. This architect of an industrial empire preferred the anonymity of a simpler existence, surrounded by what was most familiar. A voracious reader, he consumed numerous trade publications, scientific journals, and literature. With his lantern camera outfitted with a French Darlot lens, he had produced hundreds of glass slides from photographing scenes of Warren: his sisters riding bicycles in their long billowing skirts in the spring, the steamships navigating across Lake Erie in the summer, and hummocks of snow nearly obscuring Warren's streets in the winter.

Ward and his wife donated generously to civic, scientific, and educational causes, including giving $100,000 to the city of Warren for a new library. Along with his elder brother, William, better known as Will, his longtime business partner with whom he was extremely close, Ward gifted 150 acres along the Mahoning River to build a public park. But his largesse aside, Ward was not simply Warren's regal money pot.

Generally timid, Ward was nonetheless relentlessly inquisitive and enjoyed the company of like-minded individuals. An invitation to one of his salons was coveted albeit infrequent. These informal evenings featured intense brain fencing, where the inventor held forth as a kind of avuncular Socratic figure. For him, everything posed a question to be solved and every design could be improved upon. The evenings were much like Ward himself: judicious, precise, and with a strong emphasis on culture.

Ward and Elizabeth had purchased a two-acre tract of land east of downtown and were in the process of building a four-story, twenty-room estate on the bucolic stretch between Oak Knoll and Roselawn Avenues.

Elizabeth was particularly eager to move because she found the house on Park too close to the town's center and the noise insufferable as motorcars rumbled through the narrow streets, practically at her front door. Her discomfort was ironic considering her husband's considerable contributions to the auto industry.

This grand house was made of imported Scottish brick, mahogany, marble, and oak; a foyer fourteen feet wide and forty-four feet long allowed for an impressive entry from both the front and back of the house, with porte cocheres on each side. An entire wing was designed just for the servants, and there was a separate five-car garage with chauffeur's quarters and a European-style butler's pantry. Nearly an acre of the property was set aside for Elizabeth's gardens. A showpiece of cerebral elegance, the new mansion included a library on the first floor, stocked from floor to coffered ceiling with the couple's many leather-bound volumes, first editions, histories, and novels, each stamped with the Packards' personal bookplate.

They were not a couple given to unbridled displays of ostentation, but they did not reject the luxuries their wealth afforded them. The house's archways were molded to resemble the distinctive "tombstone" curves of the front grille of a Packard. The doors were fitted with cut-glass knobs, and the wall panels were hand-painted and imported from Europe. Adjacent to the sunroom was a small plumbing closet for watering plants so that the maids need not traipse across the house with water pitchers. Elizabeth's master suite had built-in fur and jewelry vaults. In an almost unheard-of luxury for an era filled with opulent distractions, plans for the marble bath called for built-in water jets.

It would be two years before the new Packard family pile was ready, but Ward, already the recipient of more than forty patents, busied himself coming up with every mechanical contrivance imaginable for the house's interior. By the time the house was finished, Ward was said to have deployed at least one thousand gadgets, each designed to replace some kind of manual labor. There was a burglar alarm and a turnstile switch that controlled all of the lighting from any room; an electric power plant was housed in the basement. So well designed were these systems that both would function perfectly well into the next century. Ward devised an elevator hidden behind the main spiral staircase and a second, secret set of winding metal stairs. The latter was meant to spirit the shy inventor up from the parlor to his second-floor bedroom without attracting the attention of the friends and visitors of his far more sociable wife.

The couple also maintained a summer estate, a thirty-two-room neo-Georgian mansion in Lakewood, where Ward built a machine shop on the second floor of the three-car garage. His continued tinkering in the workshop led to versions of a gasoline-powered lawn mower and an electric recording machine, among dozens of other prototypes and patent applications. Long fascinated with ships and boating, he devised a number of motors for his own small fleet. The DNA of several of these designs later powered American-manufactured speedboats, including every patrol boat deployed during World War II. The basement in his elegant Lakewood mansion was designed to resemble a great ship's cabin, with a nautical ladder that descended from the first floor and portholes for windows.

A visitor to Lakewood might find Ward dressed in full yachtsman regalia sailing his five-foot cruiser, his forty-eight-foot pleasure boat, or his electric launch on Lake Chautauqua. He would wear an old suit if he sailed one of his rowboats. A point of pride was his hydroplane, the first on the lake. He especially enjoyed flying it at sunset and never ceased to marvel at the odd sensation of seeing the sun dip as he ascended high enough to overtake the glowing orange ball and then watching it sink once again below the horizon line.

Ward and Elizabeth Packard's summer estate and gardens
in Lakewood, New York. Courtesy of Betsy Solis.

Shortly after his wedding, Ward began construction on the property, which cascaded all the way to the lake, where he kept a jetty and a boathouse. The couple spent the next eight years designing the property to their exacting specifications, which included constructing an eight-hole golf course, a log cabin for games, and Elizabeth's extensive formal English gardens, with neat boxed edging and endless gravel pathways connecting pools and pergolas to her groves of hardwood trees and various cutting, vegetable, and perennial gardens. Several existing cottages on the land had to be moved to make way for the new mansion. The relocation process, which took place over five years starting in 1905, was so careful that each cottage was moved without packing up its contents. As one local later described the scene, "Not a dish was broken, or an artifact damaged."

For Ward and Elizabeth, Lakewood was more than just an escape from the repressively gummy heat of Warren's summers; the town had been a favored Packard family destination stretching back to his childhood. In 1873 his father and uncle had bought up twenty-five lakeside acres along with the Lakeview Hotel and built a number of ornate cottages, creating an alternative resort colony to Newport for the growing industrial fortunes sprouting out of Ohio. The family, known for their many good works, maintained a large Gothic-style villa where they spent the summer seasons as one of Lakewood's leading and most beloved families. While just a boy Ward operated the family hotel's steam-driven electric light plant and Will ran the telegraph office.

James Ward Packard was born under a shining star. The second son of Warren and Mary (née Doud) Packard's five children, he came into the world just two weeks before President Abraham Lincoln delivered the Gettysburg Address, on November 3, 1863. A small boy with an outsized intellect, he grew up in a period of often turbulent change that offered unprecedented opportunities to bold speculators and inventive entrepreneurs. Trailblazing was something of a genetic trait.

The first Packard in America was an Englishman named Samuel Packard who sailed across the Atlantic on the *Diligence* in 1638, landing in Plymouth County, Massachusetts. Following the American Revolution the Packards pressed forward, settling in northeastern Ohio, an area known as the Connecticut Western Reserve, bordering Lake Erie on its north side and Pennsylvania on the east. Ward's forebears produced doctors, judges, and a brigadier general who served during the Civil War. In 1825 his grandfather William Packard moved to Lordstown Township, where he became its first postmaster. Some twenty years later, restless and dreaming of striking it rich, William abandoned his wife and ten children (including Ward's father, Warren) and headed to California for the gold rush. He never returned.

In 1846, during the Mexican-American War, eighteen-year-old Warren Packard crammed a rucksack with his few belongings and and traveled six miles north to the seat of Trumbull County to put down roots in a town that happened to share his name.

Mid-nineteenth-century Warren, Ohio, was a somnolent hamlet in a quilt of picturesque villages running along the Mahoning River. Ancient thick maples and elms shaded its quiet dirt roads. Horses pulled wooden carts and carriages, making soft, rhythmic thuds. Black bass filled the Mahoning and quail the surrounding woodland. The First Presbyterian Church's 225-foot spire and tower bell was the town's dominant landmark. It had been painstakingly carried to Warren by ox cart in 1832.

When the inventor's father arrived, Warren had a population of sixteen hundred people, five churches, twenty stores, three newspapers, two flour mills, one bank, and one woolen factory. On Millionaire's Row, which fanned out from the town's center at Mahoning Avenue, Victorian, Gothic, and French Empire–style mansions stood like preening peacocks. Railroad tracks were being laid across the country at a furious rate, but it would be another forty years before the full-scale construction of railroads connected Warren to the rest of Ohio. Commerce and shipping were still conducted mainly by canal, lake, and river, extending from the Ohio River to Lake Erie, which kept Warren insulated for a time from the manufacturing and industrialization that had transformed the growing steel towns of Niles and Youngstown downstream. The town developed at a languid pace into a prosperous burg of merchants and refined culture. Warren's blue skies would not be obscured by chimneys belching plumes of black smoke until the end of the century.

By the time Ward entered the picture, his father had become one of Warren's most prominent businessmen and the Packards one of its leading families. Within four years of working for a local iron merchant, Warren Packard bought the business outright, along with a competing iron foundry that manufactured axles and hammered iron. The elder Packard earned a reputation as a man of "broad vision and optimistic views . . . able to carry out to magnificent completion plans others would not dare undertake." In due course, he expanded his interests along with several partners into hardware stores and saw and lumber mills, spreading his businesses into Pennsylvania and New York. Eventually he was named a director of the Western Reserve Bank. Packard supplied the men who

were building the railroads as the railroads were building a new America. He owned the largest iron and hardware business operating between
Cleveland and Pittsburgh. It was said that his lumber business furnished
half of the wood used to build the Atlantic & Great Western Railroad that
stretched across the western tip of New York into Pennsylvania.

When Ward was a boy, the family lived in a modest brick rental house
on East Market Street and Elm Road. As Warren's fortunes improved the
family moved into a larger house on fashionable Courthouse Park. Ward
and Will had three sisters, Alaska (named in honor of the Alaska Purchase
of 1867), Carlotta, and Cornelia Olive. By the time Cornelia was born,
nearly fourteen years after Carlotta, the family had built a magnificent
French Empire–style mansion on High Street at the top of Millionaire's
Row. The Packard mansion, with its gabled roof, carriage house, fourteen bedrooms, and ballroom, was the grandest ever built in nineteenth-
century Trumbull County.

Life inside the Packard mansion was exuberantly cheerful among the
silver sets and crystal chandeliers, but it was also disciplined and purpose-
ful. Warren Packard impressed upon all of his children the importance of
education, hard work, and culture. And he set the example for his family. For fifty-one years, until he died in 1897 at sixty-nine, Warren began
the day at 5 o'clock and worked through the evening, stopping only at 9
o'clock.

Mary Packard had a softening influence on the family, and Ward was
particularly close to her. A small, bright-eyed woman, she took delight in
her menagerie of dogs and children. One family friend called Mrs. Packard a "jolly" woman who was kind and delightful, behaving more like a big
sister to her children than a parent. A standing member of Warren's aristocracy and a regular churchgoer, Mary was also politically active. Among
the first women in Warren to join the Political Equality Club, she would
attend local lectures given by the leading suffragist Harriet Upton, the first
woman to serve on the Republican National Executive Committee.

In the fall of 1874, when Ward was eleven, his parents thought it time
to expand the horizons of their intellectually curious but culturally narrow son and sent him to receive an education in refinement and worldli-
ness on a six-month excursion across Europe, beginning in Southampton,
England. He was accompanied by family friends from one of Warren's
oldest families, Dr. William and Laura Iddings, and a governess. In the
course of his business dealings, Ward's father himself had made several
trips across the Atlantic, during which he set up a pipeline to import European hardware.

From the time he could write, Ward received a pocket-size leather Excelsior diary each year. While abroad, he catalogued in his measured boy's hand thrice-weekly French and Italian lessons, visits to ancient ruins and cultural sites, and the weather. Mostly, however, he logged days of tedium and abject loneliness, noting whenever he had and had not received word from home. In the south of France on January 23, 1875, he wrote, "Weather was very hopeless the wind was blowing very hard and we did not do anything I received nothing." With attention verging on the obsessive, he made sure to assess the amount of time and distance between destinations. "Left London at quarter of three and to Liverpool at quarter of eight, 201 miles," he noted on March 16, 1875.

When tired of the art and ruins, he holed up reading *David Copperfield* or the latest copy of *Youth's Companion* magazine that he waited eagerly for his sisters or mother to send to him. Mostly he calculated the miles and hours until he would at long last return to Warren. His fancy European education did leave one indelible impression: above all else, he learned that he just wanted to go home.

Ward's universe was contained in Warren, a tight drum of little more than sixteen square miles, animated by Jules Verne's *Around the World in Eighty Days*, Nikolaus Otto's four-cycle internal combustion engine, and A. A. Pope's Columbia Hi-Wheeler. He didn't need to venture far to expand his world from the comfort of his own bedroom.

Unlike most boys, Ward wasn't given to exploring the muddy woodlands surrounding Warren. Obsessed with mechanical invention, he preferred tinkering to the exclusion of nearly all other pursuits. Every advanced device of the period—firearms, phonographs, cameras, steam engines, railroad engines, telegraphy, and bicycles—fascinated him. As his fellow students pored through detective novels attempting to puzzle together the clues, Ward developed his stock in trade: finding new ways to improve existing devices and machines and seeking to convert everyday actions into mechanical operations. Not yet twelve, he rigged up a contraption that allowed him to open and close his window without having to leave his bed.

He and Will, born exactly two years and two days apart, were inseparable. In sleepy Warren, where residents remained suspicious of innovation, wanting nothing of industry, Ward and Will became known as the brothers "who insisted on doing things and making things."

Some 240 miles away in Dayton, another pair of brothers, Orville and Wilbur Wright, began exploring flight. After their father, Bishop Milton

Wright, gave them a toy Penaud helicopter fashioned after a flying Chinese top, the two took up bicycle building, which soon led to flight machines. But the fledgling field of electricity most captured Ward's and his brother's imaginations. In the decades since the British chemist and inventor Sir Humphry Davy first demonstrated his arc lamp in 1809 by connecting two charcoal sticks to a 2,000-cell battery, a variety of electric arc lamps had been developed that created an electric current flow through an "arc" of vaporizing carbon. Since then, men all over the world sought to create a more durable and commercial version of electric illumination. Although still a boy, Ward was one of them.

One evening Ward persuaded his brother to help him concoct his own arc lamp. After the rest of the family had retired for the night, Ward and Will sawed the knob of their bedroom door in half and joined the two pieces to opposite poles of an induction coil. Their handiwork did not conduct electricity but filled their father with a mix of consternation and pride upon discovering their deed the following morning, when he came to rouse his sons from bed.

After a few weeks of intense effort, the brothers startled the gentle folks of Warren one evening with a huge arc lamp dangling out over the street in front of the Packard mansion that produced a blinding light. All of Warren emptied out of their houses and businesses to see one of the first electric lamps on a city street anywhere in the country.

As an adult, Ward was drawn to clocks and watches. It was precisely this interest that earned Dr. John Kingsley an invitation to the Packards' home on that rainy March evening in 1922. Kingsley arrived from Salem, Massachusetts, bringing a group of unique historical watches from the Lee collection, housed at the Essex Institute, a literary, scientific, and historical society. Charles Mifflin Hammond, a prominent New Englander turned California vintner who was also the brother-in-law of Theodore Roosevelt, had originally owned the timepieces, which his widow donated to the institute in 1917. The collection represented numerous examples of timekeeping from around the world. In addition to pocket watches there were 152 clocks, including some with English cone-shaped, spiral-grooved fusee movements that used pulleys to equalize the irregular pull of the mainspring, often found in clocks made during the nineteenth century. Along with several curious instruments from China and Japan there was an unusual Dutch Friesland clock that displayed a view of a harbor just below the dial where several tin ships bobbed on undulating tin waves in sync with the clock's movement. A timepiece made in 1750 by James

Ferguson of London indicated the time of twelve different world cities simultaneously.

The Institute arranged to bring a selection of the watches to Ward for his viewing at home. His enthusiasm for mechanical perfection and a lifelong passion for timepieces had earned him a reputation that nearly matched his status as an auto magnate. Over the years, he had acquired a growing collection of significant watches, becoming one of the country's most important connoisseurs.

As a boy, Ward tinkered with every clock in the family home, studying each intricate part until he understood how they worked together, like a three-dimensional game of chess. Reassembling them, he often improved their operation, a habit he did not abandon as an adult. Once, just days before Christmas, he took apart one of his in-laws' Seth Thomas shelf clocks, a popular manufacturer whose clocks were rather simple instruments that used a wooden pendulum as a swinging weight for the timekeeping mechanism. Although widely employed since the seventeenth century, such devices were prone to inaccuracies if the clock did not remain virtually motionless. On a whim, Ward removed the wooden pendulum and replaced it with an electromechanical pendulum of his own design. The new apparatus used a coil wrapped around a metallic core that produced a magnetic field. When an electric current passed through it, it provided an impulse to the pendulum, which made its timekeeping more accurate. Like the master clockmakers of previous centuries, on the inside of the wooden clock frame Ward signed and dated his handiwork in his spidery pen, *JW Packard Dec 18, 1904*, and quietly placed it back on the shelf.

Following dinner Ward led Kingsley and Judge Gillmer down to the basement, dubbed "the gun room," where he usually entertained his guests. The trio repaired to view the watches under the room's brightly painted frieze that depicted various Packard models amid a colorful backdrop of landscapes. The images were taken straight from the motor company's marketing calendars. Evidence of Ward's particular tastes was everywhere. On the mantel rested a pair of original Packard automobile lamps and an automobile clock. Small electric motors, microscopes, and various mechanical calculators and typewriting machines filled every available surface. Ward owned a copy of each make and model of these machines. Rather haphazardly, a trumpet and a Victor phonograph sat on the window seat. An ancestral bear gun hung directly over the brick fireplace, and perpendicular to it a wall rack held four U.S. Army Springfields with fixed bayonets.

Ward collected firearms much as he did timepieces, for their mechanical workmanship. One of his guns was outfitted with a diamond sight, and he was said to own one of sharpshooter Annie Oakley's pistols. A dozen years earlier he had commissioned the New York jeweler Tiffany & Co. to customize his .38 Smith & Wesson double-action perfected revolver. Tiffany engraved the gun's black barrel, adding gold, diamond, and platinum inlays, while the beautiful ivory handle was carved with scenes from the Wild West and featured a cowboy on horseback. (On close inspection, the cowboy's face looked rather similar to Ward's own visage.)

If anything, Ward was an aesthete. In his view, a piece of perfectly executed machinery should not preclude it from also being a fine piece of art.

With watches from the Lee collection spread out before him, Ward picked up one of the smooth gold pieces and turned it over in his hand. Tugging gently at the chain in his pocket, he pulled out a small penknife that dangled from the fob and deftly slid the point of the tiny blade into the case's hinge, springing it open like an oyster. The movement, the watch's inner workings, piqued his interest. Exposed, the movement revealed a mechanical tradition extending well beyond the fifteenth century. With loupe to eye, Ward magnified the minute whirling balance wheels, toothed escape wheels that clicked as they turned, and scurrying ruby pallets. The movement offered order and function in its purest, most ingenious form. Every component had a purpose that in turn sparked an entire series of events. From the outside, a polished gold watchcase was a thing of obvious beauty, but to peer beneath the dials was to view a complex universe in microcosm.

Over a twenty-four-hour period, a fine instrument ticks some 432,000 times. Every part of the mechanism strikes 18,000 blows an hour, adding up to more than 150 million beats a year. Hundreds of tiny, handcrafted moving parts are needed. The pivots measure barely .0028 inch in diameter. By one estimate, the screws are so small that it would take 20,000 of them to fill a thimble. A horologist once described the adjustment process as so delicate, "a pencil mark on a scrap of paper would make a difference in the balance." Hardly trifles, mechanical watches were constructed to last through the generations.

For an engineer dedicated to technical perfection, horology offered the ideal preoccupation. It is perhaps the only discipline where a 99 percent precision rate is considered a failure; the 1 percent variation translates into a woefully inaccurate fifteen minutes off each twenty-four-hour period. Alvan Macauley, the president of Packard Motor, would say of Ward, "Crudeness and imperfections hurt his sensibilities." As Ward studied this

Lilliputian world of screws and pivots, a new watch that he desired had already begun to take shape in his mind.

Long ago Ward had become infected with what is known in some circles as the horological virus. An obsessive fever that brought kings to their knees and lovers to ruin and turned mighty industrialists to putty, it began simply with ownership of one watch and led without reservation to the desire to own every variation and then to own something that nobody had ever held.

Watches were not just everyday objects. There was something supremely intimate about a pocket watch. Like most men of means at the time, Ward possessed more than one timepiece. They were functional, of course, but their stylish gold cases, some engraved or filigreed or enameled and set in pearls, indicated they were more than just utilitarian.

During his early collecting, Ward came to own at least eight antique and technically significant watches. An enviable inventory, they included a rare eighteen-carat gold Victor Kullberg chronometer with a minute repeater. Appointed chronometer maker to the king of Sweden in 1874, Kullberg was considered one of the most skilled horologists of the nineteenth century. On the movement of this watch were markings denoting its placing at the Paris World's Fair timing competitions: *4 Diplomas of Honour, 10 Gold Medals Award Grand Prix, Paris 1900.* Ward's early acquisitions included a beautiful late eighteenth-century enamel by the Swiss watchmaker Guex à Paris and an eighteen-carat gold verge watch by Chevalier of Paris made in 1790. Verge escapements, with their toothed wheels and a balance staff, which locked and released the mechanism's movement, had been used since the fourteenth century and were an important turning point in horology, marking the shift away from measuring time by continuous processes, such as the flow of water or the swing of pendulums, and toward the development of all-mechanical timepieces.

Since his early collecting years, however, Ward's connoisseurship had evolved sharply. By the time he entertained Dr. Kingsley, he had moved away from solely decorative pieces and lovely antiques and gravitated toward unusual instruments with the highest technological skill and precision, in particular those that incorporated a number of complicated functions outside of routine timekeeping, such as perpetual calendars and phases of the moon, called *grandes complications.* More than simply acquiring these complicated instruments, Ward took an active hand in their design, instructing the great watchmakers to create pieces with some innovative feature or to integrate an unprecedented combination of complications.

The auto magnate was the type of patron with whom the finest watch-makers hoped to curry favor. He was cultured, passionate, and fabulously wealthy. But in his approach to collecting he looked to watchmakers to transform his inventive, most florid musings into a ticking mechanical ensemble that expanded engineering beyond its known boundaries.

In 1905 he took receipt of his first watch from Patek Philippe, with the movement no. 125009. The debut timepiece, a *grande complication,* was a fine eighteen-carat gold open-face chronograph, with a minute repeater, perpetual calendar, and *grande* and *petite sonnerie* striking full and quarter hours. Unknown to both Ward and the watchmaker at the time, the 1905 chronograph was the first shot in a contest that would span decades. While Ward came to favor a handful of watchmakers, including Vacheron Constantin, he had found a kindred spirit in Patek Philippe. In Ward, Patek Philippe discovered an exquisite patron with obsessive tastes, unsparing hubris, and limitless funds to underwrite his fancies.

Beyond technology, Ward's influence extended to design, and each watch reflected his refined sense of style. Cases and rims were decorated with engravings and Art Nouveau motifs. The case backs sported his signature monogram in raised bas-relief in blue or black enamel or gold. It was a version of the same stylized emblem that he used on his own stationery.

Ward approached the creation of a new watch with the mind of an engineer and the heart of a lovesick suitor. He derived enormous pleasure from establishing some new engineering feat nearly as much as he relished discovering the answer. With each new watch commission he pursued an undiminished desire to marry superb craftsmanship with technological achievement. In doing so, he drove Patek Philippe to achieve ever greater heights of complexity and precision in ever greater combinations.

Six years earlier, Ward had received the first of what would be his two greatest *grandes complications* produced by Patek Philippe. An extraordinary eighteen-carat gold pocket watch possessed a total of sixteen complications. Ward took possession of the watch during the same year that the Packard Motor Car Company introduced its twelve-cylinder cars. The Packard Twin Six line established the company as the most popular American luxury car and was in fact called America's Rolls Royce.

The *très grande complication,* given the movement no. 174129, was the most extraordinarily complex timepiece that Patek Philippe had created in its own storied history. In addition to the perpetual calendar with retrograde date and moon phases, the watch had power reserves both for striking and movement and a sixty-minute recorder, as well as *grande* and *petite sonnerie* on three gongs, which struck the number of hours on one gong and the quarter-hours on a second, all chiming in different tones. This particular complication had been invented before electricity so that its owner might distinguish the time without actually peering at its dial.

The stunning watch's greatest technological achievement was its *seconde foudroyante* (lightning) chronometer. The chronometer measured time increments to the fraction of a second, dividing each second into five jump steps, each step indicating one-fifth of a second. At the press of a button the hand appeared to split into two, enabling the mechanism to time two separate events simultaneously, two individuals in one race, or even several successive stages of a single episode. Its manufacture sent a shock through the watchmaking world, eventually setting the stage for a horological high noon.

As the evening drew to a close Ward's thoughts had already turned to his next timekeeping masterpiece. For some time, yet another *grande complication* had been taking shape in his mind, an epic pocket watch. The instrument would stretch all horological and aesthetic limits, leaping past the handful of timepieces that had taken their place in horological history. He wanted a watch that was geographically calibrated to Warren, Ohio, with the *pièce de résistance* a celestial chart that navigated the heavens exactly as he saw them outside his bedroom window. In timekeeping's fabled history, only one other pocket watch in the modern era, the Leroy No. 1, finished in 1904 for a Portuguese coffee tycoon, featured a nocturnal map of the skies.*

Again Ward turned to Patek Philippe, who had never produced a sky chart. For this latest endeavor, Ward as usual insisted that no expense, effort, or daring be spared. For the watchmaker, with its considerable skill and imagination, this appeared to be the commission of a lifetime.

At this point, the number of watches in Ward's collection was climbing toward the hundreds. Jules Jürgensens, Agassizes, Vacheron Constantins, and Walthams, in addition to numerous French and British pieces and perhaps more than two dozen custom-designed Patek Philippe timepieces—it was a collection of masterpieces. Obsessed, he was after a watch that contained the greatest number of complications in the boldest combinations in the smallest space imaginable.

And so in the year 1922, when the Nineteenth Amendment to the

---

*Around 1780, the British clockmaker George Margetts is known to have crafted an astronomical watch with a type of celestial chart.

U.S. Constitution gave women the right to vote, the British archaeologist Howard Carter discovered the ancient tomb of the Egyptian pharaoh Tutankhamun in the Valley of the Kings, and the American-born poet and playwright T. S. Eliot published *The Waste Land*, Ward changed horological history. This latest commission would become the most important *grande complication* in his considerable collection and one of the most seminal in the world of horology.

The eighteen-carat gold pocket watch, designated with the movement no. 198023, would require five years to complete and would come to be known simply as "the Packard," one of the most celebrated timepieces in history. Knowledge of its existence would eventually reach the fabulously wealthy and immensely private collector Henry Graves, Jr., a man who made it his personal business to possess "the best of the best," a man who was engaged in his own quest to possess a complicated masterpiece of historical proportions.

For Ward, the Packard was proof that the impossible was merely a problem waiting to be solved.

CHAPTER THREE

~

A Shining Light

When Ward's train pulled into the immense dome-topped Grand Central Depot in New York City late in the summer of 1884, he confronted a city bursting out of its provincial roots. And so was Ward. Not quite twenty-one years old, he had arrived after graduating from Lehigh University, the youngest mechanical engineering student in the university's history.

Ward could have returned to Warren after graduating, settling into an affluent life of pushing along his father's business affairs, but he was much too curious about the age of machines and eager to be part of it not to move beyond that life. He still had lucid memories of his first brush with the future. In 1876, at thirteen, he had traveled to the Philadelphia Centennial Exposition with his fifteen-year-old brother, Will, where he found an Aladdin's Cave of engineering marvels. The pair had been so hell-bent on attending the exposition that they sold newspapers and operated a printing press from home in order to finance the trip. Once there, Ward gazed wide-eyed at the seventy-foot-tall Corliss steam engine at Machinery Hall. The enormous 1,400-horsepower machine weighed 1.7 million pounds and furnished power to all of the exhibits in the Hall, barely registering a vibration or whistle. Ward wandered the stalls, witnessing the latest in electric lights, elevators, fire trucks, printing presses, mining equipment, Pullman sleeping cars, drills, and lathes. He and the public were introduced to typewriters and mechanical calculators, as well as Alexander Graham Bell's telephone and Thomas Alva Edison's telegraph. Machinery Hall presented a snapshot of America's economic future.

At Lehigh University, Ward had spent his time spinning his imagination into reality. In his room at Saucon Hall, he had created a makeshift laboratory filled with a range of electrical apparatuses, all of his own design and each meant to ease the day's most prosaic activities. A magnetic device attached to his door allowed him to switch the lock on and off remotely. He constructed a large wall clock alarm using gravity batteries that he controlled from his bed; it was said that the alarm could rouse the dead. He had also rigged up a telegraph line that he operated from his room and ran through to the rooms of his friends, over which they sent and received messages. His excitement about bicycles, a relatively new mode of transport, led him to found the university's first cycling club, and the Columbia Bicycle Company enlisted him as a local agent.

"A handsome little fellow with eye glasses, a soft voice and a sardonic tongue, who knew more about the world and the practical side of mechanical engineering than the tallest of us," was how fellow student August Parker-Smith described Ward, adding, "He was anything but a grind and went through college on his nerve and natural ability supplemented by his practical knowledge of tools, machinery and electricity. It was easy to foresee, as most of us did, that he would have no trouble in making his own way in the world."

In Ward's senior year, when he wrote his thesis, "Design of a Dynamo Electric Machine," the world seemed in alignment with his intellectual velocity. The Bell Company had built the first long-distance telephone line connecting Boston and New York, and George Eastman had taken out a patent on a flexible roll of film to use in cameras. In Chicago construction began on the first steel-skeleton skyscraper, a ten-story structure, designed by William Le Baron Jenney for the Home Insurance Company. The electric lamp industry was growing rapidly, and Ward wanted to be part of it. For that, there was only one place for him to go: New York City.

Handing the station agent the claim ticket for his black leather steamer trunks neatly stenciled with his initials in white, Ward stepped off the train and straight into a city fast emerging as America's most modern, rapidly eclipsing longtime rivals Boston and Philadelphia. An entire heaving world rushed past him in a single instant. Tugs and ships cut across the harbor, arriving from ports far and wide. The unmistakable clatter of hooves and carriage wheels against cobblestone and brushed pavement signaled oncoming trolleys and hansom cabs. Steam-powered cable cars thundered across the city while elevated trains deposited passengers, soot, cinder, and noise.

As the novice engineer-inventor blinked in the sunlight, bankers and merchants scurried toward the exchanges where millions of dollars traded hands; thousands of tradesmen and workers filed into the city's warehouses and foundries. Men tipped their hats to acquaintances and women pushed babies in prams farther uptown in the calm of Central Park's open meadows. At the Fulton Market near the East River, fishmongers spilled brackish water and guts onto the mucky pavement. On barges all along the waterfront, thousands of freshly pulled oysters hung in nets awaiting delivery. At Dutch Hill on First Avenue at the foot of Forty-first and Forty-second Streets sat a shantytown inhabited by paupers, thieves, goatherds, and pigs.

As head of Lehigh's bicycling club, Ward had traveled to New York to attend events, and each time, the city held some new surprise. Then as now, it was filled with people who had come there from somewhere else. Farmers from the country, southern blacks, Greeks, Italians, and Jews from Eastern Europe poured into the city, following the waves of Irish and German immigrants, filling up squalid tenements that offered little in the way of privacy, let alone plumbing. On the Lower East Side, an entire underclass totaling some half a million lived among slaughterhouses and factories, like produce rotting in badly packed wooden crates. On Mott and Canal Streets, Chinese émigrés set up shops in cramped flats in a warren of overcrowded streets. The scruff of the Bowery, with its flophouses, beer gardens, and pawnshops, was made more sinister by the Third Avenue El train that ran above ground, blotting out the sun.

Farther uptown on Fifth Avenue a regiment of elaborate mansions transfixed passersby with their brazen grandeur. Sharing space in the rarefied air stood elegant hotels to rival Europe's. Here, afternoons were spent in gentlemen's clubs such as the Knickerbocker and the Union, and in the evenings meals were enjoyed at luxurious restaurants like Delmonico's. On Saturdays and Sundays the city's wealthiest paraded their carriages on the grand drive in Central Park; men turned out in their dandy black livery coats and buckskins while women showed off their finest French lace and fancy parasols. At the park's eastern edge, a group of American businessmen and financiers established the Metropolitan Museum of Art with 174 paintings formed out of three private European collections. In just a few years, the museum would rank as one of the finest art institutions in the world.

Having made his way from Grand Central to the piers at Fulton Street, Ward paid one and a half cents for a Fulton Ferry ticket bound for the city of Brooklyn, directly across the East River from Manhattan. Once on board, he pushed past the dense pack of commuters and leaned against the boat's worn wooden railing. Even though he was far from home, the sea immediately put him at ease. Inhaling the salty piquant air, he carefully placed one hand on the brim of his hat, keeping it safe from the gusts of wind that swirled wildly like the current. The Brooklyn Bridge, an incomparable engineering marvel completed a year earlier, cast an enormous shadow on his short journey. Spanning nearly two thousand feet, the world's longest suspension bridge was made of steel cables and featured granite neo-Gothic towers that formed arched portals. As Ward stared into the narrow strait where the tides of the Atlantic ebbed and flowed, sometimes dangerously so, a large tug surrounded by small steamers worked its way toward Wards Island and fishing smacks under sail pushed on to market. Gradually the Fulton slip on the Brooklyn side of the river came into view. He had arrived at his destination.

Brooklyn in 1884 offered Ward a colorful tableau of American city life in microcosm: unfettered urbanization replacing vast stretches of bucolic pastureland, a tangle of European immigrants attempting to grab the brass ring while a wave of Manhattan émigrés sought liberation from high rents and small apartments. Goats scavenged garbage-littered streets for lunch while children delivered pails of growlers filled with beer. On Pierrepont Place, Brooklyn's wealthiest residents lived in elegant Victorian town houses. At the Brooklyn Navy Yard, mechanical giants spit out solid iron that would later be fashioned into enormous ships. Brooklyn had vitality as well as lumbering burdens, filled with Walt Whitman's prose, forty-eight breweries, Prospect Park, and the Grays, a baseball team better known by its nickname, "the Dodgers."

Ward arrived in a thick August heat that hadn't been relieved by the searing summer rains. Work had just begun on the pedestal where the Statue of Liberty would rise up regally from New York Harbor at Fort Hood. Eight years earlier, when Ward attended the Centennial Expo in Philadelphia, visitors had paid fifty cents to climb inside the giant torch then on display. Frédéric Auguste Bartholdi, the statue's sculptor, had completed Lady Liberty the previous summer, a gift of the French people to America, but numerous financial setbacks on this side of the Atlantic delayed its unveiling until Joseph Pulitzer, the publisher of *The World*, raised funds, and on August 5, 1884, the cornerstone of the foundation was laid.

Four days earlier, with his diploma in hand, Ward had traveled from Bethlehem to the Pittsburgh offices of the Sawyer-Man Electric Company, an innovator of the incandescent lamp, where he interviewed for a job. After a brief trip to Warren, he was sitting on the Erie No. 4 Limited bound for New York to begin work at Sawyer-Man's factory at 229 Park Avenue in Brooklyn.

When Ward accepted his position at Sawyer-Man, electricity in general and incandescent lamps in particular had become rather glamorous fields. The dirty and low-luminance gas lighting that had lit the city since the

1830s had been replaced by arc lights. By the time of Ward's arrival, nearly fifteen hundred arc lamps set off New York City's streets in orbs of light.

Battles raged over the primacy of arc or incandescent lamps. Entrepreneurial businessmen produced and sold products discovered by others. Infringement lawsuits marked the industry nearly as much as new discoveries, choking the courts for years. Much was at stake as the companies raced to develop a light that was economically feasible, safe, and durable and could be used at home and at work. In short, a veritable gold rush to bring electricity to the masses was on.

Famously, on September 4, 1882, Thomas Edison installed six twenty-seven-ton dynamos (or generators) at the Pearl Street power station with an output of one hundred kilowatts, enough to power more than 1,100 lights. When Edison flipped the switch, his central station furnished lights to the buildings of the New York Stock Exchange and the *New York Times,* among others. By the following year Edison had established 246 plants that generated enough electricity to power 61,000 lamps.

As the use of electricity became widespread, a war of currents began heating up. In the first camp George Westinghouse and Nikola Tesla advocated for the use of distributing electrical power across alternating currents (AC), while Edison supported direct currents (DC). After setting the standard for electrical distribution in the country, Edison was loath to lose his patent royalties and went to extreme lengths to discourage the use of AC. Publicly he played by the rules, lobbying state legislatures, but privately he played dirty, spreading propaganda about fatal accidents resulting from the use of AC electricity. Repeatedly he used the term "Westinghoused," attempting to popularize it as common vernacular for electrocution, but without success.

By the time Ward walked the factory floor at Sawyer-Man, the company had only recently emerged from its own rather volatile and dramatic beginnings. A former Boston newspaperman named William Edward Sawyer, who was also a capricious genius, spendthrift, and heavy drinker (and an even uglier drunk), launched the Electro-Dynamic Light Company in 1878 with Albon Man, an attorney and director of the River and Rail Electric Light Company. After Sawyer's drunken behavior got him fired from his position at the United States Electric Engine Company, Man agreed to finance Sawyer's incandescent lamp experiments. Around the same time that Edison gained public acclaim, Sawyer developed some seventy-five patents, solving several problems that had long stymied others. As early as June 1878, he constructed an arc light that could be used in a controlled circuit, which enabled one light to be shut off without extinguishing the entire line. Significantly, Sawyer invented what became known as the automatic cut-off, a device that broke the circuit when the current became too strong, thus preventing overheated wires and fire. Sawyer also came up with a lighting switch that could turn on or off gradually. He claimed to have discovered a practically indestructible carbon filament no bigger than a pin and connected by wires to an electric machine that was then placed into a hermetically sealed glass globe filled with pure nitrogen. While Sawyer-Man proposed a plan similar to Edison's, establishing a central station in various parts of the city from which powerful electric generators supplied electricity, the company declared it could also provide electric light at one-fortieth the cost of gas.

As Edison's career became more celebrated, Sawyer grew to resent him, even as he stacked up one important patent after another. Sawyer complained bitterly to the newspapers that, among other things, it was *he* who had discovered the carbon filament as a conductor for the incandescent bulb, a year before the "Wizard of the Wires" unveiled his new lamp. Soon the two men conducted their feud publicly, each hurling invectives at the other via the press, filing patent infringement lawsuits along with their insults.

The public showdown turned unexpectedly grim when an angry and erratic Sawyer increasingly turned to a bottle of brandy for consolation. On April 5, 1880, doused in drink and in a particularly foul mood, he engaged his neighbor in an argument over Edison's success that ended with Sawyer shooting the poor man in front of the Rossmore Hotel on Broadway and Forty-second Street. Waiting for a conviction, Sawyer spent a year developing new lamp designs. Just one day after he received a guilty verdict, the company was reorganized into the Consolidated Electric Company with a $3 million capital infusion. Horace Little, the owner of the Eastern Electric Manufacturing Company, who also provided Sawyer's bail money, bought up Sawyer's patents.

In a strange turn of events, on April 15, 1883, shortly before his sentencing hearing, Sawyer collapsed and died. Immediately his widow announced that she was in possession of twenty of her husband's previously unknown patents. With Consolidated Electric as its parent company, this ultimately became the basis for the next iteration of Sawyer-Man.

When Ward showed up for work over a year later, the company's scandal-ridden origins had largely faded and Sawyer-Man had become a prestigious name to have on his calling card. Shortly after his arrival, the U.S. Navy awarded Sawyer-Man the contract to light the steamer the USS *Omaha*. Around the same time, Sawyer-Man installed its lighting system at the Eastern Lunatic Asylum in Virginia, "the only institution for the insane in the country in which the electric light has been adopted for universal illumination," declared the front page of the *Brooklyn Eagle*.

Ward started work at the company, toiling away as a stationary engineer in charge of a steam power plant in the dynamo room, earning one dollar a day. Within a month he was promoted to superintendent of the entire dynamo operation. He noted the milestone in his diary, writing simply, "I took charge of the dynamo room today." By November he was elevated to foreman of the mechanical department and was earning $20 a week. This too he dutifully reported in his diary with similarly understated aplomb, jotting down, "I took charge of the whole factory."

Although Ward's father had the means to support him well, following his graduation from Lehigh he was on his own. Initially he found lodging at the Berkshire, a red-brick boarding house at 204 Columbia Heights in Brooklyn, a tree-lined street of Italianate row and carriage houses overlooking the harbor. Ward paid $4.50 a week for room and board. It was hardly the Packard mansion, but the Berkshire was convenient to the ferries and bridge and featured amenities such as steam heat, an elevator, and private bath and dining, billing itself as an "ideal home for gentlemen and families."

Sawyer-Man recognized Ward as a serious young talent. As he rose inside the company, the firm also sent him out into the field. Within weeks of joining the lamp company, Ward traveled to Philadelphia to attend the International Electrical Exhibition, a "scene of bustle and brilliancy," as the *New York Times* called it, where 1,500 exhibits (including exhibits by all of the major electric light, telephone, and telegraph companies) showcased America's ingenuity. Among the items that caught Ward's eye was an instrument that measured time in the smallest possible subdivision of seconds, presented by the War Department.

In keeping with his more prosaic duties, Ward traveled to Chicago to supervise the wiring of a mill with Sawyer-Man electric lamps. All across the country, electricity was being taken up with great enthusiasm. At the time, buildings had yet to be constructed specifically for electricity, and the invention remained a mystery to consumers, who were often at a loss when something went wrong. Ward and his team of electricians erected poles and wires to distribute electric currents to power the installed lamps. Under Ward's supervision, the Chicago job went without a hitch. "Started lamps at the mill," he noted in cool triumph in his leather Excelsior diary. "Every light works."

Newly promoted, Ward received a visit from his father, whose business interests in lumber and iron stretched well beyond Ohio and often took him to New York. The financial speculator Jay Gould had once summoned Warren Packard to the city for a meeting about his iron and steel mills. Known publicly as the "Wizard of Wall Street" and privately as the "Master of Bribery," among other pejoratives, Gould at one point owned 16,000 miles of the nation's railroads. As America's business rules were being established, Gould earned a reputation for writing his own, consolidating national rail and communications systems, often with rather questionable if not underhanded methods. His attempts to corner the gold market in 1869 ran up prices, leading to a Treasury sell-off and financial panic on Wall Street.

The robber baron had offered to buy a 50 percent interest in one of Packard's enterprises, and in return Gould promised to sell him all of the scrap iron produced from one of his railways. "You can weigh the scrap yourself," Gould told the startled entrepreneur from Ohio in a bid to sweeten the deal. After careful consideration, Packard declined the offer, marking one of the rare instances when someone refused the powerful robber baron.

For their brief reunion, the two Packard men arranged to meet for a celebratory dinner at the Astor House Hotel. After finishing up at the factory, Ward returned to the Berkshire. He cleaned up, changed, and paid one cent to cross into Manhattan over the Brooklyn Bridge. Cutting across Broadway at City Hall, he rushed past St. Paul's Church, the U.S. Post Office, and the offices of the *New York Herald,* picking his way through a scrum of scruffy bootblacks, newsboys, and liveries to the four-story Greek revival hotel. John Jacob Astor, a butcher's son and former furrier who became America's wealthiest man and for whom the term *millionaire* was coined, had built the enormous hotel with more than three hundred rooms. The first in New York City with indoor plumbing, the hotel was renowned for its restaurant. Indeed the Packards dined on the Astor House's famously epic *table d'hôte,* which included frog fried in butter, calf's liver fried with pork, leg of mutton, meadow hens and currant jelly sauce, ten different kinds of game, six different vegetables, duck (black and lake), as well as grouse, oysters, stuffed tomatoes, and lima beans. For dessert there was cranberry pie, almond cakes, and rice pudding. It was a rare luxury for Ward, who had nearly lost contact with life's indulgences, although clearly not the taste for them.

His father's long face with sunken cheeks, a tangle of sideburns that brambled down past his chin, camouflaging his pencil-thin neck, was a most welcome sight for Ward. Since leaving for Lehigh four years earlier, he had scarcely been back to his beloved Warren and he eagerly received news from home. His brother Will continued to apply his more than capable management and salesmanship skills at their father's various enterprises. Carlotta was now at a finishing school for girls in Pennsylvania, and, at two years old, Cornelia Olive, whom her brothers had taken to calling "Kid Packard," had already begun to display a flair for the dramatic.

Following the lavish buffet, Ward accompanied his father on a social call to see Mr. and Mrs. John W. Peale. John had been a year ahead of Ward at Lehigh and since graduating had become a member of the coal-mining company Peale, Peacock and Kerr and the owner of nine Pennsylvania coal mines. The elder Packard had met him in the course of his own extensive business dealings. Ward had little notion how Peale might come to benefit him in the future; viewing the meeting as nothing more than a polite social call, he made a perfunctory note of it in his diary.

With their brief reunion complete, Warren headed back to Ohio and Ward returned to his room in Brooklyn and focused his attention on his new position at the factory.

When Ward became head of Sawyer-Man's mechanical department, he moved not only up the ranks but out of the factory altogether. The promotion brought him to work at the company's general offices in the newly built Mutual Life Building at 32 Nassau Street, an enormous granite and limestone edifice occupying an entire block that stretched between Cedar and Liberty Streets and across to William Street. Sawyer-Man supplied the building's extensive electric system with 1,650 incandescent lamps, requiring eight dynamos.

Just steps from the Stock Exchange, the flamboyantly hulking Drexel Building (J. P. Morgan & Co.'s headquarters, known far and wide as "Morgan's corner") and the Treasury building, whose vaults held millions in gold and silver, occupied the former site of the Middle Dutch Church and was now the epicenter of New York's financial district. The increasingly jagged skyline of moneyed cathedrals revealed the nation's new towering symbols of wealth in a soaring age. "A noble outgrowth of the century," pronounced the *New York Times* in a full-page story that also described Mutual Life's construction as "a triumph of American enterprise." The building was an eight-story dazzler, its lobby lined with onyx columns and constructed from delicately veined white marble. Considered the finest building of its day, the Mutual Life was the most costly as well.

Ward's new work address was located just two short blocks from the jewelry market at Maiden Lane. Crossing Mutual Life's richly carved portico on Nassau Street, he drifted over to the crooked, narrow, cobblestone Maiden Lane, which ran from Broadway down to the East River. Originally a brook, the picturesque passageway came by its quaint name during the period when the Dutch ruled what was then New Amsterdam and young maidens washed clothing and household linens on the spot. In the one hundred years since, the storied pathway (by then filled in by the British) had gone from a food market (with a nearby slave trade) to one of the earliest gas-lit streets, with a covered arcade for wealthy shoppers.

By the time Ward arrived, Maiden Lane was bursting with silversmiths, engravers, and skilled watchmakers as well as specialists manufacturing gold chains and lockets, lace pins, watch cases, colored gold and onyx goods, bangles and bracelets, gold hair mountings, and fancy boxes. According to one count, well over three hundred jewelers and their shops spilled out from Maiden Lane onto nearby Nassau and John Streets, considerably expanding the known boundaries of the city's first jewelry district.

A proper gentleman of the period would not be caught without a fashionable pocket watch, and a proper lady wore a stylish hair clip or locket. Maiden Lane was the place to ogle at and purchase such baubles. "The bride-to-be who could show a ring from Maiden Lane was thrice happy," trilled a popular saying from the period. On Maiden Lane Ward received a proper introduction to fine watches, in particular those from the European tradition.

In the early nineteenth century, as America was gaining its footing both as a nation and a growing financial power, Europe was in economic and political upheaval. Old empires were collapsing, new ones emerging, revolutions and military conflict continued, and the continent's established watchmakers had discovered a lucrative market in the young nation across the sea.

The venerable Geneva house of Vacheron Constantin, who took a Maltese cross as its signature mark, first began supplying America's nascent elite in the early 1800s with timepieces fashionably adorned with enamel portraits of George Washington and Lord Byron. Founded by a Genevese *cabinotier,* Jean-Marc Vacheron, in 1755, the firm had made a name for itself crafting pocket watches of the highest technical and artistic merit for the kings of Rome and Naples. Despite the market for Vacheron's distinctive timepieces, a formal distributorship system between Europe and America had yet to be set in place. Initially Vacheron's craftsmen worked the bench in Geneva and the Jura Valley in Switzerland before sending their finished pieces for export through the commercial ports of Genoa, Livorno, and Venice in Italy, the Hague in the Netherlands, and Caen on France's Normandy coast. A passel of salesmen and representatives met the ships, bringing the watches to Asia and America via dealers and agents.

In the 1820s Vacheron began taking special orders from Americans of means with very specific requests for accuracy and complications. According to the house's ledgers, many of their watches earmarked for America were more expensive than those produced for the European market. "The American pieces you referred to will probably be sold in Genoa and Livorno," wrote François Constantin, Vacheron's partner and a consummate salesman in a letter dated September 24, 1821. "Make sure they are properly regulated." While the letter suggested the need to take the American market seriously, it also revealed a subtle difference between the European and American patron. Long accustomed to extraordinary craftsmanship, the Europeans had adapted to the vagaries of inaccuracy; the American patron, who placed a premium on precision, had not.

As early as 1847, Patek & Co. began selling their wares to the forerunner of Tiffany & Co., Tiffany, Young & Ellis, a stationery and fancy goods emporium, originally located in lower Manhattan at 259 Broadway. Antoine Norbert de Patek, an exiled Polish nobleman and one of the company's founders, sized up the enormous prospects for business that lay on these shores after crisscrossing the country in 1855. Over three months Patek journeyed by steamboat and railroad, subjected to the many perils of travel in mid-nineteenth-century America, including attempted robbery and hotel fires. Yet by the time a somewhat frazzled Patek returned to Geneva, he had a sense of the American consumer. Initially he had found that a broad swath of the country's buyers demanded inexpensive yet accurate watches. But he had stumbled upon another important detail: Americans were just beginning their love affair with luxury goods. The watches of kings and queens, popes and literati were on the cusp of becoming the favored accessories of America's exploding plutocracy. As the upper class turned to Europe for its sophistication and culture, it also turned to its watchmakers to supply these portable symbols of status.

Maiden Lane was awash with watchmakers and agents representing the houses of Switzerland, France, and England. In numerous advertisements, various agents touted the world-class qualities of European timepieces. Both sides found numerous advantages. In many instances, the European houses supplied movements and dials to fit American cases. Private label relationships sprouted up whereby American retailers sold watches inscribed with the European house's name in an exclusive arrangement.

Wandering the length of Maiden Lane allowed Ward to travel horology's globe in less than a mile and surround himself with some of the finest timekeeping instruments he had ever encountered. But though he yearned to own them, he lived frugally on his Sawyer-Man salary, noting his expenditures to the last penny in his diary, where he regularly tracked his miserly living. One week he purchased a pair of overalls at a cost of $1 and in another recorded expenses totaling $7.50. At one point he borrowed the sum of $15.16 from someone named H. W. Robinson in order to pay for his board, washing, and an alarm clock and also jotted down that he faithfully repaid the amount in full in eight days.

It was during this period that Ward apparently came into possession of a handful of Swiss and French watches, including a cylinder alarm pocket watch with a stem wind in a gunmetal case of Swiss manufacture of unknown origin and a stem-winding chronometer of French manufacture, also of unknown origin. They were solid pieces, but hardly exceptional. Interestingly he also came to own an eighteen-carat gold French hunting case with an American-made Waltham movement. It is highly probable that he purchased the gold French pocket watchcase or received it as a gift and converted the movement to the Waltham mechanism, a simple engineering solution for an inventive novice horophile on a budget.

Eventually Ward found himself at Nos. 17–20 Maiden Lane. It was the site of Cross & Beguelin, a wholesale dealer of diamonds and jewelry and importer of fine Swiss and American watches. Alfred F. Cross, an American from Maine, and Henry E. Beguelin, a Swiss émigré who came from a family of watchmakers, founded the firm in 1862. The jeweler was one of the largest on the lane, with a selling floor measuring twenty-five by eighty feet and a basement used for the manufacture and production of jewelry. In a time when commerce retained a sense of chivalry, Cross & Beguelin enjoyed a prominent reputation. "No establishment of the kind in New York today maintains a higher standing in the trade and few receive a larger measure of merited recognition," as one merchants' catalogue described the firm.

Ward made a quick study of Cross & Beguelin's selection. The firm manufactured the hugely popular Centennial chronograph pocket watch in honor of the Philadelphia Expo of 1876 in a variety of sizes with nine to fifteen jewels. They sold the watches of European firms and also produced a line of their own private Swiss-label pocket watches. They also sold watch components. On his visits to the Lane, Ward apparently made the acquaintance of both proprietors as well as of their complications: chronometers, perpetual calendars, and minute repeaters.

Before long, his wanderings brought him to the elegant five-story Italianate building at 63 Nassau Street. With a cast-iron front façade and bas-relief medallions of George Washington and Benjamin Franklin, the building had finely detailed three-story-high arcades occupied by jewelers and watchmakers. Some specialized in polished cases and others in gold chains. There Ward was introduced to one of the oldest and largest watch import agents, the Abry family. In many ways, the Abrys were typical of the early import watch trade in America. Swiss-born Auguste Abry had owned a firm in the French watchmaking center of Besançon before immigrating to New York in 1832. He opened his first shop at 3 Maiden Lane. Auguste's son Jean Auguste took over before his son, Charles Leo, took up the business in the 1870s. Under Charles Leo, the family firm became Vacheron Constantin's North American agent, and he moved the business to the larger space at 63 Nassau. It was there, through Charles Abry and his successor Edmond E. Robert, that Ward reportedly first came into contact with the watches of Vacheron Constantin. As was customary, the men made pleasantries and traded calling cards, although, alas for Ward, while still under the employ of Sawyer-Man, little else. This would change soon enough.

At Lehigh University, Ward was a singular wünderkind on his way. And at Sawyer-Man, he found his true laboratory. Working on a massive scale with some of the country's best engineers, he did more than simply encounter the latest thinking in electric illumination; he seized the opportunity to advance applied electricity. With stunning prolificacy, his ideas rapidly moved from his imagination to the drawing board to prototype and finally to operational status. In 1888 alone, the year that Sawyer-Man built an enormous new six-story lamp and switch factory on West Twenty-third Street between Tenth and Eleventh Avenues in Manhattan, Ward applied for at least eleven different patents under his own name. Each patent included his signature attention to engineering refinement and mechanical detail. Among them were designs for a new lamp socket, a magnetic circuit, and a flashing apparatus to control the flow of vapor or fluid with carbon filaments.

A year later Ward devised an improved incandescent electric lamp that used a "V" stem mount. The innovative flexibility of his design enabled its use with multiple filaments (the filament being the electrical conductor heated by the passage of a current needed to create the light source). Perhaps more important, Ward claimed that his new model was less costly to manufacture than most similar lamps produced at the time.

As the decade drew to a close, Ward was now the second superintendent in charge of manufacturing at the new factory on Tenth Street, a massive building covering thirteen city lots. An article trumpeting the factory in the *Electrical World* singled out Ward and his team "for the care and judgment shown in the carrying out of every plan and detail." In the basement, where the factory's power generators were housed, several large lead conductors were connected to the fourth-floor testing room through a new vacuum pump. Ward designed the device specifically for the manufacture of lightbulbs. Powered by a Westinghouse engine, the small pump effectively extracted the air from each bulb with a ten-by-twelve-inch cylinder, creating an oxygen-free chamber where the filament was housed, which prevented combustion. The process greatly aided in creating a longer filament life and subsequently a longer lasting bulb. Ward's vacuum pump was just one of his many inventions propelling Sawyer-Man toward continued success. By 1891 all but one American light manufacturer employed Ward's vacuum pump.

As the factory churned out ten thousand lamps a day, the company became a force to reckon with. Hundreds of plants and businesses across the country burned incandescent lamps under Sawyer-Man patents. The Park Hotel in Hartford, Connecticut, Okonite's insulated wire factory in Passaic, New Jersey, a resort in Florida, and the County Jail in Brooklyn all awarded their lighting contracts to Sawyer-Man. Choosing eight hundred Sawyer-Man lamps, the Hazelton Art Gallery of Philadelphia explained that it was so disposed to light its floors with the New York company's bulbs in order to "bring out the beauties of the paintings to the greatest advantage." In April 1888 Sawyer-Man's parent, the Consolidated Electric Company, reported that 429,813 lamps were in use—a 38 percent rise in four months.

Inside Sawyer-Man, Ward's influence was everywhere; outside, the Packard name and reputation were on the rise as well. In the fall of 1888, Ward attended the tenth annual clam bake given by the American Electrical Works held in Providence, Rhode Island, a high-flying gathering under the guise of business that quickly dissolved into collegial amusements. The invitees represented some of the most prominent men in the profession, and Ward went as part of the twenty-member New York delegation of the National Electric Light Association. At twenty-five, he was one of the bold, innovative men electrifying America.

The hundreds of electric companies that burst out of nowhere to launch an industry were now fast consolidating into a few conglomerates, each snapping up patent rights like sharks after chum. In 1886 the Thomson-

Houston Electric Company purchased Sawyer-Man, and within three years the firm bought out the Brush Electric Company. Eventually Consolidated Electric sold its controlling stock in Sawyer-Man to Westinghouse.

In 1890, the year that Jacob Riis published *How the Other Half Lives,* documenting the shocking slum conditions in America's cities, Ward was earning a munificent $16 a day (approximately $383 today), working for one of the "big three" of the American lighting industry. He had moved out of the boardinghouse and into a brownstone at 107 Montague Street in Brooklyn. He studied French, consumed literature and periodicals, cycled, and dined with friends, broadening his network of colleagues. As his fortunes improved, so did his visits to Maiden Lane, with always that promise that he would return yet again. Marking his sixth year at Sawyer-Man, he carried on as always: quietly and scientifically, considering every angle from the same vantage point, always several steps ahead of his colleagues.

Public demand for more electricity continued to push engineers to come up with ever more innovative propulsion systems to deliver them. In New York City alone the population had grown to over 1.5 million. Unlike many of his contemporaries, however, Ward had not gone to New York to conquer the city; he was there to master the incandescent lamp. Once he had, he discovered that his interest in Sawyer-Man evaporated. The talented engineer and entrepreneur was simply not cut out to be a company man. At the height of his ascendancy at Sawyer-Man, longing for home and ready to strike out on his own, he decided that it was time to return to Warren, Ohio.

Back in Warren in the summer of 1890, Ward found himself a big-city success in a small town that had grown slightly larger in his absence. Four railroads now connected Warren to the rest of Ohio. There were five newspapers, seven churches, three banks, and one first-class hotel, the turreted Park Hotel hugging the edge of the central square and overlooking the courthouse. Still slow to embrace industrial developments, Warren boasted linseed oil, furniture, barrel staves, wool fabric, and carriage factories, but not an electric generating plant.

On April 22, 1890, Ward received the patent for his incandescent lamp design, the one with the "V" stem mount, conceived while still in New York. Armed with U.S. Registered Patent No. 426,055, he resolved to launch his own manufacturing plant in his hometown. The Packard High Grade Incandescent Lamp would soon transform Ward from an inventor into a wealthy industrialist and Warren from a sleepy hamlet into one of the country's largest manufacturing centers.

By most measures, the Packard Electric Company was a risky venture. With Warren's population hovering under six thousand, the electricity and incandescent light mania had not hit the city as it had others across America. Yet Ward brought more than just ambition home with him. He had a proven track record, great standing in the community, a network of contacts, and, above all, his older brother, William Doud Packard.

The two brothers were like two sides of the same coin. At just five-foot-three, Will was slightly built, like his brother, but while Ward was clean-shaven, Will sported a classic English moustache, slightly curled and pointed upward at the ends. The two oldest of the Packards' five children mirrored each other in intellect, sensibility, and industriousness. From an early age, both had demonstrated an aptitude for all things mechanical. Of the two, Ward was the indisputable mechanical genius. Will was endowed with superior managerial and marketing skills as well as a certain cunning in the art of salesmanship. During his years in New York, Ward had learned that a successful venture needed more than just a sound concept. Almost immediately he laid out his scheme to Will, convincing his brother to join him. W. D. Packard hardly needed much prodding.

As Ward had made his way in New York, his older brother had stayed close to home. As a young boy Will had helped to run the family's mills

Ward's older brother, William Doud Packard. Courtesy of the Summers family.

after school. Following his graduation in 1882 from Ohio State University, he returned to Warren, working as a bookkeeper and salesman at his father's hardware store. When the elder Packard became Warren's agent for the American Union Telegraph Company and the U.S. Express Company, Will wore two more hats: telegraph operator and express clerk. Eventually he took over management of one of the family mills, becoming a full partner in Warren Packard & Son. Joining forces, Ward and Will took up where they had left off as children. While the brothers ably exhibited their capacity for individual success, together they made for a remarkably potent partnership. But first, a skeptical Warren had to be convinced it needed the Packard Electric Company.

The brothers built a dynamo room in the family's mansion and then set up a number of lighting demonstrations for the town. In front of city hall they slung three lights over the street at central points. With a flip of the switch, electricity surged from the dynamo room to the quiet public square. Soon after their demonstration, Will's salesmanship and the family's standing in the community ultimately persuaded an apprehensive Warren to establish its first power-generating plant.

The town remained stubbornly resistant to innovation and manufacturing encroaching upon its gentle city limits, however, and few in Warren grabbed the opportunity to get in on the ground floor as shareholders of the Packard brothers' promising new enterprise. The pair managed to bring in only half a dozen local investors, including the bankers C. F. Clapp and M. B. Tayler, Jacob Perkins, a wealthy landowner and a member of one of Warren's oldest and richest families, and Juston W. Spangenberg, a successful businessman with interests in a machine shop, foundry, sawmills, and steam-traction locomotives.

The Packard Electric Company opened for business in a two-story, forty-by-eighty-foot wooden frame structure with a roof-mounted water tank on Warren's northeastern edge. A bold false-front pediment topped the factory bearing the company's name. There were ten employees. Ward took the title of superintendent and Will became secretary-treasurer. Inside the North Avenue factory, workers began manufacturing the Packard High Grade Incandescent Lamps, advertised as "lamps of all candle power and voltage to fit any socket."

The venture's success outpaced its cash flow. On October 6, 1890, Packard Electric held a board meeting to go over its strategy going forward. An initial stock offering did not attract Warrenites as the brothers had expected. Forced to raise capital or sink, Ward reached out deep and

The Packard Electric Company in Warren, Ohio, circa 1890.
Photograph by Stacy Perman. Courtesy of the National Packard Museum.

wide among the many contacts he had made at Lehigh University and expanded during his years with Sawyer-Man in New York. By January he had recruited a new set of shareholders, and the brothers formed a second concern: the New York & Ohio Company. John W. Peale, the coal magnate, was named president. Ward's nonchalant social call to Peale in New York with his father six years earlier had turned out to be rather fruitful.

The New York & Ohio Company produced incandescent lamps and transformers. All other manufacturing, including arc and incandescent electric lightbulbs, bells, gas lighting, and burglar alarms, was produced under the Packard Electric name.

Packard Electric's products gained notice immediately, earning a reputation for workmanship and quality. One year out, the *Electrical Review* described Packard as "an expert in lamp manufacture and an inventor of note." Unlike those of some competitors, Packard bulbs and lamps did not use cheaper molded glass or second-hand shells. With demanding precision, Ward focused on developing new technology, sparing little expense to ensure that every product that left the factory attained as near perfect condition "as money, long experience, and skilled labor can make it," as one trade journal hailed.

Prudently, the company offered repairs on all of its products. Although America was taking to electricity, the products were relatively new and strange. By setting up one of the earliest customer service departments, the Packards not only guaranteed their products, but they offered their clients assurance in case any problems arose and made sure they didn't have any need to shop the competition.

Having returned to Warren for good, Ward threw himself into building Packard Electric and the New York & Ohio companies. He briefly lived at the Packard mansion but soon built a three-story red-brick building on North Park Avenue adjacent to the Park Hotel. Known as the "Packard block," the building was on the site of the brick house in which his family had once lived under more modest circumstances. Ward built a top-floor apartment for himself and rented out the rest. On the frieze of the building's pediment a stonemason chiseled a lion and the words "PACKARD BLOCK" in the distinctive lettering that would come to be known as the Packard script, later used in all of Ward's business enterprises.

While Ward focused mainly on business, his thirty-year-old brother became engaged to Annie Storer of Boston and took up residence in a large Victorian house with a grand staircase on Elm Road. The two were married on June 11, 1891, and a year later the couple announced the birth of their son, Warren.

In spite of the initial public reluctance, lighting eventually became Warren, Ohio's most important industry, and it would soon manufacture more lamps than almost any other American city. Within two decades, largely due to the efforts of the Packard brothers, Warren (population 11,081) became the first city in the country to install electric incandescent lamps on its streets.

Less than fifteen months after Ward had left New York for Warren, he returned to the city in the fall of 1891 to make good on a promise he had made to himself while still with Sawyer-Man. A sunburst of orange and yellow leaves blanketed Central Park. A light rain had left a wet film of mist on the cobblestones on Maiden Lane. Ward arrived at Nos. 17–20, the jewelers Cross & Beguelin. The success of Packard Electric now allowed him to indulge his passions on his own terms. He chose an eighteen-carat gold minute-repeater chronograph pocket watch manufactured by Jules Jürgensen, one of the most famous watchmakers in the world, originally founded in Denmark in 1740, before establishing roots in Switzerland in 1836. It was said that any serious watch collector "should own at least one Jules Jürgensen." Ward's selection demonstrated his growing sophistication and interest in complicated timepieces. The minute repeater chimed the hours, quarter-hours, and minutes. The chronograph acted as a stopwatch. A sub-dial registered the minutes. Ward paid the not inconsiderable sum of $425 for the watch and an additional $27.01 for a fourteen-carat gold fob. (Together the amount would equal about $10,825 today.)

For Ward, the Jules Jürgensen represented many things. It symbolized his appreciation for technical perfection and craftsmanship, but it also marked his ability to purchase his desires. Soon he would rekindle his acquaintance with Edmond E. Robert, who would shepherd some of his earliest acquisitions of Vacheron Constantin pieces. Ward was now a collector of *grandes complications*. The silent gauntlet had dropped.

~

CHAPTER FOUR

Esse Quam Videri

On the evening of April 8, 1892, Henry Graves, Jr., son of the prominent
Wall Street financier, accompanied his father to the American Art Association's painting sale at Chickering Hall. A faint chill hung in the air that
was perfumed with tobacco and gin as the pair, dressed in black tailcoats
and glossy top hats, alighted from their carriage. A liveried coachman
delivered the men to the red-brick, four-story music hall trimmed in gray
marble on the corner of Fifth Avenue and Eighteenth Street near Union
Square. They timed their arrival to ensure their entrance came just before
the auction's first painting went under the hammer. As a general rule,
haute New York did not arrive early at such events without reason, and
the Graves family were longtime members in good standing.

Inside the music hall's grand lobby the men sliced through a sea of black
tops and tails. Aloof old New York and the latest parvenus made their way
up the marble staircase and into the large auditorium, where, fifteen years
earlier, Alexander Graham Bell had made the first interstate telephone call
over telegraph wires. A few years later, Oscar Wilde had stood on the same
stage to give his first American address, unenthusiastically received, titled
"The English Renaissance of Art." The Knickerbocker crowd clung to their
remaining old associations, which kept the growing ranks of the arrivistes,
with their bourgeois respectability, at a safe distance, but death and taxes
necessitated a civilized process by which to dispose of valued possessions.
Such high-market auctions brought the two parties together, narrowing
the social chasm between them under a tent of money and desire.

Although he preferred to maintain a lower profile, twenty-four-year-
old Henry often escorted his father, Henry Graves, Sr., to art exhibitions
and auctions, where he received an education not only in the finer things
in life but in the art and power of possessing them. Alongside his father,
Henry *fils* also learned the players in the field and how to play them. Over
the years he watched as Henry *père* amassed a well-regarded collection of

paintings that included Corots, Millets, and Rousseaus, as well as a number of watercolors, drawings, sculptures, and *objets d'art* that he stockpiled like vital materials during wartime. Earlier, at an auction held at the Fifth Avenue Art Galleries, Henry Sr. added a bronze leopard made by the acclaimed French sculptor Antoine Louis Barye to his constantly expanding collection.

The two men were quite close and shared a penchant for Old Masters and French landscapes. Among the elder Graves's accomplishments, he was a well-known orchid breeder, whose collection was exhibited around the world and written about in important horticultural journals. Orchid breeding was a rich man's hobby, and on the grounds of the family's estate in Orange, New Jersey, he maintained several conservatories where he raised his delicate, award-winning flowers. As testament to his regard for his son, he named one of his hybrid species of lady's slippers after him. The orchid, *Cypripedium Henry Graves Jr.*, had apricot-yellow petals with rosy dots and an award of merit from the Royal Horticultural Society.

The senior Graves offered his son an impeccable vantage point into the art world, for at the close of the nineteenth century he had become one of the leading collectors of Chinese porcelains in the country. The practice of amassing ceramics produced for the Imperial Court had begun with Queen Mary of England and quickly spread in the seventeenth century among the European aristocracy before being embraced by wealthy Americans. John D. Rockefeller, Jr., once wrote to his father requesting to borrow more than a million dollars to purchase Chinese porcelains, reasoning that he had never squandered money on "foolish extravagances," emphasizing, "A fondness for these porcelains is my only hobby." In 1960 Rockefeller would give the Metropolitan Museum of Art seventy-four of his four hundred pieces that dated from the sixteenth and seventeenth centuries.

Henry *père* acquired countless priceless pieces, including superb examples from the Ming Dynasty, the blue-and-white porcelains made under Emperor Kangxi, and decorated and single-color series. He obtained thirty-five specimens made in the rare peach-bloom glaze alone. Henry *fils* ardently took up the hobby, distinguishing himself by ferreting out some of the most priceless and unusual pieces, at times even upstaging his father, as when he acquired an imperial carved jade vase with dragonhead handles and the exceptionally rare set of eight Buddhist temple emblems carved in white jade.

At Chickering Hall, the sale of seventy-four canvases, largely from the Barbizon school and the Impressionists, including works by Millet and Monet, lured not only the Graveses but collectors from as far afield as

Philadelphia, Boston, and Chicago. Peter A. B. Widener, the Pennsylvania Railroad magnate, and Potter Palmer, the Chicago dry goods millionaire, came to the hall heavy in pomp, purse, and determination. Palmer had single-handedly reshaped Chicago's downtown and created the American shopping experience, establishing "bargain days," "free home delivery," and the principle that "the customer is always right." Also on the hunt was Stanford White, the preeminent architect who had designed the Fifth Avenue mansions of the Astors and the Vanderbilts. He had recently completed work on the second iteration of Madison Square Garden as well as the Washington Square Arch in Greenwich Village. Quietly dispersed among the prominent names sat several dealers with orders to spend forcefully on behalf of those "rich men who did not want their names revealed."

As was his habit, the elder Graves took a seat in the front row, under the hall's enormous electric Roosevelt organ at the edge of the stage. In its survey of the room, the *New York Times* identified the "two most conspicuous New York buyers for their own account" at the auction: Edward Julius Berwind, the coal tycoon renowned for his soirees thrown at the Elms, his Newport cottage modeled after a French château, and Henry Graves, Sr., who made both his fortune and his reputation on Wall Street.

During the previous night's auction, eighty-one paintings worth $64,180 (about $1.5 million today) had been sold, and anticipation ran high on this, the second and final day of the sale. The art world's wealthy patrons expected to be impressed, and the inveterate auctioneer of the American Art Association (the precursor to Sotheby's) Thomas E. Kirby did not let them down, coaxing spirited bidding from the attendees. When Constant Troyon's *Le Passage du Bac* went under the hammer for $27,000 ($646,683 today), the audience burst into sustained applause. Henry Graves, Sr., unburdened his wallet of $17,000 ($407,171 today), acquiring three paintings: Rousseau's *Forest of Compiègne,* an oil panel once owned by the Duc d'Orléans, and two Jules Dupré oils, *Evening Twilight* and *Early Morning,* helping to push the auction's total take to nearly $300,000 (some $7.3 million today). The following day the *New York Times* pronounced, "The prices realized mark it as the greatest art sale in New York in recent years."

The tycoons, moguls, and magnates sitting in Chickering Hall exemplified the men who put the gold in the Gilded Age, in a span of a few years having amassed the kind of enormous fortunes that Europe's aristocracy built up over generations. Europe's riches were tied up in property, but the Americans were rolling in cash, free to spend and burnish their cultural pedigrees with chattel bought from the Continent's impoverished barons. The flamboyant London-based art dealer Joseph Duveen, who had bro-

kered the sale of an astonishing number of Rembrandts and Gainsboroughs from across the Atlantic, accumulated a remarkable fortune of his
own based on the observation that "Europe had plenty of art and America
had plenty of money."

Fin de siècle America's moneyed moguls were not satisfied to acquire
railroads, steel factories, or coal mines. They were out to astonish and one-
up each other, paying enormous sums for paintings, tapestries, sculptures,
manuscripts, and mechanical pocket watches. The steel magnate Henry
Clay Frick and the railroad tycoon Collis Huntington fought over Vermeers, while the bank titan John Pierpont Morgan swept up everything
that caught his fancy. A cartoon from the satirical magazine *Puck* portrayed Morgan on the top of the world holding a giant magnet in the shape
of a dollar sign pulling all of Europe's treasures toward him. The wealthy
Boston philanthropist Isabella Stewart Gardner had amassed so many
objects that she had to build a villa to house her growing art collection.
As a boy of ten, William Randolph Hearst had asked his mother to buy
the Louvre for him; as an adult, the California media mogul procured art
in such vast quantities that he needed several warehouses and a dedicated
staff just to store and preserve his enormous collection.

In an era that would launch the luxury liner *Titanic,* these acquisitions
and their exorbitant prices were nothing less than the extension of the
outsized personalities of those who purchased them. This practice also
expressed American imperialism. Stanford White, who liberally decorated
the palaces he designed with artworks imported from Europe, defended
the acquisitiveness, saying, "In the past, dominant nations had always
plundered works of art from their predecessors. . . . America was taking
a leading place among nations and had, therefore, the right to obtain art
wherever she could." More than significant paintings and bric-a-brac, these
symbols of wealth and status represented a wholesale transfer of Old World
history to a New World whose role was currently being entirely rewritten.

An avid sportsman, Henry Graves, Jr., was an expert marksman, yachtsman, and horseman who rode competitively. His chief prerequisites for
any activity: fast and first. Henry regarded collecting with the same intensity of purpose in which he held his sporting life. A connoisseur in his
own right, he developed a keen interest in Old Master engravings, drawings, and early American and naval battle prints, eventually assembling
more than one hundred exceptional and rare works from Dürer, Rembrandt, and Whistler. All of the objects that he bought spoke to his knowledge, pedigree, and certainly taste.

Henry Graves, Jr.
Photograph courtesy of
Sotheby's, Inc. © 2012.

Henry's impressive collection enjoyed another chief distinction: only a few individuals knew of its existence.

Henry Jr. had broader interests than his father. Henry Sr. loved art and bought huge amounts of it. His purchases were often impulsive and almost always deeply personal. An "altogether extraordinary unity of taste and judgment" was how the *New York Times* once described Henry Sr.'s acquisitions. "The whole collection, in fact, is marked by a private and personal taste rather than by any tendency toward acquiring pictures clamoring for classification with the hundred best." His son's collecting came from a slightly different perspective; if he desired an object, it was because it was not simply the best, "but the best of the best."

The son of a landowning farmer and assemblyman, Henry Sr. was born in Boonville, New York, on December 11, 1838, "of distinguished Colonial ancestry." The family hailed from one of England's most ancient clans, which was said to have arrived with the Norman army. In 1635 his forebears arrived in America and settled in Concord, Massachusetts. In spite of their relatively new money, the family was included in New York society because of their pedigreed past. They possessed a crimson coat of arms decorated with a gold eagle rising out of a ducal coronet with a banner bearing the motto in Latin *Esse Quam Videri*.

In 1859 Henry Sr. went to New York City and became a clerk at the American Exchange National Bank in the financial district. In 1865, at the close of the Civil War, a period when tremendous windfalls were being made on Wall Street and the capitalists and power brokers who created them arose, he formed the brokerage firm Maxwell & Graves at 85 Cedar Street. The Maxwell on the shingle belonged to Henry W. Maxwell (along with his brother, J. Rogers Maxwell, and his son, John Rogers Maxwell). Henry Maxwell was a wealthy banker and philanthropist originally from Scotland, whose largesse was recognized with a bronze Saint-Gaudens bas-relief of his likeness at Grand Army Plaza in Brooklyn, where he lived. Maxwell & Graves were members of the New York Stock Exchange (Henry Sr. served as a governor on the Exchange for twelve years), and according to a finance and banking catalogue of the time, the firm not only afforded "every convenience to their numerous patrons, who number many of our wealthy investors and active operators," but its standing was "of the highest character."

The Gilded Age was a period of vast capital-intensive industrialization, and Maxwell & Graves was one of the shrewd power brokers that sponsored its financing. The firm created an entire series of financial institutions, giving it a reach across every engine of America's economic life. Initially it conducted a general stock commission business, buying and selling for cash, stocks, bonds, and government securities. Before long it invested in railroads in every conceivable corner of the northeastern United States.

The railways joined the distant corners of the United States: East to West and North to South. They hastened the transport of goods and with them the growth of whole towns all across the country. An entirely new and complex management structure developed, and massive labor forces mobilized around them, making the genteel and cunning men who built and owned these industries rich and powerful and propelling the country at top speed from an agrarian laggard into an affluent, industrialized nation.

For the partners of Maxwell & Graves, investing in the railroads also proved lucrative. The firm retained major positions in the Jersey Central Railroad, the Long Island Railroad and its connecting lines, the Brooklyn Trunk Line, and the Delaware, Lackawanna and Western Railroad, among others in their expanding portfolio. Through the Jersey Central the firm virtually owned the Lehigh and Wilkes-Barre Coal Company and installed J. R. Maxwell as its president and Henry Jr. its treasurer. With starting capital of $500,000, they backed the Liberty National Bank, with Henry Sr. named as the bank's vice president. Within ten years, Liberty National tallied up more than $8 million in deposits (worth nearly $200 million today).

In the process, the partners at Maxwell & Graves picked up directorships in the companies they invested in, enriching themselves as they bankrolled the exploits of others. As the firm grew more successful, it moved to bigger offices among the financial district's burgeoning skyscrapers, first at 115 Broadway and then 143 Liberty Street, eventually landing in the fifteen-story building at 30 Broad Street, an address that became synonymous with Wall Street gold.

While the press was in the habit of praising Wall Street for its Midas touch, the men of Maxwell & Graves certainly had a sharp eye for opportunities. Henry Sr. purchased capital stock in the Manhattan Beach Hotel and Land Company, established by the wealthy developer Austin Corbin, the president of the Long Island Railroad. (Corbin also happened to be a rabid anti-Semite and the founder of the American Society for the Suppression of the Jews.) The firm built luxury hotels in some of the favored summer watering holes of the rich—Manhattan Beach, Brighton Beach, and Coney Island—and also built the rail lines connecting the resorts to New York. During the summer season of 1882, the Manhattan Beach Marine Railway earned more than $16,500 (some $368,015 today), carrying 879,327 passengers to the pleasure palaces and amusement parks of Coney Island, a spit of land jutting into the Atlantic Ocean at the foot of Brooklyn. Coney Island was the most spellbinding playground the country had ever seen, with its riot of opulent hotels, freak shows, and electric attractions. There the country's first amusement park roller coaster, the Switchback Railway, based on a coal-mining train, shook the ground, and a steam-powered elevator pulled spectators up a three-hundred-foot-tall observation tower. More than simply a pleasure park, Coney Island was a showcase for the wonders of the machine age.

In the fall of 1891 a Spanish émigré named José F. de Navarro went to see Maxwell & Graves about a loan. Both Graves and the Maxwells wore many hats, and John Rogers Maxwell happened to be president of the Jersey Central Railroad. De Navarro inquired about obtaining $61,000, (roughly $1.4 million today) for his cement company from the Jersey Central. A former banker, the enterprising Spaniard had been instrumental in establishing a number of new industries; for instance, he organized the first line of steamers between New York and Brazil. While developing New

York's first elevated railroads and the famed Navarro Apartments on Central Park South, a gigantic complex of eight buildings resembling a Spanish castle, he became interested in cement, specifically the relatively new rotary kiln process of making cement.

With his two sons, de Navarro founded the Keystone Portland Cement Company. However, due to a variety of circumstances, the economic recession and the cash hemorrhage of his Central Park apartments chief among them, he was short on funds. Initially uninterested in the business of cement, Maxwell relented once he translated cement into freight that would be transported by his railroads. The loan was secured, and Maxwell became the president of the renamed Standard Portland Cement Company of New Jersey, de Navarro its vice president, and Henry Sr. its secretary. Maxwell & Graves became one of its largest shareholders after the de Navarro family. As investments go, theirs in what came to be known as magic powder turned out to be impressive. Rechristened the Atlas Portland Cement Company, it became the largest concern of its kind in the world.

By 1904 Atlas had a market capitalization of $10 million (about $240 million today) and was paying $1.5 million ($35.9 million today) in freight, increasing to $2 million ($47.8 million today) the following year. "I don't believe the Central Railroad of New Jersey has a customer of more importance," de Navarro wrote in a company history for his family. The men of Maxwell & Graves made a windfall, earning buckets of money from Atlas Portland on both the front and back ends. Atlas Portland would be used in the construction of the Empire State Building and Rockefeller Center. Contracted by the U.S. government, the company also supplied eight million barrels of cement for the Panama Canal, which many considered the "eighth wonder of the world." Without the partners of Maxwell & Graves, de Navarro would later say, Atlas Portland Cement Company almost certainly would not have become the colossus it did.

In a period of monopolies, raiders, and traders, Henry Graves, Sr., had more than just a front-row seat at America's economic ascendancy. He played capitalist poker along with men like J. P. Morgan, Andrew Carnegie, Jay Gould, Henry Havermeyer, William C. Whitney, Jay Cooke, and Leland Stanford. And he gave his seat to his namesake son. At the turn of the century, railroads and banking were a skeleton key to wealth, and the Graves family became staggeringly rich.

The Graveses existed in a happily self-contained world. Henry Sr. had married Harriet Isabella Hale, a St. Louis belle, on October 4, 1864. Three years later they left New York and moved to Orange, New Jersey, a wealthy suburban enclave four miles west of Newark in Essex County that was famous for its leafy environs and beautiful homes. "Probably no one who has lived within a radius of several hundred miles of New York City for any length of time has not heard of the Oranges of New Jersey," exclaimed the *New York Times*. Across four acres of the Graveses' homestead they built a stately brick mansion that fronted the entire length of Berkeley Avenue between Highland and Fairway Avenues, just a mile and a half from Glenmont, Edison's Llewellyn Park estate. The couple had four children. In addition to Henry Jr., there was firstborn Edward Hale, who attended the Massachusetts Institute of Technology and was a proficient musician and amateur photographer, and George Coe, an adventurer, philanthropist, and inexhaustible collector in his own right. Their only daughter, the baby of the family, Isabelle, known as Daisy, was a debutante and socialite.

Henry Graves, Jr.'s birth in Orange, New Jersey, on March 21, 1868, three years after Edward Hale's, was the last time that he came in second for anything. Henry was confident, discreet, and pathologically private. He grew up a bon vivant, entitled and cloistered, with the ability to differentiate friends from sycophants, keeping his circle tight. He knew everyone, but as one of his grandsons later described him, he was "not a back-slapper." He demonstrated courtliness in manner and appearance; regardless of the situation, he was always immaculately turned out, dressed in suits and ties, even while canoeing or on a shooting excursion in the woods.

Following his education in New York, which was remarkably undistinguished, Henry began to take up his father's business interests at Maxwell & Graves, acquiring shares and directorships in a raft of banking, commerce, and railroad businesses. As a result of his father's efforts, Henry inherited an economic princedom in the country's banking aristocracy.

The family became part of Orange's nobility and lived in a mansion of walls covered in silk and filled with luxuries and servants. It was a world of engraved calling cards and invitations to tea. There were formal dinners for thirty and elaborate dress balls for three hundred. The society pages of the *New York Times* proclaimed the Graveses' winter ball, held in January 1892, a "brilliant reception . . . one of the most successful private events that has been held in this vicinity this season."

The family dined at the Essex Country Club and regularly attended the horse shows at the Riding and Driving Club in East Orange, with its separate entrances for men, women, and children, and where Henry Jr. was an expert rider, serving on the club's executive committee along with his brother Edward. On Sundays the family attended the Brick Presbyterian Church, where Henry Sr. was an elder. Father and sons all held lifetime memberships in the New England Society of Orange. An entire *New York Times* column, "Society in the Oranges," was devoted to chronicling the goings-on of its most prominent residents. The Graves family made notable appearances, and their fashions, luncheon companions, and party decorations were keenly documented. The family marked the seasons by geography; winters and spring usually meant the Oranges, midseason was spent at Palm Beach, and summers Maine or Europe.

In June 1894 the family spent the summer season on the Continent. They sailed aboard the SS *Teutonic,* comfortably ensconced in the liner's first-class cabins, shining like jewels on the high seas. These steamer crossings had become a great transatlantic trade in money and prestige. In Europe the Graveses attended operas and art exhibitions. While Henry Sr. and his sons met with art dealers, Harriet and Daisy picked their way across Parisian couturiers. In carriages the family passed grand colonnades and manicured parks and peered into shops knowing that nearly everything displayed was theirs for the taking.

It was still a couple of decades before World War I, which would reduce the streets and rows of perfectly terraced buildings to rubble. In the summer of 1894 and into the fall, men and women crossed the boulevards with a lightness of step. Increasingly Europe's land-rich, cash-poor nobility put their heirs along with the family patrimony up for sale. In 1895 alone, nine resplendently wealthy American heiresses, known as the "dollar princesses," married members of England's impoverished aristocracy, trading their family's money for a European title. The passel of brides included Anna Gould, the youngest daughter of financier Jay Gould, who married Paul Ernest Boniface, the Comte de Castellane,* while the Marshall Field heiress Mary Leiter married Lord George Curzon, later appointed viceroy of India. The Graves family, however, hardly needed to engage in such transactions.

Upon the family's return from Europe, Henry became acquainted with Florence Isabelle Preston, the twenty-five-year-old daughter of the late William Riley Preston, a wealthy commodities broker, and his wife, Rebecca (née Duncan). They became engaged at the end of November 1895. Florence was not a great beauty, but she was attractive in the conventional Victorian style, with a soft round face, creamy complexion, and

---

*Gould and Boniface divorced in 1906, after he plowed through $10 million of his wife's money. In 1908 Boniface sought an annulment from the Vatican. Granted in 1924, it was overturned the following year, after Anna Gould appealed the decision.

large brown eyes, and she wore her soft cascade of brown curls tucked at the nape of her neck. She spoke in clipped cadences and long drawn-out vowels, as did Henry, as if their jaws had been locked into place with a gold money clip. Florence had a bright smile and a sense of fun that complemented Henry's own sharp sense of humor.

On the occasion of their engagement, a luminous miniature portrait of Florence painted on ivory was commissioned. In the oval-shaped painting, rimmed in gold, Florence wears a dreamy expression and is wrapped in a delicately embroidered, gauzy blue dress. A strand of gumball-size pearls drapes her neck, a smaller string circles her wrist, and a single pearl rests on her ring finger.

The seventh of ten children, Florence was bred to conform to decorous tradition. As a young girl she studied at Mrs. Sylvanus Reed's English, French, and German Boarding and Day School for Young Ladies in New York City, a selective and rigorous institution for the delicate daughters of the social set. At Mrs. Reed's, Florence gained "unequaled advantages" in languages while immersed in the white-gloved refinements of the day. Hers was a life of pretty dresses, servants, and tea dances.

In the winters the Prestons lived at her childhood home on Forty-eighth

Florence Preston Graves's miniature engagement portrait, painted on ivory. The couple married on January 21, 1896.
Photograph by Stacy Perman.
Courtesy of Gwendolen Graves Shupe.

Street just off Fifth Avenue. During the summers the family repaired to their country pile in Irvington-on-Hudson, an estate that stretched between the Croton Aqueduct and the Hudson River. The Millionaires' Colony, as Irvington was known, was thirty miles north of New York City and had become a favored destination of wealthy industrialists and tycoons like William Rockefeller and Charles Lewis Tiffany, who went to escape the oppressive summer heat and frequent fever outbreaks in the city. In Irvington they erected mansions overlooking the east bank of the Hudson that made even Europe's royal castles appear simple in comparison.

Like her fiancé, Florence grew up in a world of elaborate rituals of manners and concealed emotions. Their romance may have been kindled over sly glances in silk damask ballrooms, although it was more likely to have been brokered to guarantee that influence and money remained locked in the tight sphere of their social milieu. In any event, their engagement represented the enviable ideal of material and social interests. "At least 1,500 cards have been sent out," heralded the society column of the *New York Times,* reporting news of the nuptials, "and the wedding will attract to the church many people of social prominence."

A careful reading of society's hieroglyphics indicates that the Preston family's standing suffered little lasting damage from an incident that had occurred fourteen years earlier. New York had become the country's main wheat, cotton, and flour market. It was a city of merchants, and the merchants had become gamblers. At the Produce Exchange, a respectable casino, perhaps second only to the Stock Exchange in activity, some $10 million (roughly $223 million today) a day traded hands over the price of commodities like butter, cheese, hops, resin, flour, turpentine, cornmeal, and barley among men called scalpers who made "puts" and "calls" on behalf of buyers and sellers. Before his death, Preston and his partner Janvier Le Duc had speculated heavily in cotton, wheat, and lard on the Produce Exchange. In February 1882, W. R. Preston & Company, the firm that he had founded nearly two decades earlier on 66 Pearl Street, failed to make good on the Exchange's margin call on existing contracts worth some $400,000 (the equivalent of $8.9 million today). When the price dropped and the margin was called in, the company informed the Exchange that it was "unable to meet our engagements." Unfortunately for W. R. Preston & Company, even the most respectable of gamblers are capable of making a bad call.

William Preston's sagacious and stolid reputation and that of his firm became the focus of disgrace and gossip. "Few subjects have recently arisen in the Exchange concerning which so much general interest has been felt," reported the *New York Times* financial correspondent covering the fiasco. The Exchange suspended the firm, and though the deposits were eventually paid back, questions over the extent of the company's liabilities remained. The Preston family was spared from abject ruin, but the firm collapsed and its reputation was permanently damaged.

At the hour of their nuptials, however, Florence's bloodline overshadowed any whiff of scandal. The Prestons owned something the moneyed classes always desired: background. The family descended from King Edward III and counted as forebears eleven barons who served as witnesses for the Magna Carta in 1215. Despite any taint from her father's past business dealings, Florence had the kind of exalted lineage that would give even Mrs. Astor pause. America's "aristocracy" rested on the truism "The older the money, the bluer the blood," and the Prestons traced one of the oldest and grandest family trees, which stretched all the way to Emperor Charlemagne.

The wedding took place after the customary six-week engagement, on January 21, 1896, at 4 p.m. sharp. A pageant of perfectly turned-out guests gathered at St. Thomas's Protestant Episcopal Church on Fifth Avenue and Fifty-third Street. Nestled among the mansions of haute society, the church was the preferred location for society weddings. A year earlier, St. Thomas's had been the scene of Consuelo Vanderbilt's marriage to Charles Spencer-Churchill, the ninth Duke of Marlborough, certainly the most spectacular union between one of America's "dollar princesses" and a member of Europe's impoverished peerage. In exchange for her new title as duchess, Miss Vanderbilt brought a $2.5 million dowry (worth $64.6 million today) to the marriage.

Waiting at the altar, Henry stood rigid in his tailcoat, flanked by his brother Edward, serving as best man. (Within three years, the two would switch places at Edward's wedding to the socialite Jean Stevenson.) The rector, Reverend Dr. John Wesley Brown, hovered in front of the church's famed Saint-Gaudens bas-relief *Reredos*. With an imperial rustle the Graves and Preston families made their way toward their assigned pews and the six manicured ushers assembled for the well-choreographed procession of money and manners.

First came Florence's four bridesmaids, carrying white roses and wearing white mousseline de soie over white satin with little picture hats trimmed in pink velvet and black feathers. Next the two flower girls, carrying baskets of lilies of the valley and white hyacinths, walked up the aisle. Resplendent in a white silk gown trimmed with silver and embroidered gauze, the maid of honor Daisy Graves sauntered before the bride, who appeared clutching white carnations.

Florence took the arm of her elder brother, William Duncan Preston, and

a faint pink washed over her cheeks. Carrying a bouquet of lilies of the valley, she appeared to float up the entire forty-three-foot length of the nave,
where her bespoke bridegroom awaited her. In the days to come, the details
of her dress, like all of the wedding's particulars, would be remarked upon in
the charmed upper echelons of society and parsed by ordinary folks whose
only glimpse into the world of the idle rich came from the society pages.
Florence did not disappoint. She wore a gown of white satin with flounces of
point lace on the front of her skirt and clusters of orange blossoms and frills
of lace on the bodice. A point appliqué veil enveloped her court train and
was secured with a diamond crescent brooch, a gift from Henry.

The ceremony unfolded like a familiar evening at the opera. The vows
were said, the benedictions made, and the ring placed on Florence's finger.
With their arms linked and their families entwined, Mr. and Mrs. Henry
Graves, Jr. walked the length of the nave together to the Mendelssohn
march. Through the church's imposing arched doors that ushered in the
blustery winter evening, they emerged into their future.

A dull rain notwithstanding, following the ceremony the guests in all
of their finery repaired to a "small" reception at the bride's family home
on Fifty-seventh Street and Park Avenue. The newlyweds broke with tradition, most likely at Henry's urging, and chose to shun the lavish ritual of
showcasing their wedding gifts. When Florence's brother William married
Annie Fargo, the daughter of the Wells Fargo founder and president of the
American Express Company, the pair had made a great show of their silver tea sets, hammered silver salad bowls, and copper lamps. The splendid
scene was recounted in the social pages, which described their spread of
wedding gifts as "costly and numerous."

As Florence soon learned, the only thing her new husband liked to
display was a flagrant disregard for such grand spectacles. Then again, it
wasn't the material objects that Henry objected to but the hanging out in
public of his private life.

The newlyweds began their married life in the social whirl of New York
and the Oranges. Henry belonged to the right gentlemen's clubs, the
Metropolitan Club and later the Racquet & Tennis Club, while Florence
belonged to the right organizations: the Daughters of the American Revolution and the National Society of Colonial Dames. She also busied herself at formal ladies' lunches held at the city's finest hotels. Florence sat on
several charitable committees, including the American Red Cross, involving herself in a number of volunteer activities. In between she pulled linen
and silk from a small box, crafting dainty needlework. At the appointed

hour of respectability, 5 p.m., Henry drained cut-crystal glasses brimming with one of his preferred liquids: Corney & Barrow Scotch or William Penn Rye, Three Star Hennessy Cognac or Cockburn London Dry Gin.

The finest tailors and seamstresses made the couple's wardrobes, and when they desired a piece of jewelry or bauble their usual destination of choice was Tiffany & Co. Henry had long been a frequent visitor to the "palace of jewels" at 15 Union Square West. Entering through the shop's cast-iron façade painted to resemble stone, he perused the store's black walnut counters and ebony cases filled with opera glasses, pocket watches, cloisonné, silver, gold, and all manner of expensive bric-a-brac. In addition to receiving royal appointments from the queen of England, the shah of Persia, and numerous other crowned heads, Tiffany's had become the prestige supplier to America's burgeoning plutocracy, accommodating their every whim. The Graves family had long been one of the tony jeweler's regular clients. Henry generously gifted his wife with expensive trinkets and jewelry, and Florence wholly embraced these talismans of society womanhood. On her dressing table rested an array of gold-topped, cut-crystal bottles and sets of sterling silver hairbrushes. The golden lid on one of her crystal perfume bottles was set with a large ruby surrounded by a ring of five diamonds and four four-leaf clovers made of emeralds.

A devotee of Tiffany's famous monogrammed stationery, Henry Jr. regularly chose a cream-colored paper of thick stock with an embossed letterhead in blue for his private correspondence. One could quickly ascertain where they stood in Henry's affections from the way he addressed his letters. A dutiful *Dear* denoted a perfunctory relationship, while a *My Dear* expressed greater fondness for the recipient. Similarly Henry's closings articulated three degrees of fondness: *Sincerely, Sincerely Yours,* and, most warmly, *Yours Very Truly.*

For the Graveses, Tiffany & Co. offered exclusivity, something that the celebrated little shops spilling out all over Maiden Lane did not. While James Ward Packard flitted about at Cross & Beguelin's or at the Abrys' at 63 Nassau Street, poring over chronometers and minute repeaters, the Graves men turned to Tiffany's to purchase their pocket watches. The jeweler's long yellow ledgers inked their names in a sober grandiosity dating back decades alongside society luminaries, recording the numerous timepieces purchased by Henry Sr., Edward, Henry Jr., and George Coe, many of them fine utilitarian pieces, although not particularly complicated watches. In these early years, Henry was not yet an impassioned or practiced horophile. He was a young man with money, taste, and the good sense to buy his gold watch at New York's address of *tout le beau monde.*

On fashionable evenings, Henry and Florence attended the opera, not the smaller, threadbare Academy of Music where Old Money spurned the nouveau riche, but uptown at the Metropolitan Opera, built by the city's new millionaires, led by Alva Vanderbilt, on Broadway and Thirty-ninth Street. Inside the cavernous red and gold opera house, the couple attended performances of *Faust* and *Tannhäuser*. Sitting in their private box on the enviable Grand Tier (one level above the famous Diamond Horseshoe), the couple spent as much time, perhaps more, taking notice and being noticed than they did taking in the opera. On Sundays they read the Bible at church, but it was the *Social Register*, where the couple's names were enshrined, that they observed most closely during the rest of the week.

At home, Florence managed a staff of maids, cooks, butlers, gardeners, and laundresses, an entire coterie of help imported from Ireland, England, Norway, and Sweden. They tended the grounds, horses, and carriages and eventually the couple's four children.

Almost a year to the day of their wedding, Florence gave birth to their first child, Henry Graves 3rd, on February 4, 1897. A second son, Duncan, was born in 1900, and a daughter, Gwendolen, in 1903. Their last, George Coe II, named after Henry's adventurously independent younger brother, arrived in 1905.

On July 25, 1900, Henry purchased a magnificent twenty-seven-room manor with a colorful history called Shadowbrook in Irvington-on-Hudson. Spread across ten wooded acres and bordering the Croton Aqueduct, the property was once part of the homestead of Washington Irving, the grand man of American letters who wrote the classic tale *The Legend of Sleepy Hollow*. Long after his death, Irving's fanciful Dutch-style house, Sunnyside, remains standing just south of Shadowbrook, separated by a pathway and a gurgling stream.

Thomas Thornton, a wealthy Irishman of unknown origins, built Shadowbrook in 1816, along with a carriage house and separate quarters for the servants and groundskeepers. Unable to settle on one particular architectural style, Thornton chose them all, merging Renaissance, Baroque, French classical, Victorian, and Gothic Revival elements all together. Every corner of the manor overflowed with ornate details. The floor of the entryway was made of mosaic tiles, the walls in paneled oak, and the ceilings embossed in tooled leather. There were carved marble fireplaces, burled-wood paneling, bedrooms that peaked up through gabled roofs, and lacunar ceilings. Elaborate archways appeared to come straight from the Alhambra Palace. Undoubtedly the manor's *pièce de résistance* was its

Shadowbrook, the Graves family's ten-acre estate in Irvington-on-Hudson, was once part of Washington Irving's homestead. Courtesy of Cheryl Graves

Greek Revival rotunda, with a domed ceiling ringed in Corinthian columns and Tiffany windows, a lute and laurel motif, and perfect acoustics. According to local legend, Washington Irving drank his last glass of wine in the rotunda while on a neighborly visit the night he died, November 28, 1859. Returning home to Sunnyside, he reportedly said, "Well, I must arrange my pillows for another weary night! If this could only end!"*

*Shadowbrook's rotunda would go on to witness further episodes of history. In 1926 the composer and lyricist Irving Berlin, the son of Jewish Russian immigrants, arrived at the mansion with his new bride, the Roman Catholic heiress Ellin Mackay, with whom he had just eloped, to visit her mother and stepfather. Mackay was the daughter of Clarence Mackay, head of the American Posts and Telegraph Company and a known snob and anti-Semite, and the marriage left the two estranged. During the newlyweds' visit, Berlin presented a wedding gift to his wife: the sheet music to *Always*, assigning the copyright and all royalties of the song to her. Within fifty years, the song earned Mrs. Berlin some $1 million, and later daughter and father were said to have reconciled.

Forty years later, in 1966, the great tenor saxophone player Stan Getz and his wife, Monica, purchased Shadowbrook. The acoustics in the rotunda (and the showers) sold Getz on the house. "They're fantastic, they're the only place I feel inspired to sing," he said. "One of the best gifts I ever gave myself was this room."

Henry paid $125,000 (about $3.2 million today) for this castle-like mansion along the Hudson River. His great good fortune in snapping up the estate came about because of the scandalous divorce of its most recent owner, Charlotte C. Graves (no relation). The daughter of a Civil War general had flagrantly entertained her lover, J. Hamilton Jaffray of Yonkers, at the home, enraging her husband, the wallpaper manufacturer and heir Richard Graves.

Henry and Florence made the vast mansion's interior their own, stuffing it to fashionable excess with furnishings and *objets d'art*. They favored important French and English period furniture and acquired Louis XV commodes, Queen Anne desks, George III writing tables, and carved George II winged armchairs. Thick, priceless Oriental and Persian rugs covered the floors. Velvet drapes hung from the arched doorways, held back by wide cords of silk, and richly colored tapestries were drawn across elaborately carved frames. Rooms reverberated with the elegant tick-tock of Henry's antique long-case clocks. Each niche and every available space were crammed with his rare Chinese porcelains. On the walls hung his Old Masters, rare etchings, and prints in hand-carved black and gold leaf frames. The interior had the look of a very expensive, very exclusive curio shop.

As was the custom, Henry and Florence maintained separate master suites. Henry's room was outfitted in large mahogany Chippendale furniture, while his wife's bedroom suite reflected her taste: the graceful, elegant lines of William and Mary furnishings with elaborately turned legs in the shape of trumpets and spirals.

Nearly every personal item they owned carried their monogrammed initials. Henry's gold cufflinks and the family's silver tea sets were stamped with the Graves family crest. Even Florence's leather coin purse was studded in small gems and gold hinges, bearing her initials, also in gold.

A brigade of nannies and nursery maids in starched hats cared for the Graveses' four children. The boys wore dark suits and learned to bow, while their sister wore white gloves and silk dresses with bows in her hair and learned to curtsey. They ate apart from their parents until they were thirteen, when they joined their parents in the formal dining room.

Outside of the imposed formality, the children had the run of Shadowbrook's rolling manicured lawns and English gardens with yew hedges and marble fountains. Henry imported two long-haired Irish goats, one black and the other white, for his children's amusement. They put the little creatures to work, pulling a small carriage around the grounds. In the winter, when the property's lakes froze solid, the entire family skated on the ice.

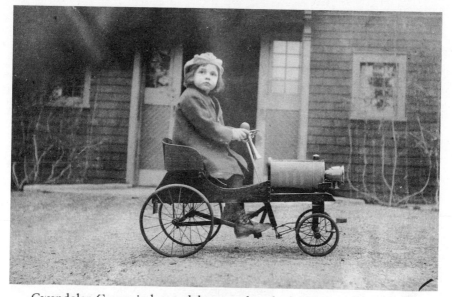

Gwendolen Graves in her pedal car, made to look like one of her father's automobiles, circa 1909. Courtesy of Cheryl Graves.

After Henry purchased a German Daimler automobile, he had miniature foot-pedal versions made for his children, which were kept alongside their father's cars in the carriage house, where the chauffeur kept them polished and oiled just like Henry's own vehicles.

The Graveses' rhythm of life in Irvington revolved around two standards: grand and grander. For Henry, that meant fast and faster. The family belonged to the Ardsley Country Club, where founding members like William Rockefeller and J. P. Morgan sailed up the Hudson from New York and slipped their yachts at the club's private dock. Its founders arranged to have the train station at Ardsley designed to match the neo-Elizabethan clubhouse. Unlike at the opera, in Ardsley the robber barons and the Knickerbockers came together over competitive sports, golf dates, and dinner dances and were convivial. In 1911, when the Sleepy Hollow Country Club opened, the Graveses were part of the elite inner circle offered membership. Its clubhouse once belonged to the mansion of Commodore Vanderbilt's granddaughter Margaret.

At the clubs Henry enjoyed a range of competitive sports, sharing his own interests with his sons. In particular, his youngest, George Coe II, displayed exceptional skill as a marksman, winning the Sleepy Hollow's trap-shooting contest at just eight years old.

The society couple:
Florence and Henry Graves,
Jr. Courtesy of Cheryl Graves.

Florence and Henry lived opulently. Their lives, like their silver and Henry's cuff links, were characterized by the motto engraved on the Graves family coat of arms: *Esse Quam Videri*—To be, rather than to seem.

CHAPTER FIVE

∾

Mr. Packard's Horseless Carriage

On Christmas morning in 1899, Will Packard's spirited seven-year-old son, Warren, bolted out of bed, flying down the stairs of the Packard mansion on High Street. His father, Uncle Ward, and his aunts joined him downstairs. In the Packard family, Warren was the object of much attention and affection. Four years earlier, his mother, Annie Storer Packard, had died unexpectedly, following a liver ulceration. The boy spent much of his time spoiled by his grandmother at the Packard mansion, while Will, who had married Catherine Bruder (a waitress he met at the Park Hotel, where he took his meals following his wife's death), traveled the globe on behalf of Packard Electric.

Outside, the cloudless winter sky pressed against the gable-roofed mansion. It was bitterly cold, and snow filled the streets. Inside, an infinitely warmer scene unfolded. Shiny, bow-wrapped boxes sat at the base of a tall and lovely Yule tree, catching the hazy morning light that streamed through the windows on tiny beads. Warren amused his grandmother, Mary, informing her that Santa Claus had brought the festively decorated tree to the house, but he knew that Uncle Ward had delivered his favorite present, a small engine that ran on alcohol.

Almost eight weeks earlier, Ward had delivered an even bigger gasoline-powered engine to the streets of Warren, Ohio: the first Packard horseless carriage.

With Packard Electric's growing success, the inventor's curiosity and interests had begun to drift. Having conquered steam engines and incandescent lamps, Ward began experimenting with internal combustion engines. Unlike a steam engine, where the fuel burns outside to produce steam, an internal combustion engine generates motion when a tiny amount of fuel is ignited within a tightly enclosed space. Considerably smaller, with fewer applications than a steam engine and more volatile, noisier, and dirtier than electric power, the internal combustion engine as

Ward the yachtsman, navigating Lake Chautauqua.
Courtesy of the Summers family.

a new power source opened up new possibilities. It was the perfect new challenge to rouse Ward's imagination, just as the electric lamp had done.

Ward's fascination first took the shape of a twenty-five-foot, four-horsepower naphtha launch. By the time he purchased the boat from the Gas Engine & Power Co. in February 1892, he was quite prosperous. The launch and its accompanying fittings depleted his accounts by $1,095.50, (more than $26,000 today). Like the rest of his personal fleet, he kept his new prize at his family's property on Chautauqua Lake, often taking his sisters out for glides on the water while dressed in his crisp yachtsman's uniform.

Soon Ward's gasoline-powered pleasure cruises turned into science experiments as the genteel inventor began tinkering with the engine to unlock any imperfections and secrets it might yield. Stealing a quiet moment in August, he slipped open his slim leather diary and aridly recorded, "Took launch engine apart and cleaned valve parts."

The idea of a horseless carriage increasingly occupied Ward's thoughts. In the drafting room at the New York & Ohio Company in Warren, Ward sat on a tall stool in front of a long wooden table. With his loose shirt-sleeves pulled up and held in place by dark armbands, he immersed himself in his latest obsession: how to build a motor powerful enough to propel a carriage across largely unpaved roads at a rate of speed faster than could be obtained with horsepower.

German and French engineers had long developed reliable gasoline-fueled carriages. The Europeans had also advanced the concept of wide-spread paved roads. Ward, the ardent tinkerer, was determined to design a motorcar of his own and in 1893 drew up blueprints for a carriage and negotiated to purchase a motor from Charles King. Originally a pneumatic toolmaker from Detroit, King had earned a measure of fame for building boat engines and some of the earliest car engines in the country. As their plans got under way, however, the economy began to falter and Ward was forced to put his plans on hold for a time.

After a period of extraordinary economic expansion, driven largely by railroad ventures and speculative financing, the economy cratered, setting off a chain reaction. Banks failed and unemployment rose. President Grover Cleveland turned to J. P. Morgan to borrow $65 million (a sum worth more than $1.5 billion today) in gold to shore up the U.S. Treasury's shrinking reserves and save the gold standard. Packard Electric became entangled in a lawsuit brought against the company over Edison patents owned by the General Electric Company, which had since acquired the Edison Company. For a time, production stopped at the Warren factories, and in their place the Packards established a Canadian manufacturing subsidiary, the Packard Lamp Company Limited, in Montreal. For the better part of nearly two years, Ward traveled frequently between Warren and a Cleveland courtroom. The legal fight dragged on in some fashion for years, ceasing only when the patents expired. While Ward's gas-powered adventures undoubtedly provided a measure of distraction from his immediate troubles, when he looked forward he saw the passing of one era and the beginning of another. America was about to enter the Age of the Automobile, and Ward had no interest in being a bystander.

Caught up in their legal woes, Ward sent his brother to the Chicago World's Fair, also known as the Columbian Exposition, which celebrated the four-hundredth anniversary of Columbus's discovery of the New World. Will arrived in Chicago on October 29, 1893. The Packards' New York & Ohio Company was exhibiting in the Great Hall of Electricity. Host to the last World's Fair of the nineteenth century, Chicago was a cultural touchstone offering more than utopian razzamatazz; it represented a celebration of the power of consumerism and America's growing business class. Technology at the World's Fair in Philadelphia, the Centennial Expo, had been a spectacle to marvel at from the sidelines, but in Chicago technology was viewed as a reflection of human progress and aspiration.

The Westinghouse Corporation had won the bid to illuminate the Exposition, making them victorious in the "war of currents." Competing against the newly formed General Electric Company, Westinghouse undercut its rival's million-dollar bid by half. On the evening of the Fair's opening, President Cleveland pushed a button, and more than 200,000 bulbs transmitted by Nikola Tesla's alternating current system lit up the fairgrounds of the White City.

Packard Electric stood in the middle of it all. The company announced itself with an illuminated sign on which *PACKARD LAMPS* burned from five hundred Packard Six Candle–Power bulbs. No longer the wide-eyed spectators they had been seventeen years earlier in Philadelphia, the Packard brothers were now successful exhibitors at the world's first all-electric fair. Packard Electric's lamps, cables, fuse boxes, and transformers stood alongside the world's best examples of power plants and dynamos, as well as the Edison Tower, a stunning column made from thousands of colored lamps and topped by a huge incandescent bulb jutting up from the General Electric exhibit.

Inside the Great Hall of Electricity, a movable sidewalk ferried people about the six-acre structure. Among the many pleasures on display was an entire house outfitted with electric appliances, including lamps, irons, stoves, burglar alarms, and fans. There were machines for sewing and machines to do the laundry. As one of the Fair's guidebooks noted, "There will be no occasion for lighting a match in it for any purpose whatsoever." Edison had unveiled the phonograph at the Paris World's Fair in 1889 and promised an even more impressive spectacle in Chicago four years later. The Wizard of Menlo Park may have lost the "war of currents," but he still knew how to cause a stir: he introduced the modern world to his Kinetoscope, a motion picture projector.

Over 184 days, 28 million people—roughly a third of the American population—braved the blistering heat and torrential rains to make their way through the fairgrounds' two hundred buildings. Exhibits included the public debuts of Cream of Wheat, Juicy Fruit gum, and Aunt Jemima's maple syrup. In the Manufacture Building, Patek Philippe was invited to sit on the jury for horology for the first time, but it was the machine-made "American method" that set new standards in watchmaking. For five cents, fairgoers could use one of the thirty-two "better class" bathrooms installed by the Clow Sanitary Company. And visitors gladly paid what was popularly described as the criminal price of fifty cents—the cost of the entry ticket to the fair—for a whirl on an astounding new recreational conveyance, the Ferris wheel. Running alongside the fairgrounds, a mile-long avenue known as the Midway Plaisance featured a model Eiffel Tower and reproductions of Ireland's Blarney Castle and a Javanese village. Outside the fairgrounds, a young piano player named Scott Joplin entertained the crowds, toying with a new syncopated rhythm called ragtime.

Will's visit coincided with the final two days of the Fair. He was on something of a mission. With precious little time to view the Expo's many wonders, he headed straight to the massive Transportation Building at Jackson Park. The striking building designed by Louis Sullivan, with its polychrome exterior and fantastically detailed arched entry, had a 165-foot cupola reached by one of eight elevators that also carried passengers to the various galleries. Entering the ornately carved "Golden Door," the diminutive Ohioan found his bearings. Great Britain's pavilion stood to his immediate right, France's to his left. Will picked his way through the hall's interior naves and window arcades, which flooded the building with light, giving it the look of a Roman basilica. While most of the visitors crowded into the wildly popular show of bicycles, Will strode in the opposite direction, angling away from the saddlery exhibits and toward the small space set aside at the building's rear. There he found the horseless carriages.

Four years earlier, the Universal Exposition in Paris had put on an impressive show of steam- and gas-powered vehicles. A three-wheeled French tricar, from the firm Serpollet-Peugeot, and the German engineer Gottlieb Daimler's lightweight four-wheeler fueled by petrol had caused a sensation, spurring French automotive engineers to greater innovations. Yet in 1893 such mechanized contrivances elicited little excitement in America. Nearly every town possessed stables and blacksmiths; hundreds of companies engaged in the manufacture of carts, carriages, and wagons. Bicycles were the latest transportation fad to capture America's attention, and it was bicycle owners who agitated legislators for better roads. Inside the Transportation Hall, enormous American-made steam locomotives loomed above the fairgoers like the Egyptian pharaohs at Aswan. Going forward, the Exposition presaged a world transported by trains, American-made trains.

The Packard brothers saw things differently. Will hurriedly grabbed the opportunity to view the first and only gas-powered motorcar on American soil: the 1.8-horsepower Daimler. Much at the display was crude and

unreliable, with some models no more than a two-seated buckboard powered by an electric motor. Will gave little of his attention to the electric
vehicles, among them products of three American companies, including
William Morrison's battery-powered surrey. It was the shiny tiller-steered
German Quadricycle with wire wheels that had lured him into the building. For those paying attention, the vehicle proved to be quite a sight.
Yet the Daimler hadn't even rated a mention in the exhibition's catalogue,
and few newspapers had reported its existence in their extensive coverage
of the monumental Exposition. Will made a close study of the machine,
relaying his observations to his brother upon his return to Warren. It
would not be the last time he performed such a role on Ward's behalf.

Then again, Will was not the only visitor who studied the Daimler.
Long before the Columbian Exposition had closed its gates and receded
into history, Henry Ford, an engineer from Dearborn, Michigan, who
worked at the Edison Illumination Company, had come to Chicago. So
had Charles Duryea, an innovative mechanic from Springfield, Massachusetts. Like Will, they also left Chicago utterly fascinated.

Within months of his return to Massachusetts, Charles Duryea and his
brother Frank bought a second-hand horse carriage and outfitted it with
a one-cylinder engine and three-speed transmission. On September 22,
1893, Frank drove what he called his Buggyaut at top speed, five miles
per hour, down six hundred yards of a Springfield street. While Charles
remarked that the car "ran no faster than an old man could walk," the Buggyaut was America's first gasoline-fueled automobile. Two years later the
brothers launched America's first car company, the Duryea Motor Wagon
Company. Other American mechanics and engineers quickly followed
suit, and an industry was born. In November 1895 a periodical called the
Horseless Age made its debut, "published in the interests of the motor vehicle industry." By then the country had just four automobiles capable of
making any kind of a successful run, and one of them was imported from
Germany.

For American automakers 1895 would prove to be a watershed year. The
Chicago Times-Herald provided the catalyst, sponsoring the first American auto competition. Billed as the "Race of the Century," the newspaper
offered $5,000 in prize money, generating a great deal of publicity for the
fledgling industry while boosting newspaper sales. Many viewed the automobile, unlike the phenomenally popular bicycle, as nothing more than
a passing fad.

The race almost didn't get off the ground when an early blizzard hit Chicago in November. Several participants never showed up. Others among the ninety registered contestants missed the deadline for entry because their hand-built cars were not completed in time. Chicago police stopped two competitors en route, forcing them to hire horses to pull their vehicles, telling the gentlemen that they had no right to drive such conveyances on city streets. After several delays, the race commenced on a severely cold Thanksgiving Day. Originally planned to cover a ninety-two-mile round trip from Chicago to Waukegan, a ten-inch snowfall the morning of the race forced organizers to shorten the route to nearly half that.

In the end, just six participants made it to the starting line. Chugging through the wet and icy course at an average speed of 7.5 miles per hour, the Duryeas' motorized wagon took first place. Its driver, Frank Duryea, clocked in at ten hours and twenty-three minutes. The race galvanized the public and established the automobile as a technology to reckon with. Countless mechanics began their own inventions, and before the year was over more than five hundred various motor vehicle patents had been filed.

Following the great race, Ward revived his own motorized escapades in earnest. Pulling out his old drawings, he returned to his drafting table, added a subscription to the *Horseless Age* to his favored reading material, and renewed his research. On January 15, 1896, Ward and Will journeyed fifty-five miles southeast to New Brighton, Pennsylvania, to the Pierce Crouch Engine Company to have a look at a three-horsepower motorcar owned by Dr. Carlos Booth, a Youngstown, Ohio, doctor.

Dr. Booth had helped design the vehicle (Crouch built its motor) and claimed it was capable of running up to sixteen miles per hour on level surfaces, climbing hills of 15 percent grades, and driving through six inches of mud at five miles per hour. Booth had hoped to enter the *Chicago Times-Herald* race but had not completed the machine in time. Instead he used his "cab," as he referred to the vehicle, to make house calls, earning him the distinction of being the country's first physician to use an automobile to make his rounds. He reluctantly gave up the practice, however, after receiving an avalanche of complaints from Youngstown residents who protested that his car "made a commotion among the horses." Visiting New Brighton, the doctor graciously invited all comers to see his car and experience the pleasures of "motoring with driving."

Around the same time that he examined Dr. Booth's "cab," Ward hired a draftsman and pattern maker named Edward P. Cowles to work at the factory in Warren. While employees at Packard Electric and the New York & Ohio Company churned out bulbs and dynamos, Cowles was given a salary of $12 a week to work on a new motor wagon.

In Europe on Packard Electric business, Will couldn't help but notice that the automobile was in wide use there. Particularly in France, motorized vehicles jostled with horse-drawn carriages. Europe was not Will's favored destination. Over the course of his many trips there he had found the Continent to be rather uncivilized. On an earlier visit, he wrote to his mother complaining, "The 'Great River' Thames is nothing more than a canal." When it came to European automobiles, Will found them to be "rather crude cars." Yet despite his censorious appraisal, upon his return to Ohio he arranged to have one of Europe's best horseless carriages, the French De Dion-Bouton motorized tricycle, sent home to Ward.

An odd little machine mass-produced in Paris, the De Dion had a single-cylinder, .75-horsepower gasoline engine mounted above the front wheel. The vehicle's main mechanics also consisted of a carburetor, a high-voltage ignition with a spark plug, and a timing wheel. In a road race against four-wheeled vehicles between Paris and Marseille, the De Dion, bearing Michelin tires, performed rather heroically, coming in fourth.

Although initially excited, Ward quickly turned on the troublesome little De Dion. It was hard to start, made a racket, was hard-riding, and, among a litany of technical faults that he leveled at the machine, it had poor lubrication, a bucking carburetor, and an unforgivably faulty ignition. After thoroughly picking apart and dissecting the motorbike, Ward made several considerable refinements. First, he transformed the tricycle into a quadricycle and added a new passenger seat, making it something of a hybrid between a motorcycle sidecar and a dentist's chair on two wheels. He overhauled the motor, building a single-cylinder, air-cooled vertical engine geared directly to the back axle. Much to the distress of Warren's residents, the quadricycle became a familiar sight as Ward piloted the machine on hell-raising runs for miles without stopping.

Over the course of nearly a year, Ward wrung every last bit of mechanical information out of the De Dion. Once he had made every possible improvement to its performance, he discarded the French motorbike like a pair of old shoes, giving it to his nephew Warren. The quadricycle was just the latest toy in a swelling list of his uncle's mechanical castoffs that Warren would come to inherit.

In Michigan, Henry Ford had built his own motorized, four-horsepower quadricycle. On June 4, 1896, Ford took the machine out for its maiden voyage on the streets of Detroit and not long afterward sold it for $200 ($5,170 today). That same year, a twenty-two-year-old inventor named Ransom E. Olds terrorized his hometown of Lansing, Michigan, driving the first Oldsmobile, which fumed and screamed down the street and "suggested infernal origin," at 3 o'clock in the morning. Over in Flint, Michigan, a high school dropout and one-time cigar salesman named William C. Durant bought a road cart business with $2,000 ($51,707 today) in 1896 and within four years made the Durant-Dort Carriage Company the leading manufacturer of horse-drawn carriages. This brought Durant to the attention of a Scottish-born engineer named David Dunbar Buick, and by 1904 Durant was manager of the Buick Motor Company. Within another four years Durant had founded General Motors. Many others went on to try their luck in the motor business.

In the summer of 1898 Ward traveled to Cleveland, where his brother had gone for several long stretches. The older Packard's eyesight had been failing him, and he sought treatment at Caulfield Hospital. Sometime after the death of his first wife, Will had contracted syphilis, which was said to have caused him severe vision problems. To help himself cope, he built an exact replica of his Warren mansion, down to the placement of the doorknobs, in Lakewood, allowing him to get around his country estate with ease. As his eyesight worsened, he hoped to find some medical treatment.

During Will's stays in Cleveland, both of the brothers had become acquainted with Alexander Winton, the owner of the Winton Bicycle Company and a pioneering automobile manufacturer. A year earlier, Winton, a Scottish émigré with an infamous temper, had produced two single-cylinder automobiles and shortly thereafter incorporated his Winton Motor Carriage Company. Built by hand and brightly painted, the automobile had padded seats, a leather roof, and gas lamps and rode on B. F. Goodrich tires. In 1897 Winton set the world record for the mile. The following year, he tested his ten-horsepower model around a Cleveland horse track, attaining the astounding speed of 33.64 miles per hour. Quashing any lingering skepticism about his vehicle's speed and reliability, that same year Winton took his motorcar on an 800-mile endurance run from Cleveland to New York, making the trip in ten days.

The Packard brothers had already discussed Winton's successes before Ward arrived in Cleveland the morning of July 22, 1898, to investigate the car for himself. Alexander Winton personally took Ward out for a test drive, which left him positively disposed toward the vehicle. He wrote in his diary, "To Cleveland tried Winton Motor carriage—fine price $1000.00, [equal to nearly $26,000 today]—immediate delivery."

After several weeks in which he was confined to his hospital room, Will finally took his own turn at the tiller on August 4. George Weiss, a principal investor in Winton and the organizer of Cleveland's first automobile club, accompanied him. Already Winton Motor Carriage was well on its way to becoming the largest manufacturer of gas-powered automobiles in the country. In advertisements Winton touted its cars' many attractions: speeds of three to twenty miles per hour, handsome styling, and construction. Most of all it proposed an entirely new road to the future. "Dispense with a horse and save the expense, care and anxiety of keeping it," read one ad. "To run a motor carriage costs about ½ cent a mile."

On August 13, Ward returned to Cleveland and took delivery of his brand-new car. Of the twenty-one Wintons sold in 1898, Ward's car was number thirteen.

As it turned out, it was unlucky number thirteen. The car took nearly eleven grueling hours to cross sixty-five miles, the return journey proving not only grindingly slow but also treacherous. The open-topped vehicle hobbled sluggishly through the clammy Ohio summer, broke down frequently, and, just three miles from Warren, needed to be towed in by horses. Upon hearing of the episode, Alexander Winton was so unsettled that he dispatched his own foreman, William "Bert" Hatcher, to Warren the following Monday. Hatcher stayed nearly a week repairing Ward's car until it was up and running again.

Afterward Ward's faith in his Winton was restored, and he decided to show off his new horseless carriage, taking one of his Packard Electric shareholders on a twenty-mile joyride to Bloomfield. In his diary he pronounced the drive "great." Whenever the "motor," as Ward took to calling the machine, was idle, he parked it in the stable behind the Packard mansion, arranging to have Mr. Burns, the stable master, look after it as he did the family's horses. Throughout the fall Ward enjoyed his Winton, thrilling local businessmen with rides. In the summer he expertly piloted the vehicle to Lakewood, where he took his mother and sister Carlotta for drives to nearby Jamestown.

During his visits to the Winton factory in Cleveland, Ward had taken an interest in George Weiss, whose intelligence and confidence he admired. Not only did Weiss possess a successful track record in the fledgling motorcar business, but he had demonstrated his ability to handle the volatile Scotsman Winton. After purchasing his Winton, Ward struck up a correspondence with Weiss. In November Weiss and his wife drove to Warren and paid Ward a personal visit. In his letters to Weiss, Ward alluded to his interest in entering the motor carriage business and his desire to partner with Weiss in the future enterprise. "I must go up sometime this week on business and hope that you will be at home," he wrote in one. "I am anxious to talk 'horse' with you."

As the New Year approached, however, Ward's enthusiasm toward his "motor" began to dim considerably. A temperamental machine, the Winton continued to present a variety of problems both big and small, and Ward made numerous trips to Cleveland to confront its inventor over them. On several occasions, he was forced to leave the car behind and return to Warren by train. Between November 1898 and May 1899 he wrote up a veritable charge sheet on the Winton: the batteries needed replacing, the engine was noisy, the ignition was unreliable, and the radiator could barely contain water. Then again the expanse of crude roads posed their own challenges to the car's sturdiness.

On June 10, 1899, with months of disaffection behind him, Ward once again charged into Cleveland with his troubled Winton. During this visit to the factory, which would be his last, he is said to have challenged the tetchy automaker with a catalogue of the motorcar's difficulties and weaknesses and to have offered a list of suggested improvements. Annoyed to the point of rage, Winton turned to the slight fellow standing before him and rather infamously fumed, "If you are so smart, maybe you can build a machine of your own." To which Ward swiftly replied, "I will."

~

CHAPTER SIX

Ask the Man Who Owns One

Returning to Warren following his confrontation with Winton, Ward immersed himself in the New York & Ohio Company. Inside the factory he carved out a twenty-square-foot area for the designing and building of a horseless carriage, with a small corner set aside as an "experimental space" for assembly and testing. The Winton had given him the final push to make a serious attempt on his own, convincing him that he could build a much better car than any available.

His mother agreed, and in February 1899 Mary Packard invested $600 (about $15,500 today) in her son's new venture. In March Ward sent his brother on a four-month excursion to Europe "to make a special investigation" of the continent's automobiles. Tasked with looking into the mechanical traction of the latest Daimler and Benz automobiles in Germany and the graceful chassis techniques of the French manufacturer Panhard et Levassor, Will was expected to write up a detailed report on his findings. Awaiting word from his brother, Ward wrote to George Weiss in April, "I believe that this report would be of some value to anyone contemplating starting into the business here. It is a branch of work which has a very great fascination for me and it is not impossible that I may go into it someday."

After Ward formally asked Weiss to work for him, things moved swiftly. Within weeks the two men had hashed out their plans. On June 17 Weiss sold his 170 Winton shares for $12,000 (some $310,247 today), raking in a profit of $7,000 (about $180,977 today), and traveled to Warren. Nearly two weeks later, Ward and Weiss agreed to enter into a partnership, Packard & Weiss, and convinced Winton's top mechanic, Bert Hatcher, to work for them. Born in South Bend, Indiana, Hatcher had reached only the eighth grade, but his lack of formal education did not prevent him from quickly resolving technical problems. Hatcher's work on Ward's troubled Winton nearly a year earlier had made a lasting impression. The pair offered Hatcher a salary of $100 a month and $5,000 worth of stock, along with the opportunity to purchase more shares if the nascent company moved forward in business.

Hatcher proved a gifted draftsman too. On July 7 Hatcher submitted his first drawings for the new Packard carriage and Ward promptly put him in charge of the drafting room. Ward and Hatcher ran the mechanical operations, while Will Packard and George Weiss managed the venture's financial and business dealings.

Ward applied himself with the rigor and discipline of a scientist and the eye of an artist. His mechanics in their coveralls and heavy boots gathered in the corner space set aside at the New York & Ohio Company factory, and Ward, dressed in a suit, oversaw them, working long and punishing hours on Job 357, the first prototype: the Model A. Ward's experience with his Winton and the De Dion-Bouton had allowed him to see how to make numerous points of improvement, but his main preoccupations remained speed, durability, and beautiful lines. Influenced by the graceful French carriages he had studied, Ward envisioned an elegant automobile. Above all, he was adamant that his Packard did not appear to be "a carriage in want of a horse."

With Ward at the helm, ideas were floated back and forth in the musty air, and then bit by bit they were interpreted on paper. The men considered the results from design, adaptability, and manufacturing perspectives, with every detail offered up to severe scrutiny. Ward's own experience in manufacturing transformers and generators at Packard Electric provided great understanding of how to insulate a high-tension electrical current. Still, the Model A team faced a daunting set of engineering challenges, from steering and braking to transmitting power from the engine to the rear set of wheels. Even petrol was difficult to come by. Only grocers sold gasoline, and most kept little more than five gallons on hand. The Packards built a fifty-gallon tank in a small house near the barn to store their own supply.

Experimental designs were put forward as the men unfurled drawing after drawing. Eventually they agreed upon an initial prototype based on a ram's horn–shaped front frame supported by leaf springs. A one-cylinder, nine-horsepower engine mounted horizontally under the driver's seat would power the vehicle. In the factory corner, the automobile that gradually had evolved in Ward's mind was built.

Ward insisted on using only the highest-grade materials that would stand up to the rigors of the road, but finding suitable ones became a perpetual struggle for the engineer. "Early manufacturing," he would later recall, "was one damn thing after another." Allergic to the idea of cutting corners and convinced that French foundries were superior, Ward imported gray iron cylinder castings for his engines, as well as leather upholstery. In the Packard shops fiery sparks blew off metal as bolt, nut, and curve was each uniquely quenched, bored, grinded, and welded to make the first parts and tools that would build the prototype motorcar. Exhaustive tests followed, then technical modifications, and then more tests.

Dissatisfied with the kind of steel that was commercially available to make his gears, Ward bought costly armor plate steel sheets. The steel proved so hard that it regularly ruined the tools required to turn out the gear wheels, and it was increasingly difficult to find matching grades. To remedy the situation, Ward and his team intensively researched the heat treatment of metals and created an entirely new process. Adapted broadly, it was credited as one of the "greatest contributing influences in the present mechanical age."

In November Ward constructed a new shop, separate from the New York & Ohio Company, specifically for the motorcar. After several tests and adjustments, the first engine measured in at an underwhelming 7.1 horsepower, well under Ward's desired nine horsepower. More tests and adjustments were ordered until Packard's "No. 1 Carriage" met Ward's exacting criteria.

The first Packard, the Model A, took her first run on November 6, 1899, in Warren, Ohio. Courtesy of Special Collections, Lehigh University.

• • •

The maiden voyage took place on the chilly morning of November 6, 1899, the day after Ward's thirty-sixth birthday. The sky hung low and gray over Warren, and the ground resembled a quilt of snow, slush, and chunks of hard-packed ice. Made of varnished black wood, the vehicle dubbed the Ohio Model A came equipped with an impeccably fitted single-wide seat of tufted, buffed leather and a fine leather dash, although it had no doors. It deployed a tiller lever for steering and a set of heavy flywheels that gave it an even torque. Ward had to consider all eventualities, and just in case the car might not perform under its own power and required a tow-in by horses, the automobile featured a whip socket.

Born of Ward's imagination and the sweat of his mechanics, the Model A was unlike any horseless carriage yet built. For all of the car's elegance and style, the first Packard came with a number of inventive features that would become industry standards. It boasted two forward speeds and one reverse gear made possible by sliding a belt drive. An automatic spark advance transmitted to the rear wheels through a chain, and another mechanism provided a constant spark, regardless of the engine's speed. An accelerator foot pedal, one of the first, controlled engine speed. A cooling system was built with a recirculating pump, and the radiator was capable of holding four gallons of water. Ward designed the impressive H-slot gearshift, and until the automatic transmission was introduced in 1939, every car manufacturer deployed this ingeniously simple design.

Over the next twenty years the automaker's various models would earn some 454 patents and forty-one trademarks under the Packard marque, introducing numerous innovative firsts, including the steering wheel, front and rear bumpers, and double windshield wipers. The Packard would also feature the first sun visors, glove compartments, and air-conditioning and would go on to make four-wheel brakes standard equipment.

As the mechanics wheeled out the Model A, their exertion blasted puffs of breath into the biting air, and the frozen ground crunched beneath their work boots. It was said that Bert Hatcher began the ritual of pushing switch buttons and cranks, and the engine crackled to life with a sputter. He thrust the clutch into the first gear, and the vehicle lurched forward and began smoking and rattling its way toward 2 High Street, reaching a top speed of about twenty miles per hour. The sight of the high-sloping auto caused quite a sensation on Warren's quiet streets, and residents rushed out to see the commotion. By one reckoning there were at least a dozen runaway horses. Red-cheeked from the cold, Ward observed the event from the sidelines; his nephew Warren also watched as the Packard chugged past, proving its mettle, ably hauling through slush and ice and stopping only when Hatcher threw the switch that signaled the engine to stop.

Following the Packard's victory lap, Ward pithily wrote in his diary, "First carriage got running." The event blazed across the front page of the *Warren Tribune* with more excitement and some exaggeration. The machine, reported the paper, "proved satisfactory in every particular. It was expected the car would make 30 miles an hour and it can easily go 35 miles. The successful completion of the machine will probably mean a factory for the automobiles in this city."

Indeed, following the test run, Packard & Weiss officially became the Automobile Department of the New York & Ohio Company. Five days after Christmas, Will opened the tidy ledger where he had recorded all expenditures originating from the Packard & Weiss partnership over the previous six months and tallied up Job 357. Excluding Hatcher's salary, the direct costs to build the first Packard came to $2,189.72 (equivalent to more than $56,000 today), nearly double the annual income of the average American working in manufacturing at the time.

Within months, the newly established marque turned out five additional vehicles with improvements in design and functionality. Will took delivery of the official "first" Packard, after the prototype; numbered A-21, it had a lower front axle and a stylish dashboard with a brass rail and a compartment to hold the crank, tools, and the driver's personal items. George Weiss received A-22. On February 3, 1900, George D. Kirkham, an owner of the Harris Automatic Press Company in nearby Niles, Ohio, paid $1,250 (a sum worth more than $32,000 today) for Packard A-23. In May three of Kirkham's associates bought the remaining Packards. By September Packard & Weiss had sold a total of eleven automobiles, six of which were the new Model B's.

At the turn of the century, America had more blacksmiths than doctors and more bathtubs than telephones. Only 30 percent of homes and businesses had running water, and barely 15 percent of all households had flush toilets. There were roughly eight thousand cars, 144 miles of paved roads, and a competitive auto industry taking shape. Between 1895 and 1902, 128 pioneering auto firms emerged on the scene, the majority of them clustered in the Midwest. A fervid trade-in market appeared as rival automakers exchanged just about anything for an automobile, taking in horses, horse-drawn buggies, even saddles and harnesses. A Packard competitor once told Ward, "We could allow a couple of thousand dollars on a second hand wheelbarrow as a trade-in and still make a profit on our car."

In the beginning these woefully undercapitalized automobile companies vied for a limited pool of funding and an even more select group of clients. Financial tycoons from Huntington to Vanderbilt continued to put their money in railroads, and the bankers and investors followed suit. The automobile remained a greatly misunderstood creation, rarely considered anything other than a racing vehicle. Automobiles were a rich man's game, and newspapers relegated coverage of them to the sports pages. An automobile within economic reach of the average American wouldn't appear until 1908, when Henry Ford debuted his Model T, which, thanks to the assembly line, reduced the sticker price to an affordable $825 (about $19,764 today).

Few had yet to divine a world in which the automobile would be a viable form of popular transportation. As the first Packards rolled out of Warren in the year 1900, the U.S. Census Bureau listed the manufacture of automobiles under the category "miscellaneous." Slightly more accommodating, the august *Literary Digest* presaged a limited future for the motorcar, predicting that, while the cost of the horseless carriage would inevitably fall, "it will never, of course, come into as common use as the bicycle."

Forty-five years earlier a segment of the population had not seen a future for the railroad either. In 1845 a Lancaster, Ohio, school board refused to lend its building for a debate over the feasibility of the railroad and telegraph, announcing, "If God had designed that His intelligent creatures should travel at the frightful speed of fifteen miles an hour, He would have clearly foretold it through His holy prophets. It is a device of Satan to lead mortal souls down to Hell."

In the early trying days of manufacturing, Ward faced numerous obstacles, many of which were raised in his own hometown. At the close of the nineteenth century, the proudly residential burg demonstrated no more enthusiasm for the automobile industry than it had for steel at the century's start. Residents made it difficult for mechanics hired to work for the Packards' fledgling automobile manufacture to find housing. The town council passed a law that forbade workmen from carrying their lunch pails down Mahoning Avenue. Any and all entreaties to improve Warren's roads were met with genteel outrage. If they were good enough for wagons and carriages, expressed the civic minders, then they were good enough for automobiles.

Hazards were everywhere. Scornful farmers routinely placed wooden boards with nails and broken glass on the roads. Mud holes were regularly nurtured. On a drive, any man at the tiller had to remain vigilant around low-hanging tree branches. Packs of barking dogs bolted after the cars in hot pursuit, nipping at their wheels—basically oversized bicycle tires made of a single tube filled with a mixture of glue and feathers to prevent punctures. During test runs Ward sent his nephew Warren to fend them off, armed with small glass balls filled with ammonia. When the dogs scrambled for an attack, Warren threw a ball onto the ground, releasing a stinging, pungent odor into the air, which discouraged the animals from advancing any farther.

In fact the boy had a variety of functions. At times, while the Packard team tried out new models, he had to jump out of the cars as they descended a hill and hurry to place a brick under the rear wheel. As Warren later recalled, on many occasions when the single chain-drive to the rear axle broke while going up a hill, he scampered out and became the vehicle's human emergency brake while it was repaired.

As the Packard evolved, so did its testing facilities. A system of pulleys and shafts tested the cars under a variety of loads, speeds, and periods of time. Before a vehicle left the factory, a tachometer was attached to its engine to measure its distance through speed and rate of revolutions. Hydraulic dynamometers gauged horsepower.

Ward insisted on putting his vehicles through relentless tests. Mindful of the poor state of America's roads, he believed it was important to ensure that his cars could withstand the punishing conditions that the average driver encountered every day. On a specially built, rough-and-tumble test track, each Packard was subjected to a series of arduous trials. Cars were expected to climb a steep dirt incline covered with loose gravel, and then descend it, traversing ruts, detours, and sharp turns. Test drives were done under every conceivable condition, tracked across sand traps and water puddles, and during frigid winters when ice accumulated under the fenders. Packards slogged through thick mud puddles and pushed on even as hail and rain thumped angrily against the cars' metal body. After they were subjected to all manner of abuse, the vehicles were studied to examine the impact on transmission, rear axle, and drive shaft. The cars were then dismantled and the interior parts studied. A Packard wasn't just built; it was built to last.

The invention of the steering wheel resulted from a nasty spill that Will and Warren took when Will lost his grip on the unsteady tiller while motoring to Lakewood. The pair was sent airborne and dropped in a cornfield.

At a time when dependability was still a goal to reach for most automobiles, the *Horseless Age* bestowed high praise on Ward's strenuous efforts with the Model B. "Solidly built to endure high speeds on rough roads," the journal reported in its May 16, 1900, issue, "and workmanship is thorough and first class." Pleased, Ward and his team were now ready for even greater recognition. They had much to prove.

On the afternoon of May 18, 1900, Ward left Cleveland in his Model B, piloting a rough and raw route through Cuyahoga County in just four hours (not including a rest break for one hour and fifteen minutes). According to the *Warren Daily Tribune,* the 1,600-pound cruiser "stood the test in fine shape." Rather remarkably, the car consumed only three and a half gallons of gasoline. "Mr. Packard carried along an extra can of gasoline," reported the paper, "but it was not needed."

Eight days later, as dawn broke on May 26, George Weiss joined Ward in Cleveland, and the two traveled 125 miles east to Buffalo, New York. Reaching speeds of up to twenty-two miles an hour, the men arrived in thirteen and a half hours in the Queen City, where Lake Erie meets the head of the Niagara River. More important, they managed to make the entire

Ward takes one of his Packards for a spin with its "new-fangled foreign contraption," the steering wheel. Courtesy of Special Collections, Lehigh University.

journey without a single accident, breakdown, or confrontation with any of the horse-drawn carriages they passed along the way. The only episode occurred at their hotel, where the two men, hair matted and coated with grime from head to toe, gave the hotel staff a fright. The receptionist took one look at the pair and told them there were no "suitable rooms to let."

Not all of Ward's driving trials were as successful. Once, while he was en route to Niagara Falls, a torrential rainstorm turned the roads into a sea of mud. The automobile came home in a boxcar. This, however, was the exception rather than the rule. Satisfied with his motorcars' progress, he believed the Packard was now ready for a wider public.

Preferring to leave the salesmanship to others, Ward remained buried in Warren with his research and experimentation and sent George Weiss to New York City to represent the Packard motorcar at the first National Automobile Show, held at Madison Square Garden during the first week of November 1900. Ambitiously energetic, Weiss had shipped four Packards ahead of him: two standard Model B's and two of the latest special edition Model C's. The Packard's "new-fangled foreign contraption," as the steering wheel was called, made its national debut at the Garden.

The weeklong car show presented the latest in automotive technology from electric-, steam-, and gasoline-powered vehicles, which the *New York Times* called an "uninterrupted whirr of horseless vehicles in great variety." Packard joined thirty-one motorcar companies making a showing, among them Winton, Baker, Duryea, Riker, the American Automobile Company, Oldsmobile, and Daimler.

The steep ticket price, fifty cents, did not appear to keep the crowds away. Forty-eight thousand spectators watched in their bunting-draped seating galleries as the parade of cars engaged in feats on the specially built oval track. A martial band played. During the timing competitions, motorists vied to see which engine could ignite the fastest. There were obstacle courses and contests for accelerating and braking. A wooden hill built on the Garden's roof caused quite a commotion as visitors gasped at the remarkable sight of a gas-powered motorcar traversing a graded hummock. In a bit of egalitarian showmanship, on ladies' night the fairer sex was allowed to try their hand at a lap around the oval.

New York's upper class came out in force. Men like John Jacob Astor IV, William Rockefeller, and William K. Vanderbilt turned out in white shirtfronts and tuxedo coats to view these "curious contrivances." The New York audience warmly greeted the arrival of an alternative to hansom cabs if only to help alleviate the estimated 450,000 tons of horse manure and 15,000 horse carcasses that required removal from the city's streets each year.

While the Packard made a respectable showing in the brake contest, the company shone most brightly in demonstrations outside of the official events. Shrewdly, Weiss displayed two of the Packards outside of the Garden and talked up reporters, telling them, "We have no million dollar factory, but are turning out thoroughly practical road vehicles for delivery, which is more than many who are making big claims are doing." William Rockefeller happened to pass by during one of Weiss's presentations. Impressed by what he saw, the Standard Oil cofounder bought two models straight off the exhibition floor, the first of many Packards that the tycoon would own. Hollis Honeywell, a Boston Brahmin, purchased the third Packard. All in all, Packard made a spectacular showing at the Garden.

Four days after the Auto Show closed, Packard established a dealership in New York City at 114 Fifth Avenue. By January the company had orders for thirty-three cars worth $40,000 (more than $1 million today) in sales. Eight months later, Ward, Will, and Weiss reorganized their partnership into the Ohio Automobile Company, completely separate from Packard Electric and the New York & Ohio Company.

Leveraging his momentum and recognizing the importance of broadening recognition of the Packard marque, Ward participated in a series of endurance runs. At the time, lack of reliability came to define the automobile in the public's mind. Wanting to demonstrate the superiority of his motorcars on a wider national platform, Ward entered a six-day contest sponsored by the Automobile Club of America. During the second week of September 1901, eighty vehicles would drive five hundred miles of seldom traveled and notoriously ruinous roads between New York City and Buffalo, where the Pan American Exposition was under way. With the finish line at the Expo, the contest was primed for maximum publicity.

A total of five Packards entered the race. Ward drove the first, along with George Weiss; Bert Hatcher and Charles Chaffee, a Packard family chauffeur, drove the second; A. L. McMurtry, owner of the New York City Packard dealership, piloted the third car; and two private Packard owners each drove the final two. Of the eighty vehicles entered, none were electric, twenty-six were steamers, and only five were European.

On September 9 the drivers cranked their engines at the Fifth Avenue starting line to raucous cheers. At 8 a.m. sharp, the starting pistol was shot and the vehicles began a thrilling run. Once they cleared the relative luxury of New York's paved streets, however, the drivers faced a host of difficulties. Motorists got lost and cars broke down regularly. Just thirty miles out, in

Tarrytown, the sand was ten inches deep and concealed dangerous rocks that badly jolted the vehicles. The drivers also had to contend with pelting rain, steep grades, and debris while navigating around many of the stalled and abandoned cars that cluttered their path like animal carcasses in the desert. On day 3, while freewheeling down a hill, Hatcher turned abruptly and pulled his rear tires off their rims in the process. By day 4, one of the race's most famous entrants, John Jacob Astor IV, decided he'd had enough; he relinquished his car to his chauffeur and fled to the comforts of home.

As Ward and the other participants hurtled through the tiring final leg, President William McKinley was assassinated. An anarchist named Leon Czolgosz shot him in the chest and abdomen as he attended a reception at the Pan American Exposition's Temple of Music on September 6. The previous day McKinley had delivered a speech in which he said, "Expositions are the timekeepers of progress. They record the world's advancements. They stimulate the energy, enterprise, and intellect of the people, and quicken human genius." Eight days later he died from his injuries.

Owing to the tragic event, the organizers chose to cut the race short, ending it in Rochester. The Packard cars had all turned in remarkable performances. Even Hatcher had been able to quickly recover his rims and press on. Overall Ward turned in the second fastest average speed; Hatcher came in third. Four Packards earned first-class certificates for averaging twelve to fifteen miles per hour. Of the eighty vehicles that started the race, only half survived, but all five Packards made it to the finish line.

Standing head and shoulders above all of the endurance runs, it was the first transcontinental race, from San Francisco to New York, that Packard foreman and test engineer E. T. "Tom" Fetch piloted in 1903 that most captured the nation's imagination. At the time, the entire country had less than two hundred miles of paved roads; there were no gas stations or road maps, no road signs, and no passable roads anywhere west of Chicago. No one had taken such a journey in a horseless carriage.

The race began on May 23, 1903, when Dr. Horatio Nelson Jackson impulsively took on a $50 bet to drive his twenty-horsepower Winton on an odyssey that would take him three thousand miles. Just weeks into his trip, Jackson's whim became a national sensation. Fetch convinced Ward that he should get in on the run, arguing that a trip like this could prove that "self-propelled carriages" were no longer experimental toys but utilitarian vehicles. Sharing the young foreman's views, Ward had a Packard Model F, a 1902 Runabout, shipped by rail to California. Fetch nicknamed the twelve-horsepower car "Old Pacific."

On June 20, 1903, Fetch peeled out of the Cliff House in San Francisco, following an old Union Pacific railroad map, the only published guide of the period. He rode on thirty-four-by-four tires inflated to eighty pounds of pressure and topped twenty miles per hour. Given their late start and the number of other marques jumping in, Packard chose to gain maximum publicity for Fetch's adventure by having him take the most demanding route that included the steepest passes in the Sierra Nevada. As his passenger, Fetch brought Marius C. Krarup, a reporter for the magazine *Automobile,* to document and photograph every treacherous mile.

At points, Fetch had to unfasten and then refasten fence lines to continue on. He lay down canvas strips over muddy paths in order to get traction. They reached Carson City, Nevada, a silver-mining town, shortly after a murder had been committed, and so shocked the town that everyone, the sheriff included, left the crime scene to gape at the first horseless carriage they had ever seen.

After sixty-three days out, Jackson arrived in New York on July 26. Fetch, who had left three weeks after Jackson, cruised into the city on August 21, making the physically taxing journey in just sixty-one days. Two hundred cars met Fetch and Krarup on the outskirts of New York and escorted the exhausted duo to the Astor House Hotel for a celebration. Asked to address the excited crowd assembled outside of the famous hotel on Broadway, Fetch eyed the throng before spitting out, "Thank God, it's over."

As the first Packards began making their appearance across the country and Ward grappled with issues of speed and endurance and the assembly of intricate moving parts, he also acquired the first of several complicated timepieces. He had commissioned the London watchmaker E. Dent & Company, who, among its many distinctions, had built Big Ben, to make the first of three special pocket watches for him. Ward took a hand in his watches' designs, making specific requests. Each piece was said to cost the equivalent of a row of houses in London.

Ward's inaugural piece, received in 1902 with the movement no. 32544, was considered one of the most complicated watches produced in English history. It featured a tourbillon escapement; an exclusive, highly complicated, and fragile mechanism on its own, this tiny device made of some seventy different even tinier pieces was combined with five other complications that included a split-second chronograph. By depressing a button on the rim of the eighteen-carat gold watch, Ward could time two events simultaneously with one split-second hand over the other, noting accurately the times of each. It was just the kind of watch that came in handy when he tested the speed and endurance of his early model Packards.

Ward's interest in watches seemed a natural evolution of his intellectual curiosity, just as the automobile was a natural progression of his technical wizardry. Until the mid-nineteenth century, the most talented mechanical engineers were watchmakers. A marriage of artistry and technology, a *grande complication* provided more than just a timekeeper: it represented a masterpiece of mechanical engineering. And the principles of watchmaking applied to the craft of auto making. The intricate moving parts, design, and production all required knowledge of advanced and applied sciences. Both disciplines involved orchestrating inventive solutions to myriad challenges. For Ward, the automobile, like the mechanical watch, was engineering in its most beautiful form—an object of utilitarian artistry.

While Dent crafted its masterpieces for Ward, another important English house, Smith & Sons, put the finishing touches on the first of three customized, highly complicated pocket watches for him as well. The first was a minute repeater enclosed in a polished eighteen-carat gold, partial hunter's case with a small round crystal window surrounded by Roman numerals. The piece, with the movement no. 13159, had nine different hands indicating various functions. The movement had won a Class A certificate award from the Kew Observatory, connoting the watch's superior merit at the trials at the National Physical Laboratory of Great Britain. Along with the Geneva Observatory trials, the Kew trials were considered a kind of Olympics for watchmakers.

Ward's days wandering Maiden Lane to take vicarious enjoyment in the timepieces there were long behind him. Not only did he have the genius to know what he wanted, but he now had the money to acquire it. He also had an ally in his growing horological pursuits.

Back when Ward first began working on his naphtha launch in Chautauqua, he made the acquaintance of Fred E. Armitage. It is not clear whether the two men had met previously, but they had much in common, including an obsession with all things mechanical. Born in 1864 in Jamestown, New York, Armitage was something of a polymath. Having grown up on Lake Chautauqua, he was a first-rate rower and sailor who used his talents to manufacture some of the earliest American power launches—a vocation that also likely brought him to Ward's attention. Armitage was a jeweler by trade, and before going into business for himself in 1890, he co-owned the jewelry shop Phillips & Armitage in Jamestown, about five miles from Ward's summer home on Lake Chautauqua. Armitage's singular establishment catered to the area's growing wealthy summer resort clients and visitors to the esteemed Chautauqua Institute.

Like Ward, Armitage gravitated toward timepieces and meticulous precision. He would earn a measure of local fame when tasked to overhaul Jamestown's clock. He designed a system coordinating a master clock with thirteen others across town, all synchronized to the hour.

The two men had a personal friendship as well as a business relationship, and Ward's diary soon became riddled with references to visits from "F. E. Armitage." There were dinners in Warren and tête-à-têtes in Lakewood. More than just a skilled jeweler and technician, Armitage had become an American agent for Patek Philippe and Vacheron Constantin. Far from the bustle of fashionable Maiden Lane and the grandeur of the big-name shops sprouting up on New York's Fifth Avenue, Armitage operated quietly near the shores of Lake Chautauqua and Ward came to depend on him for many of his most important timepieces. Armitage would broker Ward's unique Patek Philippe walking stick watch and express Ward's desires to Europe's finest watchmaking houses as he built his splendid collection of complicated and custom-designed commissions.

In the first decades of the twentieth century the Ohio Automobile Company increased its output. Packards became sleeker, with a longer wheelbase and larger tires. Red became the standard color. Ward had built more than a stylish and dependable automobile. At the dawn of the new moneyed century, the Packard automobile was *the* symbol of the glittering new plutocracy.

The auto industry itself marked a new defining line in America's fast industrialization. There were men like Ward who built automobiles and men like Henry Graves, Jr., who bought them. "The Packard is a car for a patrician," wrote Elbert Hubbard, the prolific writer-philosopher. "It belongs to the nobility and is of the royal line. It is conceded to be without a superior in reliability, workmanship, and luxurious and artistic finish. To own a Packard is the mark of being one of fortune's favorites. It satisfies ambition, soothes aspiration, and gives a peace which religion cannot lend." The company's slogan, extracted from Ward's confident response to a Pittsburgh man inquiring about the newest model, said it all: "Ask the man who owns one."

Packard's automobiles, like the watches he acquired, had the finest in styling and quality, beautiful lines and craftsmanship that housed the most technologically advanced, highest endeavors in mechanical engineering. And they were built to last.

CHAPTER SEVEN

~

Eagle Island

In early July 1911, Henry Graves, Jr., arrived in the Adirondacks just as the high season began. Once again the mountains swelled with some of New York City's wealthiest families as they made the summer pilgrimage, exchanging their brick and marble mansions for the rustic decadence of their Great Camps in the woods. It was a change of place but not of splendor.

In the *Gwen,* the sleek motorboat named for their only daughter, Henry, Florence, and their four children sliced through a light mist hanging over the tranquil cobalt waters of Upper Saranac Lake heading toward Eagle Island. Their retinue of servants trailed behind in a separate boat carrying the family's steamer trunks as well as the family's pets, a Mastiff and a Cavalier King Charles spaniel, and enough supplies to last through the two-month summer season. They were the very same cooks, nannies, and housekeepers who attended to the family's every need in Irvington and Manhattan. A second set of local caretakers, guides, and woodsmen also tended to them on the island.

Like the chimerical Brigadoon, Eagle Island Camp appeared for a brief period each summer, enchanting its visitors before retreating into the frigid snow drifts of winter. Accessible only by boat, it was inhabitable at the tail end of spring, after the lake had finally thawed, allowing for safe passage. Eagle Island, named for the bald eagles that once nested there, sat on thirty-two forested acres commanding a view that stretched across Upper Saranac Lake to the Narrows and included two smaller islands called Big Watch and Little Watch.

Despite being surrounded by six million acres of unspoiled wilderness on their journey north, the family did not replace their opulent clothing, couture hats, and jewels. Henry traveled in a long coat, starched shirt and tie, and a short-brimmed hat. Florence wore a fashionable hobble skirt and Chantilly lace blouse, her hair swirled into a flattened sundae under a wide plumed hat; and a fur wrap around her shoulders fended off the morning chill. The boys wore Fauntleroy suits, including fourteen-year-old Henry, whom everybody called Harry; little Gwendolen wore a frilled dress with smock skirt and silk ribbons in her hair. As they made the short journey to their camp, great blue herons sailed overhead and loons sang their plaintive wail. The incessant buzzing of the ghastly black flies that made life painfully unpleasant during May and June had mercifully disappeared by the time the Graveses arrived.

William Meagher, Eagle Island's longtime caretaker and family friend, met the group, offering a firm, warm handshake to Henry, and accompanied them to their private island sanctuary. As usual, the New York newspapers' social columns announced the arrival of "Mr. and Mrs. Henry Graves, Jr., and their party."

The papers gave no indication of the anticipation Henry felt for the annual motorboat regatta, a great spectacle that brought the curtain down on the season each summer. This year, determined to win at any cost, he had come armed with a secret weapon, a 200-horsepower engine that he furtively installed in his treasured fifty-foot speedboat, a gloriously agile craft named the *Eagle*. But first he and his family had several busy weeks to navigate.

Life at the Great Camp was one of perpetual motion, with games of croquet and tennis, sailing, hunting, and fishing trips. Guests arrived at

Henry pilots his family in the *Gwen* to Eagle Island, Upper Saranac Lake.
Courtesy of Cheryl Graves.

Eagle Island during carefully timed jaunts. Among the barrage of activities, a raucous costume party at the main lodge was on the social calendar, where, crackling with brio, Henry and Florence dressed as "Orientals" in long silk Chinese tunics. Henry's dear friend Edwin Gould, the son of Jay Gould, showed up in an impossibly tall makeshift paper hat and a sandwich board that announced, "This Way to Eagle Island, Water Sports, Girls Swimming!!!, No Admittance for the Motley."

Eagle Island dinners took place in the octagonal dining room, its ceiling designed to resemble the underside of an open umbrella, under which any number of lovely events settled breezily. There was dancing to the Victrola or alongside Florence as she played the ebony upright piano in the great room, the same room where the servants lined up Henry's issues of *Collier's* and *Scribner's* in perfectly measured rows on the front table and kept the large Tiffany & Co. clock on the mantel regularly wound. At dusk came the blissful ritual of cocktails served on the veranda as the sun disappeared into the water, cigarettes tipped from long black holders, gossiping over recent events: the recently crowned king of England George V or, say, the end of Ty Cobb's forty-game winning streak. Among less delicate ears the tittle-tattle about which camp owners might have sneaked off with their lakeside servants traveled across a network of whispers.

During sun-dappled afternoons, Henry's full-time guides rowed the family and their friends in long, narrow wooden paddle boats between the lakes and streams, delivering them on quaint shores where their servants laid out lavish picnics served on gold-rimmed china with linens and silver monogrammed with Henry's initials, *HGJR*.

Summers at Eagle Island provided the Graves children with a luminous Never Never Land where they shed their white-glove restrictions under a canopy of birch, spruce, and pine trees. Athletic and outdoorsy, they delighted in the months they spent energetically exploring the woodland, swimming, and canoeing. Duncan, known as "Bud," competed in the annual swimming races. At summer's end he would take first place in the twenty-yard contest. With his Graflex camera, Henry 3rd, called "Harry," became a keen photographer, documenting the family's summers through his accordion lens and pasting the photos in leather albums emblazoned with his initials in gold; his parents and siblings posed rigidly on bark-trimmed benches and at play, canoeing, lying on hammocks, swimming in the cove. Even more than his siblings, George Coe II, whom everybody called "Toot," took to hunting and exploring the wider Adirondacks, filled with wildlife and enveloped in peaked mountains that revealed lakes, glaciers, marshes, and thickets of woodland in infinite shades of green.

This was the Graveses' fourth summer at Eagle Island, the second since Henry had bought the Great Camp, which he now officially listed as his summer residence in the *Social Register*. In 1903 Levi P. Morton, the former vice president of the United States under Benjamin Harrison, and his wife, the inimitable hostess Anna Livingston Street, built the camp as their second Adirondack retreat. Morton erected his first, Camp Pinebrook, on the west side of the lake in 1898. A one-time governor of New York, Morton had also served as the country's ambassador to France, where he was best remembered as the man who drove the first rivet in the Statue of Liberty (into the big toe of her left foot) when she was under construction in 1881. Outside politics, Morton had made a fortune on Wall Street, but at eighty-six he had decided to slow down. Now residing full time at his gentleman's farm in Rhinecliff-on-Hudson, he sold his Morton Trust Company to his sometime friend and frequent rival J. P. Morgan (which would later become part of Morgan Guaranty Trust) and leased out Eagle Island for several seasons.

Henry leaped at the opportunity to own this rare jewel, purchasing it in the spring of 1910, the camp and all of its contents, including the Gustav Stickley furniture and its vibrant, worn Oriental rugs. Among the glittering Adirondack trophies, Eagle Island wasn't the biggest of the Great Camps, but it was an exceptional piece of property.

The architect William Coulter, pioneer of the Adirondack style, designed the sprawling Great Camp as a compound that included several lodges with thirty-foot ceilings made of peeled spruce, stone, and paned windows, all connected by walkways, creating the impression that one was suspended amid the treetops. Not long after the camp was completed, Eagle Island became the fawning subject of a *New York Times* article describing how the wealthy residents of the Adirondacks continued to incline toward luxurious appointments at their mountain retreats. "Of the many camps clustered about the Wawbeek on Upper Saranac Lake," the paper observed, "none attracts more favorable comment than the rustic camp built upon Eagle Island."*

Even in an age of excess, the Great Camps stirred the imagination. Alfred Gwynne Vanderbilt's colossal Camp Sagamore required two hundred men and two years to build. Its massive fieldstone fireplace in the main lodge was so large that the rest of the structure had to be built around it. On Vanderbilt's instructions, Sagamore featured a large casino and a two-story wigwam where the prince of American aristocracy hung his Renoirs and imported women from New York City to amuse his male

---

*In 2004 Eagle Island Camp was listed on the National Historic Register.

guests. On fifty thousand acres, William Avery Rockefeller erected his own private Adirondack reserve, where he also constructed a private railroad station. "The great forest is to him a playground," quipped the *New York Times*, "not as a hunter or a fisherman, but a spot where he may forget some of the cares of his strenuous business concerns."

Proving that it was possible to bring the opulence of their city lives into the forest while conquering this unforgiving topography where the Iroquois and Algonquin had fished and trapped a century earlier, society gilded their camps with bowling alleys, golf courses, game rooms, themed cottages, movie theaters, and Japanese tea pavilions. The historian Maitland De Sormo wrote of the Great Camps, with their stunning array of architectural features and indulgent diversions, "[They] represent the most conspicuous measuring stick in the mountains as well as elsewhere."

For Henry Graves, Jr., an oligarch of a less flamboyant temperament although no less expansive tastes, Eagle Island offered seclusion, privacy, and the luxury of home. At a time when many American cities still lacked such amenities, Eagle Island had hot and cold indoor plumbing and telephone lines. A power plant furnished electricity to the entire island. Among the eleven buildings there were seventeen bedrooms and separate servants quarters with a laundry, refrigerated pantries, and entire rooms

A season staple of Eagle Island, the Graves family and friends swim at their private bathing beach in front of the boathouse. Courtesy of Cheryl Graves.

just for linen, china, and ironing. For amusements, the island had its own bathing beach and sailing cove, a tennis court, guest cottages, and two boathouses. Henry built the second one, a stunner that matched the main lodge, with its own veranda and sleeping quarters for the boatmen who maintained his fleet of four motorboats (including the *Gwen* and a twenty-seven-foot speedway runabout), four canoes, four guide boats, and several smaller boats set aside for the servants' use.

The Graveses rubbed elbows with their neighbors on the southern half of Upper Saranac: the beer heiress Isabel Ballantine, who owned Moss Ledge, with its teahouse overlooking the lake, and a group of wealthy German-born Jews who established their private Great Camps here, a move spurred in no small part by the accepted social anti-Semitism of the upper classes that barred them from most private resorts and social clubs. Across from Eagle Island, Adolph Lewisohn, a banker and philanthropist known as "the Copper King," retreated to Camp Prospect Point, a series of remarkable Swiss-style chalets on a bluff overlooking the northern reaches of Upper Saranac that towered over the shoreline. The Wall Street tycoon and Goldman Sachs scion Henry Goldman owned Bull Point, modeled after English Tudor country mansions and originally built for the investment banker and arts patron Otto Kahn.

During the previous summer, just a month before the regatta, Henry casually piloted his speedboat, the *Eagle,* which had earned the moniker "the Queen of the Adirondacks," on the lake. He had motored up to the Saranac Inn to pick up his guests Edwin Gould, his wife, Sarah, and her mother, Mrs. George F. Schrady, for their stay at Eagle Island. Among the status-conscious camping colony, the appearance of the *Eagle* was tantamount to revealing a straight flush. This season, however, Henry kept his fastest boat under wraps, refusing to let anyone near it except for his camp mechanic. His main rival, the champion swimmer Charles Daniels, an eight-time Olympic medalist and the inventor of the "American crawl," had secretly purchased one of the fastest boats in the world the previous summer, the thirty-three-foot, four-cylinder, seventy-five-horsepower *Arab III.* He redecorated it, renamed it the *Ketchup,* and quietly shipped it to Upper Saranac. This year Henry was not about to take any chances.

Behind the wheel of the *Ketchup* Daniels had won the previous regatta by mere seconds. He bested the *Perhaps,* piloted by Jules Bache, founder of the brokerage firm J. S. Bache & Co., a bon vivant and collector of millions of dollars' worth of Goyas, Raphaels, Rembrandts, Botticellis, and Vermeers. As September 2, the date of the new season's race, approached, it became widely known that many had secretly installed new engines in a bid to outdo the competition and win the championship.

On the summer afternoon of the showdown, reporters from all of the major New York City papers descended on the lake to cover the highly anticipated motorboat race in their society pages. Henry finally unveiled his souped-up *Eagle,* as Bache, Daniels, Lewisohn, and several other representatives of the camping colony also appeared with their supercharged motorboats.

At the pop of the starting pistol, Henry threw the full power of his engines into gear, jolting the *Eagle* forward, missing the *Perhaps* by just a foot. Taking control of his craft, Henry straightened and shot down the lake. Together with his competitors, he clipped and charged, turning the waters, as one observer described, into "the fury of a hundred storms." Covered in foam, he shifted gears and pushed his 200-horsepower engine to its maximum threshold, lifting the *Eagle* into the lead. By the time he reached the first mile mark at Ward's Point, he had gained five hundred feet over the *Ketchup.* Smiling to the roaring crowd, Henry approached the first of many sharp turns, thrusting his engine down to navigate it. He had raced through the first round at thirty-five miles per hour and now sprinted to the finish. He had put a mile between him and the *Ketchup.* It was over. Stepping from his boat, Henry was received by a standing ovation. He ended the season in first place, just as he had intended.

In business, Henry fared similarly, although he had earned his position by default. Five years earlier, his sixty-seven-year-old father had died while spending the month of August with his wife at their Kineo summer cottage on Moosehead Lake in Maine. Upon his death he left a personal estate worth $5,660,000, "of which only $345,622, was in the State of New York," reported the *New York Times.* While this was certainly a fortune (equal to $135.5 million today), it represented only a portion of the considerable web of interests he held in commerce, railroads, and banking through his brokerage firm, Maxwell & Graves, a sum that conservatively elevated his wealth by many times that amount.

Henry's two brothers played some part, although largely titular, in the Wall Street firm that bore the family name, each reaping generous rewards. The eldest, Edward Hale, appeared the least active in business affairs. He lived with his wife, Jean (née Stevenson), and daughter, splitting their time between mansions in Orange, New Jersey, and Tuxedo Park, New York, engaged in music and the social calendar. The never-married George Coe lived in a villa on the family's Orange estate, engaged in philanthropy (preferring to donate to a host of organizations anonymously) and travel, exploring exotic ports of call. He filled his summer estate, Sylmaris, in Osterville, Massachusetts, near Cape Cod, with thousands of priceless objects he collected over time: antique bejeweled Indian daggers forged in gold, French Charles-Honoré Lannuier armchairs, English and Irish glass from the eighteenth and nineteenth century, and white jade snuff-boxes decorated with emeralds and rubies. George also shared his brother Henry's taste for rare prints. The youngest, their sister Daisy, married well, twice. With her first husband, Lyman F. Goff, the son of one of the most prominent industrialist families in New England, Daisy had one daughter, Isabella, and after being widowed at age thirty-three, she married Ernest Atherton Smith, a wealthy Orange resident.

Two years before his death on December 20, 1904, Henry Graves, Sr., drew up his will, establishing a trust from the bulk of his estate in order to provide for his descendants for generations to come. He bequeathed his household furniture, personal effects, and works of art equally among his wife and children, providing his widow with an annual income of $20,000 (roughly half a million dollars today). Harriet Isabella Graves also received the family's mansion in Orange. From the documents filed it seems that Henry Sr. did not leave a nickel to anyone outside of his family, to any charity, or to any institution of art or botany or any church that he had graciously patronized during his lifetime. In death, his sense of noblesse oblige ran inward toward his family and outward to its future generations.

The terms of his will called for the core of his millions to be put in trust and its residual income to be divided equally, a portion paid to every lineal grandchild and great-grandchild, ad infinitum, on an annual basis throughout their lifetime. Henry and his younger brother, George Coe, were assigned as trustees, with Henry given final authority over the legal instrument. Under its terms, if one bloodline died out, the core would be divided among the surviving descendants. Having already guaranteed his children's fortunes, it appeared that Henry Sr.'s singular desire was to wield financial authority over his family like royalty, ensuring the wealth of his blood heirs in perpetuity. His will stated, "Inasmuch as my sons have already received from me sufficient capital to make them independent during their lifetime, it is my wish that the principal of my estate should remain intact, without depletion or distribution for the benefit of my grand children and their families."

Henry's personal portfolio resembled an alphabet of American industry. On paper he owned millions in stocks, bonds, and, later, U.S. Treasury bills. Following his father's death, he was entrusted to manage 9,100 shares of the Delaware & Western Lackawanna Railroad worth $455,000 (some $10.4 million today), making him the eleventh largest shareholder in the company, behind George Baker and William K. Vanderbilt. He also managed 1,975 shares of the Lackawanna Coal Co., worth $98,750 (about $2.3 million today). This was in addition to his own 1,850 shares in the two entities, worth $92,500 ($2.1 million today). This placed Henry in a golden circle of twenty-five men with controlling ownership of both the coal company and the railroad that it serviced. The money rained down on Henry Graves, Jr.

By the time Henry Sr. died, the chasm between Old Money and New Money had practically disappeared. There was just money. And a great deal of it at that. Marriages took place between America's haute bourgeoisie and the established line of blue bloods. Old Money families rubbed up against the self-made, reshaping the contours of class and society. These mergers helped balance snobbery and cash but did little to recast the broader inequities. Between 1870 and 1900 the country's wealth quadrupled. But the swell of riches did not spread very far, as 80 percent of Americans earned less than $500 a year, leaving the remainder of the country's considerable wealth lining the very large pockets of a very small minority.

In New York City arrivistes stormed the gates of the upper echelons. Wealthy self-made midwesterners and highly successful immigrants arrived and multiplied, and just behind them a new middle class stirred. At the bottom of the pile, a great mass of humanity lived in the margins; poor immigrants who lived in fetid tenements and commuted uptown to serve the wealthy, the workers who labored on railroads and in coal mines, factories, and sweatshops, all for low wages and punishing hours, grew increasingly agitated.

Not that society noticed. The Diamond Horseshoe set demonstrated few reservations about their affluence amid the growing swells of want. The party glittered on. One sumptuous affair followed another, each requiring copious reserves of money. The same could not be said for taste. Mamie Fish, the infamous wife of the Illinois Central Railroad magnate Stuyvesant Fish, once threw a dinner party in honor of her dog; the canine arrived wearing a $15,000 diamond collar. The Chicago tycoon Cornelius Kingsley Garrison Billings put on a notorious dinner to celebrate the completion of his 25,000-square-foot stables. At forty years old the shameless eccentric had just retired and moved to Manhattan. For his soiree, held at Sherry's on Fifth Avenue and Forty-fourth Street, imitation grass and dirt covered the ballroom floor and Billings's thirty-six guests were served food and drink while mounted atop horses rented specifically for the event, by waiters dressed as hunting grooms. The celebrants ate from trays fastened to their saddles and drank champagne from special tubes attached to their saddlebags. The old guard was aghast at the display of unfettered vulgarity, while everybody else was just aghast.

Four months before their father's death, Edward Hale Graves and his wife threw an "intimate" St. Valentine's Day *bal poudré* for seventy in the ballroom of their Orange mansion on Scotland Road, decorated to resemble Louis XVI's court at Versailles. The guests summoned their tailors and seamstresses to conjure up ornately adorned eighteenth-century costumes with white powdered wigs for thousands of dollars. Jean Graves wore a pale blue brocade gown with a white, flowered petticoat, blue ribbons suggestively braided into her full white wig. Edward wore an extravagant court costume in plum. When the clock struck half-past ten, Edward and Jean and seven other couples performed a choreographed minuet.

The public grew more resentful, troubled by the moguls' wealth and power and by the duplicitous means by which many of these fortunes had been made. Many leaders also began to question the industrial order of America and its monopolistic practices, whereby the companies' founders and top executives continued to enrich themselves while everybody else got poorer. Attempting to corral support for a political offensive to break the financial stranglehold of the few on the many, the lawyer and future Supreme Court Justice Louis Brandeis wrote in 1913, "They control the people through the people's own money."

Four months after the death of Henry Graves, Sr., President Theodore Roosevelt gave his State of the Union Address to Congress on December 3, 1906, and proposed the introduction of a federal inheritance tax. "The man of great wealth," he asserted, "owes a peculiar obligation to the State because he derives special advantages from the mere existence of government." Roosevelt reasoned that the government should "put a constantly increasing burden on the inheritance of those swollen fortunes which it is certainly of no benefit to this country to perpetuate."

The plutocracy did not share this perspective and had little interest in parting with their vast fortunes, viewing any shift of the economic scales as an attack not just on their money but on the system they profited from greatly. Established plutocrats and those of more recent vintage were as ruthless in quelling dissent as they had been in amassing their riches. On September 17, 1913, nine thousand workers went on strike at John D. Rockefeller's Colorado Fuel and Iron Company, sparking one of the most severe clashes between labor and capital in American history. The miners and their families had moved into a tent colony after being evicted from their company-owned houses. On April 20, 1914, state militia, company thugs, and guards opened fire on the camp during a fourteen-hour stand-off that left nineteen people dead.

The tide of revulsion at America's wealthy classes and their excesses reached a tipping point. In 1909 Congress passed the Sixteenth Amendment to the Constitution, allowing for the taxation of individual incomes; the states ratified it four years later, levying taxes on personal income above $3,000 (about $72,000 today). The public outrage directed at the "squillionaires," the term that the art consultant and Renaissance historian Bernard Berenson coined somewhat derisively for the American industrial barons whose castles he helped hang with priceless Old Masters, won a battle if not the war on May 15, 1911. That was the day the Supreme Court upheld a lower court's ruling that Standard Oil was an "unreasonable monopoly" under the Sherman Anti-trust Act and ordered its dissolution. The inheritance tax that Roosevelt had fought for, however, would not be enacted until 1916, eight years after he left office.

The hue and cry against the plutocrats helped spur an elitist drift to good works, unveiling an era when massive acts of wealthy exhibitionism turned to great philanthropic spectacles. The public face of conspicuous consumption became swept up in a tidal wave of conspicuous philanthropy. Parthenons of music, medicine, and literature were erected. John D. Rockefeller helped found the University of Chicago and the Rockefeller Foundation. Andrew Carnegie spent millions during the last years of his life establishing libraries and funding academia and the arts and sciences. In California, Henry E. Huntington announced the creation of a trust to make public his colossal private library, which held some of the finest precious manuscripts and first editions in the world along with his vast art collection. It also contained one of the most distinguished assemblies of seventeenth- and eighteenth-century British paintings outside of the United Kingdom, including Thomas Gainsborough's *Blue Boy*.

Among the Graves brothers, it was George Coe who wore the title of philanthropist and made numerous significant gifts during his lifetime. He would donate the sixth largest organ in the world to the Grace Episcopal Church in Orange, New Jersey. After giving the Metropolitan Museum of Art thirty rare Old Master prints, including several by Rembrandt and Van Dyck, he was named one of the Museum's benefactors. Later he emptied almost the entire contents of his Osterville estate into the Museum. His collection of furniture helped establish the Met's American Wing.

Henry's philanthropic dabbling equaled perhaps less than a year's expenditures on his shaves. He gave unremarkable donations in the hundreds of dollars to organizations such as the American Red Cross. When a group of Irvington locals led by the composer Jennie Prince Black, the wife of the printing magnate Harry Van Deventer Black, attempted to erect a memorial for Washington Irving and was beset by delays and financial troubles, Henry did step up, donating a triangular wedge of land on his Irvington property for the effort. Eventually a bronze bust of Irving flanked by two of his most famous literary characters, Rip Van Winkle and King Boabdil, was constructed on a six-foot marble pedestal. Outside these rather small gestures Henry did not engage in the erection of grand institutions or the willing of personal artworks to them. He demonstrated little desire to see his private tastes become public property.

Four months after his father died, Henry dissolved Maxwell & Graves, which had propelled the family's enormous fortune over the previous four decades. (Henry W. Maxwell had died in 1903.) Rather than take up his father's seat on the New York Stock Exchange, he "transferred" it to America's top-ranked male tennis champion, William A. Larned, for $80,000 (about $1.9 million today). The family kept the expansive Maxwell & Graves offices that managed a host of businesses in which they maintained a foothold. From 30 Broad Street Henry administered his father's trust, running the family's extensive network of assets, but mostly he occupied his time with collecting.

CHAPTER EIGHT

~

The Collector

On the evening of February 27, 1909, Henry Graves, Jr., waited patiently as a jostling crowd of more than a thousand agents, buyers, and millionaire speculators stuffed to overflowing the cavernous white and gold auditorium of Mendelssohn Hall on Fortieth Street, just east of Broadway. Henry did not come to bid. Rather, from the sidelines he oversaw forty-nine of his father's paintings go under the hammer in a prestigious single-sale auction conducted by the American Art Galleries. The celebrated auctioneer Thomas E. Kirby, who had dropped the gavel on many of Graves's own purchases, opened the sale with the observation that the group of paintings possessed "the highest average and the most beautiful quality of any collection of pictures ever sold in this country." As Kirby rapped his gavel on the evening's last lot, an Anton Mauve watercolor, the crowd broke into robust cheers. The auction brought in a total of $233,250 (an amount worth over $5.5 million today), with Jean-François Millet's *Sheep Shearing* topping the lot, fetching $27,500 (about $658,829 today).

Over the previous two days, Kirby had orchestrated the dispersal of 650 of Henry Sr.'s Chinese porcelains, jades, and lacquers. The auction elicited a great deal of interest among the Diamond Horseshoe set. A widows and wives gallery of New York's richest—Mrs. John Jacob Astor, Mrs. Ogden Goelet, and Mrs. Potter Palmer—along with Charles P. Taft, brother of the president-elect, arrived eager to bid, prompting the *New York Times* to proclaim, "The number of prominent persons who were purchasers made it a social event."

Very few people, certainly among those sitting in Mendelssohn Hall, had much of a clue that Graves's son was a prolific collector in his own right. As the buyers and dealers attended the remarkable auction—the entire three-day sale brought some of the highest prices for art, realizing $281,452 (a tally worth nearly $7 million today)—Henry had slipped into the evening a collector in his own right.

• • •

Henry's acquisitiveness and that of his peers displayed a curious mix of vanity, ignorance, and greed, which for many frequently resulted in miscalculations about the authenticity of their purchases. Of the sixteen Goyas purchased by the avaricious sugar magnate Henry Osborne Havemeyer, only four were originals. Isabella Stewart Gardner bought two thirteenth-century French medieval sculptures whose legs and torsos were fakes. Even the redoubtable J. P. Morgan was fooled a number of times, in one instance buying what he thought were bronze doors of a cathedral in Bologna.

Henry needed to surround himself with the best. He did not become a scholar or write critiques on the art and objects that drained his accounts and filled his homes. He did not discover unknown artists and promote them to great heights. He did not become a patron or build a marble edifice to display his collections. His sole interest was personal satisfaction. Henry derived enormous happiness from the very private knowledge that he had accumulated nothing less than the world's most superlative treasures.

Henry shows who's boss
to one of his dogs.
Courtesy of Cheryl Graves.

Stepping out from his father's shadow, Henry demonstrated a streak of enterprise in his collecting. A ferocious predator, Henry delighted in the chase. Procuring seemingly unobtainable objects gave him the most gratification, and he exhibited an uncanny ability to acquire choice specimens no longer on the market, frequently seeking those without rival. Under rather opaque circumstances he came to own two eighteenth-century French color prints from the Hermitage in Russia. From the Friederich von Nagler and Berlin Museum collection he purchased one of the most famous of all Rembrandt etchings, the *Hundred Guilder Print*. From other cloaked and sundry sources, he scooped up a brilliant selection of drawings by Whistler and Dürer.

Henry's naval prints were among the finest anywhere. Without fanfare, he purchased Charles Willson Peale's 1778 mezzotint portrait *His Excellency George Washington Esq.*, one of three known copies, as well as significant portraits of Lady Washington, Alexander Hamilton, and James Madison. Albert Reese, one of the few who came into contact with Henry's collection, exclaimed, "Mr. Graves aimed for the best and whenever possible exchanged a good impression for one still better." As director of the Print Department at the Kennedy Art Galleries, Reese would later quip, "That his acumen was seldom at fault is attested by the extraordinarily high level of his acquisitions."

Henry took a keen interest in French glass paperweights, something of a niche preoccupation. Produced in Italy and France during the mid-nineteenth century, paperweights did not receive the same recognition as most *objets d'art*, although many collectors believed that their precise artistry ranked alongside illuminated manuscripts, portrait miniatures, and Fabergé eggs. For Henry, the appreciation appeared to come from their delicate perfection of precision, frozen for all eternity inside of a solid glass dome that both protected and magnified the design. He acquired examples made by Baccarat, Saint Louis, and Clichy from the classic period, the brief sliver of time between 1845 and 1860. Despite their relative size, much like his naval prints, they represented power, with cameos in miniature of General Lafayette in silver and another with the profile of Queen Victoria. Henry owned nearly two hundred paperweights, an impressively large assembly of some rather small items.

Throughout his lifetime, Henry also accumulated an extraordinarily comprehensive collection of rare coins, pieces of historical significance and in particular thousands that had been in brief circulation. Among them, he owned Colonials, half-dimes, half-dollars, silver dollars from 1883, Quarter Eagles, and the most notable of his collection: an 1804 silver dollar, also known as the "King of Coins." Curiously, he kept his paperweights and coins in a carved Chippendale mahogany cabinet in his master bathroom.

Henry's collecting went beyond the typical rich man's acquisition for his times. When buying at auction he slipped from view and most of the time engaged in private transactions. During this heady period of ambitious patronage among the wealthy, few collectors of Henry's magnitude remained unknown to others and to the public. As his grandson Reginald Fullerton, Jr., would later remark, "People were largely unaware of these collections. He never talked about them."

Over the years Henry had developed his collecting philosophy: "If you're not going to buy the best, don't bother." He applied this reasoning to mechanical timepieces, but his interest in them settled rather slowly and somewhat later than his other collections.

Along with his English and French period furniture, his Irvington mansion contained a number of long-case clocks. The rooms were filled with the distinctive ticking of the masters: Thomas Tompion, George Graham, and Joseph Knibb. Despite such rarities, however, Henry's connoisseurship remained precocious in most respects.

The Graves men's regular patronage of Tiffany & Co. initially brought Henry to Patek Philippe, soon enough his chief watchmaker. In 1906 the grand dame of American jewelers followed society's coattails uptown, relocating from Union Square to a grand Stanford White–designed marble emporium in the style of a sixteenth-century Venetian palazzo, across the street from the Waldorf-Astoria Hotel, at Fifth Avenue and Thirty-seventh Street. Here, Reginald Claypoole Vanderbilt would select a 16.25-carat heart-shaped diamond and platinum ring for his fiancée, Gloria Morgan. And it was at Tiffany that Patek Philippe seduced New York's plutocracy.

In the opening decades of the twentieth century, Patek Philippe's presence in America expanded greatly. Alfred G. Stein, a watchmaker who first worked the trade on Maiden Lane in 1885, joined Tiffany in 1889 and then operated his own distributorship selling Patek Philippe at 68 Nassau Street. Stein played a crucial role between Geneva and the American market. He had served as the only American juror at the horology trials at the Paris Exposition in 1900 and was awarded the Order of the Palm from the French government for his work. In 1901 Patek Philippe rewarded him with a seat on its board of directors. He did much to nurture American interest in the watchmaker, touting its technological achievements particularly among a market keen for complicated pieces, and he translated American tastes to Geneva. It was Stein who urged his retail network to continuously apprise customers of Patek Philippe's most recent endeavors.

In particular, he encouraged the spread of results from the Observatory competitions and the significant horological trials at the World's Fairs.

Backed by a clever, widespread marketing campaign, Patek Philippe earned a reputation as "The World's Best Watch." As one newspaper advertisement read, "There is a watch which just about reaches perfection in its time keeping qualities. This is the Patek Philippe. . . . It practically lasts forever."

Patek Philippe's Observatory prizewinners excited Henry's attention, affection, and competitiveness. The Geneva Observatory contests began in 1866; they were a six-week endurance test of positional error, temperature compensation, and the average daily rate. Watchmakers presented only a fraction of their most finely regulated watches at the timing tournaments, and the winners were known to be the most precise timekeepers in existence. Given Henry's penchant for speed and winning, it is not surprising that Patek Philippe's chronometers, those awarded first prize and top honors at Geneva, captivated him.

According to one expert, Henry acquired a few high-performance Patek Philippe pocket watches, including a minute repeater, as early as 1903. Eight years later he purchased a timekeeper that was awarded first prize at the Geneva Observatory competition. As World War I engulfed Europe, he snapped up several more timepieces, a great many that were specially commissioned for him but also a selection that was not, including some prizewinning Observatory pieces.

Nearly all of Henry's timepieces were engraved with the Graves family coat of arms. Photograph courtesy of Sotheby's, Inc. © 2012.

High-performance timekeepers were small golden reflections of Henry Graves, Jr., himself: dedicated to perfection and exclusion, at once discreet and transcendentally grand. Each commission reflected his "taste for purity of design as well as complications." He sought cases of eighteen-carat gold or platinum. Nearly all were engraved in Geneva with the Graves family arms bearing the motto *Esse Quam Videri*. Before long, his fanatical interest in award-winning accuracy brought him to desire chronometers evolving to *grandes complications*.

Obsessed with accuracy, Henry could express just how rich and powerful he was through his private acquisition of the world's best timepieces. His interest in precision had little to do with plebian needs for punctuality, for he could set his own pace and demand those around him to follow; instead it was a manifestation of his desire to inhabit a world populated with the best. A pocket watch of this caliber defined Henry, with its whirl of mechanisms providing uncontestable certainty under his control. It was nothing less than an article of faith, of absolute perfection. As his interest in *grandes complications* advanced, so too did his need to possess the most finely regulated, complicated, and precise instrument ever manufactured.

In 1916 Henry acquired yet another prize. The same year that President Woodrow Wilson was reelected with the campaign slogan "He kept us out of the war," Henry took a lease on a magnificent fifteen-room apartment at 420 Park Avenue. The brand-new seventeen-story double building took up the entire block between East Fifty-fifth and Fifty-sixth Streets. The lavish winter flat contained a forty-foot living room, a salon, library, and dining room as well as five maids' rooms. The apartments at 420 Park were millionaire's rentals. At a time when the average American's annual income was roughly $6,300, Henry paid $1,333 a month in rent.

The broad street, originally called Fourth Avenue, became elite New York's latest destination once the railway tracks leading to Grand Central had been placed underground. Renamed the much more elegant-sounding Park Avenue, it soon became an ideal location for developers who built towering apartment blocks that resembled stacks of mansions in the sky. Situated in close proximity to the toniest shops, hotels, and restaurants, the leafy, perfumed thoroughfare transformed into an appealing address for the swelling ranks of Fifth Avenue exiles and America's bright, shiny new millionaires. "Croesus's Sixty Acres" was how the economic chronicler Maurice Merney described the avenue in an article for the *North American Review*. Within a decade, four times as many millionaires would inhabit this "plumed, pearled lady with the beauty spot, much to be desired and costly to possess," than in all of Great Britain.

This grand movement toward apartment living among society, lamented the columnist R. L. Duffus, was a sure sign that Fifth Avenue's "Millionaire's Row was doomed to disappear," and with it the great line between those who had earned their wealth and those who had inherited it. This tidal wave of wealth produced unintended consequences, Duffus explained, "the pressure of the population which has made land so expensive that even the fabulously rich are moved to economize in the use of it."

In 1916 more violent changes were taking place. In Europe the Great War continued to rage nearly two years after a Serbian nationalist had assassinated Archduke Franz Ferdinand in Sarajevo on June 28, 1914. The Allies—primarily Great Britain, France, and Russia—were pitted against the Central Powers of Germany, Austria-Hungary, and Turkey. While America continued to beg off, wanting nothing to do with the bloody turmoil, British conscription laws went into effect. More than one million of the king's soldiers were deployed across Belgium and France.

Despite President Wilson's campaign promises, America soon became caught up in the Great War. A year earlier, on May 7, 1915, a German U-boat had fired a single torpedo, sinking the RMS *Lusitania* as she made her way to Liverpool from New York, killing 1,195 passengers. Alfred Vanderbilt was among the casualties. The sinking enraged American public opinion and helped sow the seeds of animosity that would push the country to declare war two years later.

The year 1916 marked another battle line. Poring over Observatory reports and horological studies, Henry Graves, Jr., found himself in the uncomfortable and altogether unfamiliar position of being number two. Among elite horological circles, news had spread that Patek Philippe had manufactured a stunning eighteen-carat gold pocket watch with sixteen complications. The number of complications alone instantly put the piece into the highest level of *haute horlogerie*. Patek Philippe's remarkable technical feat was one for the history books. Equipped with *foudroyante* (flying seconds), chronograph with split second, perpetual calendar, and *grande* and *petite sonnerie,* the piece, with the movement no. 174129, was such an extraordinary example of mechanical engineering and precision timekeeping that the United States Naval Observatory was said to have asked for permission to examine it.

The pocket watch had been delivered on January 31 to its fifty-three-year-old commissioner, James Ward Packard. For Henry, this was undoubtedly a bit of unsettling news.

The two men had never met, although Henry, a car enthusiast, most certainly knew of the great automaker. It would have been nearly impossible for him not to have known of Ward's estimable watch collection, for theirs was a small circle, and the specialized world of horophiles was smaller still.

In this age of the machine, these two men came to represent the two faces of success in a quickly evolving America: the self-made midwesterner, Ward, a brilliant spark of industriousness whose mind was stuffed with learning and curiosity, and Henry, the descendent of an illustrious line that stretched across the sea, whose pockets were lined with a great inheritance originally forged by his father in the wood-paneled rooms of Wall Street.

Although both men were intensely private, their insular environment and upbringing had created a pair of opposites: Ward was a builder of exquisite objects and Henry a buyer of them. Yet the lives of the automobile tycoon and the son of railroad riches eventually came to intersect over a shared fervor for mechanical watches, the most nearly perfect machine that man had ever produced. Ward's passion began at the internal mechanism and gravitated outward, while Henry's desire commenced at the watch's understated surface, its hallmark of power, sophistication, and status, before moving inward. This mix of opposing traits eventually brought the pair to the same endgame and to the salons of Switzerland's finest watchmakers. In an era of boundless excess, when industry and fortunes soared, two men with resolve and great wealth found themselves in pursuit of the grandest of all *grandes complications*.

The Graves-Packard steeplechase took place when the obsession with glittering trophies fueled all manner of gentlemen's rivalries. Andrew Carnegie, known for, among other things, his own gargantuan limestone and brick mansion, set out to build a garage for his automobiles and made sure that it too was equally impressive. Carnegie's imposing three-story Georgian-style garage had space for five electric cars, three charging panels, a lift where batteries could be removed and replaced, and living space for eight people, including his footman and four chauffeurs. The banker Otto Kahn answered Carnegie with his own opulent "automobile house" on East Eighty-ninth Street, which housed ten cars and twelve men, including a groom and four chauffeurs.

The horological arms race between Henry Graves, Jr., and James Ward Packard was a gentlemen's contest fought on velvet chairs in hushed salons, aided by discreet Swiss watchmakers, with requests propelled forward in formal letters on richly embossed stationery. Desires were subtly refined and encouraged in part by the watchmakers themselves from one collector to the other. According to one of Geneva's insiders, there were always commercial matters to consider. Graves's and Packard's independent occupations collided into a great collecting challenge, which Patek Philippe, shaking loose its Swiss discretion, would come to describe as "one of the most remarkable and fascinating duels in horological history."

From the beginning, Ward held the upper hand.

CHAPTER NINE

~

Acceleration

The automobile magnate had no idea that he was headed into a horological battle even as he emerged out front. In 1917, the year after Ward took possession of his extraordinary tabernacle of time, his Patek Philippe with its sixteen complications, he received yet another wondrous little timepiece from the watchmaker. Encased in eighteen-carat gold, this pocket watch, with the movement no. 174623, featured a perpetual calendar and indicated the phases of the moon. Mechanically flawless, the watch was also deeply personal. With Warren, Ohio, at the center of his universe, Ward had requested an additional mechanism to indicate the exact hour that the sun rose and set each day over his beloved hometown.

In Geneva, Patek Philippe's artisans and technicians consulted with scientists and astronomers. While the astronomical event occurred daily across the globe, the exact time varied not only throughout the year but every twenty-four hours due to a variety of factors: the Earth's axial tilt, the movement of the planets in their elliptical rotation around the sun, and one's position on the planet. In order to craft the mechanism, the watchmakers needed to calculate Warren's exact latitude and longitude. After working out its many particulars, with each data point computed manually, Patek Philippe produced a satisfyingly beautiful instrument.

Under the dial sat two hand-crafted, toothed sections embedded with minuscule differentiated gears—just a few millimeters apart—placed delicately upon an extension axle where the minute and hour hands pointed to when the sun rose and set. The specially designed cams were calibrated to rotate once a day over 365 (and one-fourth) days annually. They indicated both sunrise and sunset within five minutes. (Had it not been for local atmospheric influences, the watchmakers could have shaved off even more time.) When Ward commissioned this singular watch, he was traveling frequently to Detroit, where his automobile company had relocated and where he was expected to spend a great deal of time. He didn't enjoy extended stints outside his familiar sphere, and removing the timepiece from his pocket to gauge the sun's ritual path over Warren undoubtedly gave him a connection to home.

In the nearly two decades since Ward had built his very first horse-less carriage, the Model A, the Packard Motor Company had gone from a cottage industry to America's first luxury brand. While Ward had nearly everything to do with making Packard the marque of engineering and artistic perfection, the company's phenomenal growth was due in no small part to the massive efforts of Henry Bourne Joy.

The son of James Frederick Joy, a Michigan railroad magnate who had hired the attorney Abraham Lincoln for his first major case, involving the Illinois Central, Henry Bourne Joy was born into privilege. One of the "Princes of Griswold Street," Joy had graduated from Yale, married into one of Detroit's wealthiest families (the Newberrys), and invested in steamship lines, sugar refineries, real estate, and railroad companies. A veteran of the Spanish-American War, he had become intensely interested in technology in general and the automobile in particular.

In 1900 Joy traveled to New York City with his brother-in-law Truman Newberry to take a look at horseless carriages. While they were inspecting a steam-driven vehicle, the water-pressure gauge exploded in their faces. Suitably turned off steam-propelled cars, they went to visit the Adams & McMurtry car showroom and came upon two Packards idling at the curb just as a fire engine charged past. When Joy inquired whether the cars might start—a considerable issue with motorcars at the time—he was invited to "jump in." As the story goes, not only did the Packard start on command, but the men raced off in hot pursuit of the fire engine. The episode so astonished Joy that he and Newberry both bought Packards on the spot. Returning to Detroit, Joy ordered a second Packard, the new Model F, with a blue body, red wheels, and polished brass work, for $2,500 ($64,634 today), an event deemed so newsworthy that the *Detroit Journal* ran the headline "New Automobile Which Harry Joy Has Ordered" on November 7, 1901.

Beyond impressed, Joy not only drove his Packard to Warren the fol-lowing year, but he sank $25,000 (roughly $621,489 today) into the Ohio Automobile Company. At the time Ward and his brother, like most auto-makers of the day, were quite anxious for outside capital. In the process of expanding, they began building a new 32,000-square-foot factory in anticipation of greater capacity, but were quickly burning through their cash reserves at a time when bankers still remained unconvinced of the industry and viewed its financing as a fool's game. As a result, many of the early pioneers had already begun to disappear. Joy's cash infusion sealed the Packard's survival. Coolly, Ward noted in his diary on July 1, 1902, "Joy here."

Very quickly Joy began visiting regularly with Ward, providing his views on vehicle models and manufacturing, and he soon ratcheted up his investment in the firm to $100,000 (some $2.5 million today). Not long after the Ohio Automobile Company incorporated as the Packard Motor Car Company in September 1902, Joy brought in a number of his wealthy Detroit friends and family as investors. Total capitalization of the new venture was now $400,000 (roughly $9.9 million today), and the classic shift between the quixotic founder and the visionary money managers began. Joy and his associates now owned a controlling $250,000 (about $6.2 million today) interest, while Ward, his family, and original shareholders like George Weiss held on to the remaining $150,000 (about $3.7 million today). Ward and Will plowed a great chunk of their money back into Packard Electric.

Joy provided more than just working capital (the company increased its output to produce one motorcar per day); he maneuvered a reluctant Ward to move Packard Motor from Warren to Detroit with the expectation that Ward, as president, would relocate as well, continuing his role in the operations of the company. As Joy wrote to Ward in January 1903, "I want you and we all want you to be the whole thing in the Packard Co."

Until October 1903, when the Packard Motor Car Company officially moved to Detroit, nearly every component of the Packard automobile was manufactured in Warren. Initially Ward gave his assurances to Joy and the other Packard investors that he would indeed relocate to the new Detroit factory. In fact he began the process of selling Packard Electric for $200,000 (some $4.7 million today) to the Lamp Trust, freeing himself up for the move.

The pioneering industrial architect Albert Kahn was contracted to design the new factory on East Grand Boulevard. Unlike his contemporaries, Kahn, the son of an itinerant rabbi who had emigrated to Detroit from Germany in 1884, believed that meeting the functional needs of industrial design was every bit as innovative and honorable as erecting mansions and monuments. At the time, early auto plants, like the early Victorian mills, were dirty, cramped, dark, and inefficient. Kahn's Packard factory was unlike anything ever seen. Built of reinforced concrete and glass, it was impressively clean, open, well-proportioned, and naturally bright. The sunlight poured through the factory in fat beams. Considered to be the most modern automobile manufacturing facility in the world, it caught the attention of Henry Ford. Soon the Flivver King commissioned Kahn to design his Highland Park factory, a gargantuan facility nearly two-thirds glass, where the Model T was perfected and the assembly line introduced.

Once the Packard Factory was completed and the first cars readied for manufacture, Ward announced that he had had a change of heart. Joy had been in constant contact with him over the factory, selecting its superintendent, and along with engineer Charles Schmidt received Ward's outlines for the first automobile to come out of the new factory. Initially he told Joy that he needed to stay in Warren to oversee the close of his deal with the Lamp Trust. According to Joy, Ward explained that, unless he agreed to remain behind to manage the electric factory, the Lamp Trust would not only forfeit the down payment they had made on the deal but would withdraw the offer. Since he could not run two manufacturing factories in two different cities simultaneously, Ward chose home.

Both Ward and Joy shared a belief in quality and an insistence on continued research and testing, but a subtle friction over philosophy had surfaced between the two men. Besides uprooting the company to Detroit, Joy directed the move toward bigger, multicylinder engines and a strategy of building one luxury model annually. Ward, the classic engineer, passionate about complicated pocket watches, held that the simplest solution was the best for car design. He once told a reporter, "More than one cylinder on a Packard would be like two tails on a cat—you just don't need it."

Other issues arose between Ward and the moneymen from Detroit. Since Joy's glittering circle had invested in Packard, Ward's authority had waned considerably. A brittle tension between the two principals erupted as Packard began its transformation into a great company, forcing Ward to reconsider his role. By July 1918 Ward, who had already reduced his presence in the firm, had decided to withdraw from the company altogether, but Joy prevailed upon him to remain.

It was an uneasy rapprochement. Much to Joy's dismay, Ward did not move to Detroit as Joy had hoped and traveled there only for board meetings and other events. As the years wore on, Ward's "inborn shyness" got the better of him and he journeyed there less and less, finding the meetings excruciatingly painful. Mistakenly, Joy and other company executives interpreted Ward's reluctance to visit Detroit and take part in the company's management as indifference to company affairs, but his wife, Elizabeth, later explained, "He worried himself sick whenever he had to go to Detroit. His shyness caused him more suffering than most people realized." In 1909 Ward stepped down as president, staying on as a major

Ward and the new president of Packard Motor Company, Alvan Macauley
(right), in front of the factory on East Grand Avenue, Detroit, 1915.
Courtesy of Betsy Solis.

shareholder and chairman of the board. Joy took over as president.* In
1915 Ward formally resigned from active participation in Packard affairs.
Alvan Macauley, a Lehigh graduate and former head of the American Ari-
thometer Company, stepped in as Packard's new president. Three years
later Ward had put Detroit behind him.

Where Ward was dispirited by the public engagement his role had
required as Packard grew, Henry B. Joy grabbed the opportunity and
became a devoted promoter of the Packard. He spoke to Congress on
behalf of the industry, forcefully lobbying for tariffs on foreign cars to pro-
tect domestic companies whose labor was said to cost two and a half times
that of Europeans. Personally demonstrating that the Packard was quite
simply "one of the best in the country," he drove from a Detroit showroom
to one in New York in three days flat.

---

*Disgruntled with the changes at Packard, in 1905 George Weiss broke all ties to the
company and exchanged his three hundred Packard shares for twelve Model L's, which he
later sold from his Cleveland home.

Within ten years of moving to Detroit, the Packard factory had grown to 3.5 million square feet, with its own water tower, lumberyards, research laboratories, canteens, and seven thousand workers. It had become the largest manufacturer of luxury cars on the planet. In 1902, the year before the company moved, the Warren factory had produced 192 cars and posted a loss of $200,000 ($4.9 million today). By 1912 the company had orders worth $14 million (more than $312 million in today's terms), and the Packard Motor Car Company would not experience financial woes for the next twenty-five years.

Detroit was now the undisputed capital of America's automobile industry. Packard's move, wrote a journalist nine years after the event, had been "the connecting link between the romance of the geniuses and the romance of the industry." The days of cranking out horseless vehicles in wooden sheds were long past; now sleek automobiles were rolling out of gleaming factories. The auto industry had gone from a speculator's game and a toy for the rich to become a generator of wealth accepted at the pinnacle of American society. At the twelfth annual Automobile Club of America banquet held at the Waldorf-Astoria on December 20, 1911, the guest of honor was none other than President Howard Taft.

While James Ward Packard had done much to engineer the automobile, Henry Ford made it possible for the greatest number of people to afford them. Ford priced his cars reasonably and in 1914 announced that he was raising the minimum wage at his factories to five dollars a day—well over twice the the pay of most—enabling his employees to purchase the cars they built. As wages soared, so did automobile sales. The increase in mobility and incomes helped push home ownership. Automobiles and trucks completely altered the production and distribution of the entire spectrum of consumer goods, and prices dropped considerably. Variety and quality increased, as did purchasing power among a growing middle class. From houses to food to clothing to the cars themselves, the automobile enabled millions of American consumers to buy products and services that had once been luxuries for the wealthy.

In 1908, when General Motors' founder William C. Durant pronounced that 500,000 cars would eventually run across the country's roads, most believed these were the musings of a crank. The naysayers found themselves on the wrong side of history. The impact the automobile had on social mobility and the effect that mobility had on society was astonishing. No longer tied to train schedules or the restrictions of a horse-drawn carriage, individuals could now travel where they wanted and when they wanted. Before the car, most people had never traveled beyond a fifty-mile radius. The limitations of transportation meant that most communities had developed within fifteen miles of a railroad or waterway. The car narrowed the geographical divide between city and rural life and paved the way for the suburbs.

American automobile manufacturers catered to every spectrum of society. General Motors would spearhead the concept of aspirational consumption, creating several tiers of automobiles based on price and style categories, from the low-priced volume car, the Chevrolet, up to the Oldsmobile and Buick and finally to the luxury lines of a Cadillac.

As for the Packard, its reputation was peerless. In a *Fortune* magazine article on the automobile industry, the writer exclaimed that, next to the Ford Motor Company, Packard "was the most valuable name in the auto industry." Although removed from operational management, Ward had begun nearly all of the guiding principles that brought the company to preeminence. Beginning with the steering wheel and the accelerator pedal, his numerous innovations all became industry standards. He established a corporate genetic code based on technological and aesthetic exceptionalism (the Packard was said to be a gentleman's car built by gentlemen); he implemented innovative and rigorous testing and research facilities.

The Packard's elegant Goddess of Speed hood ornament remains one of the most enduring symbols in automobile history. Photograph by Stacy Perman.

Ward gave consumers cars that elevated their status, products that they hadn't known they needed or wanted, and he understood the importance of nurturing their comfort level as they became accustomed to new and sweeping phenomena. Just as he had reassured his lightbulb customers at Packard Electric, Ward had made automobile expertise and repairs available through strong dealer relationships as well as pioneering sales and advertising strategies. A network of elegant Packard dealerships sprouted up all over the country, many of them resembling industrial palaces or movie sets, with glamorous showrooms displaying the cars and backrooms where servicing and repairs took place. The company's marketing and advertising department produced perhaps some of the most beautiful corporate literature ever generated; there were leather booklets and owner manuals, and the German artist Henry A. Thiede illustrated calendars, catalogues, and countless editions of *The Packard* magazine. All played off Ward's now unmistakable utterance, "Ask the man who owns one."

Packard executives well understood the importance their customers played in promoting the Packard brand, and the marketing department kept comprehensive lists filled with copious details on them: William K. Vanderbilt bought his first Packard in 1911, and William Randolph Hearst in 1917. W. F. R. Murie, head of the Hershey Chocolate Company, bought his first Packard around 1910 and in the course of thirty-five years purchased forty-five in all, even custom-painting one to match the color of a chocolate bar. The company also kept a list titled "Prominent Packard Owners of Long Standing," which catalogued every head of state, monarch, and potentate who owned a Packard; the list would come to include Chiang Kai-shek, the shah of Iran, the queens of England and Spain, the emperor of Japan, Czar Nicholas II of Russia, the Aga Khan, General George Patton, Presidents Franklin Delano Roosevelt and Harry Truman, and Hollywood royalty such as Clark Gable.

More than merely documenting its patrons, Packard publicly traded on the people who owned one. As America's plutocracy broadened, Packard's marketers cleverly paid a none too subtle homage to Mrs. Astor's famously exclusive ballroom with the advertisement, "The Packard Four Hundred," a beautifully produced roster celebrating those families who had owned Packard cars for twenty-one years or more. Similarly, the company promoted its "Famous Packard Gateways," an attractive catalogue of the stately front gates of famous Packard owners across the country.

In the years immediately following his decision to step down at Packard Motor Company, Ward accelerated his watch collecting. Between 1914 and 1920 he acquired dozens of watches, filling his desire for engineering perfection with complicated mechanical timepieces. His diary, functioning more like a ledger, became dotted with his minimalist entries, noting his visits to Wittenaur's, the Swiss firm on West Thirty-sixth Street in New York City, meetings with his longtime dealer F. E. Armitage, and random notations on a price, feature, or movement number. In 1915 he acquired an eighteen-carat gold tourbillon escapement pocket watch, movement no. 500–2, from S. Smith & Son. The following year, the British house delivered the automaker an eighteen-carat gold timepiece, movement no. 211682, with a split-second stopwatch, made of nonmagnetizable materials, considered a highly unusual request for the day.

Ward's *grande complication* had done little to sate his passion. The more watches he acquired, the more interested he became in discovering novel approaches to timekeeping, to see just how far he could push the labyrinthine inner workings of a complicated watch. The more watches he commissioned, the more obsessed he became.

The man who believed that simplicity stood at the crux of car design thrilled at combining the greatest number of complications within the very limited case of a pocket watch. Throughout history the most important watchmakers pursued their craft the same way that Ward approached mechanical engineering: obsessed with perfection, tirelessly developing new complications, and all the while relentlessly refining existing solutions.

Although he enjoyed the classic complications such as perpetual calendars, moon phase indicators, and split-second chronographs, Ward was particularly enchanted with minute repeaters, the chiming mechanisms that allowed the wearer to hear the time. The complication dated back to the seventeenth century, when man lived by candlelight. Watchmakers produced tiny hammers that struck small bells to mimic a church carillon, converting the beats of time into music within an accuracy of sixty seconds. At the push of a slide button on the edge of the case, mechanical sensors measured the time based on the position of the gears that drove the hands and then, in succession, chimed on the hour, the quarter-hour, and the minute—all in a different tone to designate each interval of time.

The core of Ward's collection came from Europe, and the best pieces from Switzerland. Paul Moore, secretary of the Horological Institute of America, called these "matters of selection." From the Ancienne Maison of Auguste Agassiz, the Swiss family firm established in St. Imier in 1832 and the forebears of the watchmaker Longines, Ward commissioned an eighteen-carat gold pocket watch with an enamel dial and Roman numerals. The sophisticated instrument featured an eight-day up-and-down power reserve indicator. Originally called a *réserve de marche,* this complication was designed to show the amount of remaining stored energy by indicating the tension on the mainspring at any given moment. When fully wound, the steel hand pointed to the 8, descending as the power wound down each day.

Shortly after the turn of the century, a new kind of timekeeper, the wristwatch, slowly made its way into the affections of fashionable gentlemen. First introduced in the late nineteenth century, the wristwatch's precursor was primarily worn by ladies, usually on a neck chain or as a pendant, before becoming a bracelet. Men had initially rebuffed such timepieces as the "mark of sissies." In 1904 the Brazilian aviation pioneer Alberto Santos-Dumont asked Louis Cartier to design a watch that left his hands free as he piloted his "flying machines," and the perception of men's wristwatches began to shift. In 1911 the Parisian jeweler began selling its Santos model wristwatch to the public. But it was the Great War that really changed the fashion, after soldiers' "trench watches"—bulky pocket watches affixed to straps—provided hands-free ease of function during combat. Off the battlefield, they were quickly renamed "officer's wristwatches" and gradually became symbols of masculinity, particularly after other pilots, equestrians, sportsmen, and automobile enthusiasts found them appealing.

In 1918, three years after Patek Philippe began offering its first leather-strap "officer's watch," Ward acquired what appears to have been his first wristwatch, an engine-turned (*guilloché*) cushion-cut Patek Philippe, made of eighteen-carat gold with iridium hands, an enamel dial, and a sixty-second sub-dial. On the back he had the cartouche inscribed, *Fabriqué pour James Ward Packard, Warren, Ohio, 1918, par Patek, Philippe & Cie, Genève.*

Ward viewed the wristwatch as a novelty. Even as the pocket watch waned in popularity, his affections remained tethered to those timepieces that, within ten years, would be called old-fashioned "turnips."

During his early collecting days, Ward had commissioned select pieces from Vacheron Constantin, working through Edmond E. Robert at Maiden Lane. He now dealt almost exclusively through Armitage in acquiring his timepieces, all designed as homages to art and science. A beautiful Vacheron pocket watch unusually forged in twenty-carat gold with a delicate floral bas-relief case, no. 233573, had a pair of dials engraved with the image of a house in a lush garden, one in fourteen-carat gold and the second made of enamel. Ordered in 1917 and finished a year later, the watch had a subsidiary dial indicating the seconds and was delivered to Armit-age for Ward on March 11, 1919. Ward acquired another in 1918, for his wife Elizabeth, a silver, square-shaped pocket watch with an elaborately engraved case embedded with a round enamel dial and its sub-dial for seconds. On the back was a scene in bas-relief of winged archers on horse-back. (Ward had also gifted his wife a Patek Philippe pendant watch made of platinum and enamel and encrusted in diamonds.) Earlier Ward had acquired a Vacheron Constantin pocket watch, received in 1912; it was a beautiful although relatively utilitarian piece of eighteen-carat gold, an enamel dial with Arabic numerals, and a sixty-second sub-dial.

Ward's fourth piece made by Vacheron Constantin, a chronograph made of twenty-carat gold, with the movement no. 375551, measured both elapsed and conventional time. It was the most finely complex watch that the engineer had requested of the watchmaker. Acquired in 1919, it was also one of the most singular examples that Vacheron Constantin would craft in its own history. Among the chronograph's complications

Ward acquired his rare twenty-carat gold Vacheron Constantin pocket watch chronometer with trip repeater, *grande* and *petite sonnerie,* and half-quarter repeater in 1919. The monogrammed case matched the automaker's personal stationery. Photographs by Stacy Perman.

were a trip repeater, *grande* and *petite sonnerie,* and, most unusually, a half-quarter repeater that struck at seven and a half minutes as well as on the hour and quarter-hour, all in different tones. It was unusual and rare. No other Vacheron chronograph with a half-repeater was known to have been made before it or since, a particularly precise timekeeper for a particularly exacting man.

Although Ward maintained his affair with Vacheron Constantin, his affections toward Patek Philippe never wavered, particularly when it came to constructing his most complicated pieces. At times, he communicated his desires directly to Geneva without Armitage brokering the transaction. Each piece, produced to his requirements, was built after he approved blueprints and designs, down to the color of the dial, the type of hands, and almost always the engraving of his name on the cuvette or movement itself.

In 1918, a year in which Ward hotly pursued a great number of complicated pocket watches from Patek Philippe, he made an exclusive order for a silver and gold open-face watch with a dual time zone on a subsidiary dial and oversized luminous numerals, with the movement no. 190757. It also indicated mean time, set to the international clock in Greenwich, England, which reflected world time on a twenty-four-hour scale—an interesting choice for a man who had reduced his world to Warren, Ohio, and Lakewood, New York, with a few side trips in between. The same year, he received another eighteen-carat gold Patek Philippe pocket watch, movement no. 174907, with a gold dial, an up-and-down reserve indicator, contained in an exceptional Murat-style case.

Curiously, Ward could sometimes seem detached about his collection. No matter how much he may have paid for a watch, he never considered it a rare treasure but wore it regularly. And yet once his watches were manufactured and in his possession, he was already thinking about his next commission. A great number of his watches sat in their fitted boxes untouched and idle for years. Even so, his interest in watches went beyond their engineering. He also made sure to embed in them a part of his soul.

In 1920 Ward received a highly unusual, open-face, eighteen-carat gold minute repeater with *grande* and *petite sonnerie* and an up-and-down indicator, given the movement no. 174876. Fascinated with the sea, a boating enthusiast and inventor of naval motors, he had tasked Patek Philippe to craft for him a pocket watch with a ship's bell striking. This ancient nautical timekeeping instrument had never previously been miniaturized to fit inside a pocket watch.

In the days before such a striking was included in instruments for ships' rather large clocks, a sailor on watch would strike a bell after a sandglass emptied into its bottom chamber during six shifts divided into six hours each. Ward's watch was built with tiny gongs that followed the bell system of a ship's clock that struck on each half-hour up to eight bells; for instance, sounding six strokes on a single gong at 6:15 followed by one *ting-tang* double strike. At 6:30 the watch struck on a single gong followed by two *ting-tang* double notes. At 6:45 it sounded six strokes on a single striking followed by three *ting-tang* double notes. Levers on the dial allowed Ward to silence one or all of the double notes of the *petite sonnerie* on the quarter-hour, the gong of the *grande sonnerie* on the hour, or the ship's striking. It was a particularly difficult undertaking for the watchmaker. Of the watch's thirty-seven jewels, the small donut-shaped rubies placed into pivot points to reduce friction within the movement, twenty-one were for timekeeping functions and sixteen were used for the striking sections.

As Ward challenged Patek Philippe with one ingenious complication after another, Henry Graves, Jr., stepped up his game and entered the realm of *grandes complications*. During Ward's most prolific period of commissions, between 1912 and 1918, Henry purchased at least five pocket watches from Tiffany & Co. Certainly Henry had the money and the drive to go toe to toe with the automobile magnate over complicated pieces, but on his own he didn't have the kind of technical and creative gifts that Ward possessed to push Patek Philippe or any watchmaker to produce artistically inspired timepieces. He could only chase a step behind the imaginative genius.

In the middle of 1919 Henry sauntered confidently through the doors of Tiffany & Co. directly under the great clock held aloft on the giant shoulders of Atlas. He was expected and recognized. Tiffany maintained an interest in Henry, along with the rest of society's elite who passed through the store's Corinthian columns, weighted down in ropes of jeweled baubles. It was said that Charles Tiffany himself had initially employed a small staff dedicated solely to keeping exhaustive files on its wealthy clientele, filling the pages with details regarding their financial status, newspaper stories, and photographs. When these well-heeled patrons arrived, they were identified immediately, addressed by name, and permitted to take away their distinctive boxes in robin's-egg blue first and pay later.

The Graves family file stretched back to Henry Graves, Sr. In 1886 the jeweler had reserved a leather impression of Henry Sr.'s name with detailed flourishes for embossing his briefcases and bookbindings (at the time, a practice considered quite a special request). The jeweler had long become accustomed to Henry Jr.'s frequent visits and now acted as the middle man between him and Geneva, providing blueprints and delivering pricing, scheduling, and finally the finished product. As with Ward's pieces, these watches took years to finish. Usually Tiffany handled the details of the engravings, sending the pieces to Geneva to be stamped with the Graves coat of arms and motto.

A trip to Tiffany to buy a watch had become something of a pastime with the Graves family. A year earlier, Henry's son Duncan had purchased a pocket watch of his own at the jeweler's, and his eldest son, Harry, acquired four watches, at least one of which he gifted to his soon-to-be fiancée.

Something of a ne'er-do-well, Harry had left Princeton University after spending an unremarkable semester largely inside the local cinema. He returned home to Irvington and to Margaret Dickson, the dark-haired beauty he had become smitten with while summering with his family in the Adirondacks. The New Jersey debutante daughter of Joseph and Mary Dickson was a superb match for Harry, as the affluent Dicksons had major interests in the Delaware and Hudson Canal Company.

On April 18, 1918, with the drums of war in the air, Harry somewhat impetuously ended his life of unblinking privilege and signed up for Britain's Royal Air Force. America had yet to officially declare war on Germany, and Harry, apparently rebelling against the dull respectability of his life, jumped into the fray, joining up with the RAF in Canada. Just before leaving for training in Toronto, however, he married Margaret, on July 6, 1918. The pair wed in a lavish ceremony at Manhattan's St. Regis Hotel ballroom, which was decorated in peonies and palms. Harry was twenty-one and his bride nineteen. They gave each other Patek Philippe watches from Tiffany inscribed with their initials. Harry's watch was also inscribed, *Royal Air Force's England MDG to HG.*

Before Harry had left his Canadian barracks, however, the Great War ended on November 11, 1918. He and his new bride honeymooned in Palm Beach and then settled into an apartment at the St. Regis Hotel. Harry assumed a position at the New York Trust Company, which would soon merge with Liberty National Bank, where his father had held a board seat and the family a sizable tranche of stock. As the couple's second wedding anniversary approached, Henry laid out $80,000 (about $921,516 today) for a colonial mansion for them spread across two acres in Ardsley-on-Hudson. Like their parents, the newlyweds easily slipped into the high life of their social set, with servants and nannies taking care of

Henry Graves 3rd.
Courtesy of Cheryl Graves.

their houses and children. Their daughter Mary Dickson was born barely nine months after their wedding, and within two years she was joined by Florence Barbara and then Henry Dickson.

Despite his gentle carriage, Henry had a powerful ability to translate his desires into possessions. With the receipt of a spectacular pocket watch, James Ward Packard had unknowingly placed Henry into the position of having to reassess what exactly he was made of.

It was during this period that Henry commissioned the first of his *grandes complications,* an open-face eighteen-carat gold pocket watch with minute-repeating *grande* and *petite sonnerie* with split-second chronograph, thirty-minute register, perpetual calendar, and ages and phases of the moon. In all, it had twelve complications. He would not receive the timekeeper for six years.

In many respects the year 1919 marked the shot across the bow. Over the next ten years, both Henry and Ward would order a majority of their pocket watches from Patek Philippe with ever more combinations of complications. In their passion for collecting, it appeared that both men felt each other's gravitational pull. And Patek Philippe shrewdly rode the orbits of two of its most important patrons in America, if not the world. The firm dedicated its top watchmaking resources to developing the finest watches for them.

Before the Great War a locomotive of art, literature, food, fashion, science, and philosophy had roared out of Europe's academies, cafés, palaces, and streets. After the war Europe's streets were scarred, its cafés empty, its palaces charred, its houses turned to rubble, and its citizens deeply wounded. The once great Continent had cratered, its economies as blasted as its cities and countrysides. More than an ocean separated the once dominant Europe and the now ascendant America.

From Patek Philippe's workshops on the Grand Quai in Geneva, the collectors' quest appeared to be an example of hubris in an exuberant new era of American world influence. For the watchmaker, Graves and Packard soon represented a new kind of prized patron in a history populated with kings and queens, statesmen and the celebrated, who would come to include Tolstoy, Tchaikovsky, Einstein, and Curie. Henry Graves, Jr., and James Ward Packard changed the game. These gentlemen rivals had bottomless resources, they knew without hesitation what they wanted, and they were out to win. And they came to desire the same thing: to own the grandest of the *grandes complications*.

In many respects, the contest to build the most complicated watch in the history of mankind was no different from the race to circumnavigate the globe, to scrape the sky with the world's tallest building, or to discover the vast riches of the Egyptian pharaohs buried under the ancient sands of the Valley of the Kings. All such undertakings required determination and a touch of obsession. A *grande complication* also required a small army of brilliant and skilled artisans, technicians, and watchmakers. The competition between the two men began like most pursuits of this sort: with an innate desire to be first, to be bigger and better, to go where nobody else had gone before—to bend history to their own terms.

Since the beginning of timekeeping, the most important and complicated watches were the result of one contest or another. Indeed the men were following the tradition of kings stretching back five centuries.

CHAPTER TEN

Vallée de Joux

In the beginning there was the sun.

At first light, early man understood that the heavenly line between dawn and dusk had begun again. As darkness fell, he observed the invisible hand at work, replacing daytime with night. Mankind's relation to the Earth was defined by the seasons and the cycles of the celestial bodies: the sun, the moon, the planets, and the stars that measure its passage. Slowly he began to capture time's ceaseless motion. To experience time was one thing, but to use it was quite another. At the turn of every epoch, the instruments devised to calculate time represented more than the evolution of science; they symbolized the apogee of civilization itself.

The clock marked the starting point of the modern world.

Over twenty thousand years ago, Ice Age hunters in Europe scored lines and bored holes into sticks and bones to count the days between the phases of the moon. The Sumerians, living in the Tigris-Euphrates Valley five thousand years ago, developed a calendar dividing the day into twelve periods, each split into thirty equal parts. In pre-Columbian Mesoamerica, the Mayans created two calendars: the first, 260 days, and the second, 365 days, were both based on the sun, the moon, and the planet Venus.

The ancients followed the path of the sun. Three thousand years before Christ, the Chaldeans invented a bowl-shaped sundial with a pointer that cast a shadow across its dial, marking twelve hours of the day. Built by the Egyptians, who also constructed one of the oldest scientific clocks, the narrow, towering pyramid-shaped obelisk, the sundial was perfected by the Greeks. Its use quickly spread to Rome in the third century BC after the emperor waged a naval battle against Egypt and Roman soldiers returned to the Republic with a sundial in tow. Later the Egyptians designed water clocks, known as a clepsydra, from the Greek meaning "water thieves." These stone vessels made with sloping sides enabled water to drip at a nearly constant rate from a small hole. Although these devices attained high levels of refinement, they also remained decidedly inaccurate.

Astronomers were the first to note more exacting measurements of the Earth's tilted whirl on its axis in minutes, seconds, and fractions of a second. More than a thousand years after the Egyptians marked the shadow of the sun, the Greek mathematician, physicist, and astronomer Archimedes further advanced the clepsydra, building an instrument of complicated mechanisms incorporating one of the earliest sets of gears. As the eminent American scholar and economist David S. Landes noted, "The clock did not create an interest in time measurement, the interest in time measurement led to the invention of the clock."

As Europe fumbled over the heavens, the Chinese became the most advanced timekeepers of their day. Chinese horologists had developed a series of remarkable mechanized astronomical clocks, and in the year AD 1086 the Chinese emperor tasked one of his diplomats, Su Sung, an administrator and scientist, to build an astronomical clock to surpass all others. The result was an elaborate thirty-foot-tall clock tower designed to mimic the movements of the sun, moon, and stars based on the Chinese calendar and powered by a water wheel. The astonishing machine had a brass armillary sphere that indicated sunrise and sunset. Five front panels opened to a parade of changing figurines ringing bells or gongs, each holding tablets indicating the hour or other special times of the day.

A marvel of the era, Su Sung's masterpiece took eight years to complete, but its reign was short-lived. In 1126 invading Tartars from the north stole the magnificent clock. Despite their best efforts, succeeding Chinese emperors failed to restore the great instrument or even to duplicate it. The tragic loss of Su Sung's water clock marked the end of the high art of Chinese horology. Knowledge of it remained lost until the nineteenth century, when Su Sung's illustrated monograph, originally prepared for his emperor, was rediscovered.

During the Middle Ages, a scrim of darkness fell across Europe, and virtually all technological advancement ceased. In the Islamic world, however, science, medicine, mathematics, literature, philosophy, and astronomy flourished, as did masters of the sundial and clepsydras. The caliph of Baghdad honored the Christian emperor Charlemagne by gifting him with a water clock. Islam considered timekeeping a sacred obligation in order to maintain the correct prayer times and to determine the direction of Mecca. The days and months were ruled by the lunar calendar. The Muslims also greatly refined the astrolabe, an ingenious little device that tracked how the sky appeared over a specific place at a given time, which they re-created from translated Greek texts during the eighth and ninth centuries. In the twelfth century, Islamic Spain introduced the astrolabe to Europe, where it became the most popular astronomical instrument for some three hundred years.

The evolution of the mechanical clock, however, followed a long, slow path. Mechanical clocks first appeared in fourteenth-century churches and monasteries in Europe, where the well-ordered calls to worship obeyed a stringent prescription of time. As early as 1321, the clock tower at Norwich Cathedral in England summoned the monks and nuns to prayer six times a day. These primitive clocks conducted the energy of falling weights through a gear train made of wheels and pinions that indicated time by the striking of bells. (The word *clock* originally comes from the Latin *clocca*, meaning "bell.") This weight-driven clock offered many improvements over earlier mechanisms; unlike sundials or water clocks, it was largely impervious to adverse weather.

After the discovery that time could be divided into small measureable intervals, several breakthroughs in timekeeping became possible, including the invention of the escapement, which provided the connection between the regulating device and the gear wheels. A power source such as weights or a coiled spring caused the gear wheels to turn, while the escapement controlled the rotation, regularly releasing the teeth on the gear wheel, counting the beats of time. It was the escapement that enabled the ticking element at the heart of the entire mechanism, and with it the mechanical clock was born.

The invention of the mechanical clock is one of the seminal events of humankind, ranking alongside the invention of movable type in advancing civilization. The ability to measure time was the major driver in transforming Europe from a torpid, remote quilt of tribes and polities into a world power that amassed wealth and power through knowledge and technology. The clock liberated time from nature. Emerging from Europe's church towers, the mechanical clock soon held sway over economic, political, and social life. The sound of the chimes during Vespers gave way to the peal of the village clock announcing the opening and close of markets. Time no longer solely marked a schedule of religious practice; it became the standard measure of productivity. A gauge of the value of man's time was now possible.

The rulers of Italy, Germany, France, England, and the Netherlands quickly seized on this new technology, each vying to take the lead in this emerging craft that commanded great skill and greater fortunes. These towering spectacles with their astronomical indications, automated figures, and carillons quickly became symbols of affluence. The majority of clocks were made for public use, and time became the business of the state. The powerful bells inside Europe's belfries denoted regal supremacy. In 1370 King Charles V of France commanded that all of Paris's clocks be synchronized with that of the Palais Royal to prove his dominion over time. While the clock had emancipated man from nature, the royals had determined that time belonged to the realm of kings.

Almost as soon as clocks began showing up on church towers so did the desire for smaller timekeepers. But miniaturization demanded more advanced technology, which meant higher costs. Individual ownership was confined largely to royalty, courtiers, and the richest of merchants. As clocks diminished in size, the prestige conferred on their owners increased exponentially.

When clocks moved from the turret to the table, they also became more portable and could reside in private residences and carriages. As they gradually shrank in dimension, clocks became fashionable personal accessories. Indeed they became an early sign of the individualism that would pervade the Renaissance period within the ranks of the privileged classes. Along with fashion and convenience, miniaturization led to greater precision; since miniaturization required a greater degree of technical perfection, opening the door to greater errors, watchmakers by necessity were forced to create more precise timepieces. In other words, the mechanical clock still left much to be desired.

During the first 150 years of their existence, mechanical clocks were crude, inexact, and clumsy. It would be another four centuries before precision timepieces started to appear. The word for *hour* attained its value as a particular interlude of time only in the fifteenth century, while the words *minute* and *second* followed a hundred years later. Given the dearth of technical innovation, watchmakers gave scant attention to accuracy and instead spent their time competing with one another to create the most ingeniously elaborate pieces to delight their captivated patrons.

An entire trade industry developed around watchmaking, which soon organized into highly dedicated guilds, with each country developing specializations based on taste, demand, and production. Artisans working with gold, precious stones, and enamel turned out timepieces that were sumptuously designed and grandiose in appearance. For watchmakers, until reliable instruments eventually absorbed their energies, virtuosity trumped accuracy. These first clocks were jewels, symbols of wealth, power, and status. Their timekeeping ability was ancillary. Very quickly these novelties became some of the most coveted objects.

Watchmakers fought for the honor to craft showpieces and gain favor with society's wealthiest, most influential patrons. In 1518 François I of France dispatched a slice of the French treasury to pay for two miniature gilded clocks embedded in the handles of a pair of daggers. In England craftsmen hoping to impress Elizabeth I devised a tiny alarm watch set inside a ring; at the appointed hour a small prong extended and softly grazed the queen's imperial finger. While these enchanting works of genius captivated and charmed their royal wearers, they did not have the same miraculous effect on the telling of time.

At the end of the fifteenth century, the spring-driven coil made its debut, ushering in the first important development in horology in over a century. Unlike the weight-driven clock that needed to remain stationary, the coil spring distinctly benefited transportability. However, it also posed a new challenge: while weights displace equal heft as they drop, the force of a coil decreases as it unwinds. The solution came in the form of the fusee wheel, a conical wheel placed in between the mainspring and the wheel train that, when pulled, compensated for the retreating force of the spring. The spring-driven clock introduced the truly movable timepiece and, with further innovations in miniaturization, the object that came to be called the watch.

Around the year 1500, eight years after Christopher Columbus sailed to America, a German locksmith named Peter Henlein invented the drum watch using a ribbon of steel closely wound around a central spindle to maintain the motion of the mechanism. It was housed in a cylindrical case with a hinged lid that concealed the dial. Henlein lived in Nuremberg, a grimy city of red-tile roofs that, along with Augsburg, was a major center for clock making, metalworking, trade, and finance. While some accounts dispute Henlein as its inventor, the drum-shaped forebear, called the Nuremberg egg, of the watch with a mainspring-activated drive came out of southern Germany during the first half of the sixteenth century. It is the earliest known portable timepiece. Made of iron and constructed with pins and rivets, it functioned for forty hours before it required rewinding. However, the device had only one hand, to indicate the hour.

While portable, these drum watches were hardly compact. Heavy and bulky, they were not much different from table clocks, yet they brought timekeeping one great leap forward to the pocket. The making of them spread among watchmakers, who added components such as alarms and striking mechanisms. People no longer had to depend on the church bell or town square clock, and they began organizing their lives around the rhythms of their own day. Time was now within the individual's grasp.

With the greater emphasis on time came a deeper awareness of the brevity of life. The Black Death, the devastating pandemic that killed nearly a third of Europe's population between 1347 and 1350, had left an indelible mark on the continent. The ability to keep time changed commerce and politics and also became entangled in society's views of virtue, salvation, and diligence. With the unalterable notion that time was precious and death always close by, it became quite fashionable to carry small skull-shaped watches called *memento mori* as reminders of mortality.

Mary, Queen of Scots commissioned several of these death's head watches during her lifetime. Among her notable collection of *memento mori,* the intricately designed silver gilt skull that French watchmaker J. Moysan of Blois fashioned for the doomed queen is perhaps her most darkly impressive. The watch is elaborately engraved with scenes depicting the Garden of Eden, the salvation of man, the Crucifixion, and time devouring all things, each tableau accompanied by verses from the poet Horace. On the skull's forehead is the figure of Death, with his scythe and an hourglass. A hinge on the jaw opens to reveal the watch's movement occupying the hollow of the skull, where a small hammer strikes a silver bell on the hour. On the flat where the roof of the mouth lies under the base of the brain is a silver dial plate rimmed in gold and richly carved in a scroll pattern with the hours marked in large Roman numerals. Before the queen was executed for treason in 1587, she gave the watch to one of her ladies-in-waiting.

During this age of clocks, the aristocracy delighted in luxuries, and *montres d'horloge,* as watches were called, were eagerly sought. Men wore the engraved, embossed, and bejeweled pieces on thickly braided chains around their neck or hanging from their belt. Hans Holbein's famous portrait of Henry VIII circa 1536, now hanging in the Thyssen-Bornemisza Museum in Madrid, shows the king wearing what appears to be an elaborately designed drum watch hanging from a thick gold chain.

Not to be outdone, women paraded their fanciful, sumptuous watches as highly decorative accessories to match their costumes. According to an inventory compiled in 1523 of the jewels owned by the regent of the Netherlands, Margaret of Austria possessed a watch "in the likeness of an apple, with a little golden chain, with a ring at the end of said chain."

During Queen Elizabeth's forty-five-year reign, she received numerous horological gifts: clocks as pendants, staffs, rings, hanging from chains, and fashioned into whimsical shapes, such as a horse bearing a globe with a crown, a frog, flowers, and even the head of Elizabeth herself in high relief. They were forged in gold, set in mother-of-pearl, or gilt brass, all garnished with precious stones. One inventory of the queen's possessions included more than thirty clocks and watches, only a partial accounting of her ultimate collection.

The dawn of precision mechanical timekeeping coincided with the Age of Discovery. From the second half of the fifteenth century through the seventeenth, Europe's explorers sailed the seas in search of new trading routes and lands, claiming magnificent bounties of gold and silver, coffee, spices, sugar, tea, and tobacco on behalf of king and country. Explorers such as Columbus, Sir Francis Drake, Vasco da Gama, Bartolomeu Dias, and Ferdinand Magellan triggered a global shift, leading to the rise of colonial empires and contact between the New and Old Worlds. The economic center of the world swung from the Islamic Mediterranean to Western Europe as the scientific ferment sweeping across Europe made the discovery and colonization of far-off lands possible. As the compass, sundial, and numerous advancements in cartography and navigational instruments aided the explorers, these years also saw a prodigious volume of developments in the arts and sciences.

One hundred years after the Gutenberg Bible was printed, Nicholas Copernicus published his theory on a heliocentric universe, *On the Revolutions of the Heavenly Spheres*. Leonardo Da Vinci filled his notebooks with scientific observations, including studies on anatomy, the laws of motion, the flight of birds, and the actions of water. Pope Gregory II introduced the Gregorian calendar, and Galileo helped to found modern science. Nearly one hundred years before the Dutch mathematician and scientist Christiaan Huygens revolutionized horology in 1656 with the invention of the pendulum as a means for measuring time, it was Galileo who first discerned the use of a system based on natural oscillations.

Hungry for new technology, monarchs, nobles, and popes assembled great collections of timepieces representing the latest in innovation, including expensive clocks. Kings appointed royal astronomers, mathematicians, and clockmakers to their courts. A great patron of astronomical research, William of Hesse supported Tycho Brahe, himself the son of a Danish nobleman and the most famous astronomer in Europe. Brahe made the most precise instruments and observations of the heavens before the advent of the telescope, cataloguing one thousand stars. William also brought the finest horologist of the period, the Swiss-born Jost Bürgi, to his kingdom in Kassel, Germany. Considered the most advanced clockmaker, Bürgi invented mechanisms that radically improved the accuracy of clocks, transforming them from mere whimsy into powerful scientific instruments. For William, Bürgi built a clock that marked hours as well as minutes and seconds and, as William boasted, fluctuated no more than an extraordinary one minute a day.

Bürgi's talents were so highly desired that for a period he traveled regularly between Kassel and Prague, where he worked for Rudolph II, king of Hungary and Bohemia and Holy Roman emperor. Rudolph raised court patronage to heights of extravagance never before seen, turning his royal castle, the Hradčany, into an extensive cabinet of wonders. Packed inside were paintings, decorative arts, and every contrivance of the day: maps of the heavens, globes of the Earth, rhinoceros horns and leopard claws, clocks and scientific instruments. The entire city of Prague was reengineered as a center of art and science. In addition to objects that portrayed the world in its entirety, the emperor brought the best minds in Europe to his court. Brahe was named imperial mathematician, followed by Johannes Kepler; both established observatories in Prague. Among his possessions, the emperor also kept his own precious *memento mori,* a skull watch with a movable lower jaw that struck the hours against the upper one.

Amid this flowering of knowledge in Europe, clock making quickly became an extension of the rivalries and one-upmanship that existed on the continent. In the seventeenth century, to rule the seas was to reign supreme. As the great seafaring powers pressed for dominance, celestial navigation gave way to better timekeeping instruments. Sailors had to calculate both latitude and longitude to navigate accurately. While latitude could be determined by measuring the sun's position or that of the stars above the horizon, longitude was trickier. Keeping time at sea meant confronting a host of uncertainties: the ship's motion, gravity in different latitudes, variations in humidity, and temperature changes.

As the numerous wrecks lying at the bottom of the ocean made painfully clear, the need to determine longitude was absolutely necessary. Longitude posed the greatest scientific problem of the day, and by the seventeenth century all of the great maritime nations were offering royal sums to the individual who came up with an accurate and reliable way to measure it. King Charles II founded the Royal Observatory in 1675 to solve the problem of finding longitude at sea, setting the stage for the Longitude Prize. With an Act of Parliament, in 1714 the British government offered a purse of 20,000 pounds (worth nearly $20 million today) for a solution that could provide longitude to within half a degree (two minutes of time). Sir Isaac Newton enhanced navigation with the 1687 publication of his *Principia,* enabling sailors to predict the movement of the moon and other celestial objects based on his theories of motion. However, Newton declared that defining longitude was a feat of impossible proportions. Indeed "finding longitude" soon became a term for the pursuit of fools.

Yet Europe's greatest thinkers rose to the challenge to unravel this ancient mystery that had baffled seamen, confounded scientists, and held statesmen hostage. In the end, the prize went to John Harrison, a carpenter and self-taught clock maker from the English village of Hull, who challenged the entire scientific and academic establishment. After nearly thirty years and numerous trials and political machinations, Harrison won for his fourth attempt at building a marine timekeeper by accurately fixing a ship's position at sea from celestial observations. In 1764 the device known as the H4 correctly measured the stars at sea during a forty-six-day journey between London and Jamaica. It was found to be in error by only five seconds.

This transformative contest did more than aid safe passage across the oceans. Harrison's achievement spurred others to improve upon his success, men like the French clock maker Pierre Le Roy, who invented the detent escapement; the inventive Swiss clock maker Ferdinand Berthoud, builder of marine clocks; and the British horologist Thomas Mudge, who devised the lever escapement. All brilliant rivals, they set the course for the modern chronometer, an extremely precise timekeeper designed for accuracy under the vagaries of temperature and pressure.

Previously timekeeping had followed two tracks: lavish items for the aristocracy and high-precision navigation instruments. From this point forward, clock making was elevated to a marriage of science and art in equal measure. And it was small, landlocked Switzerland that emerged to challenge the 200-year-old grip that Italy, Germany, France, and the Netherlands had long held over timekeeping.

In the middle of the sixteenth century, Geneva was a speck of a republic in a universe of warring kingdoms and principalities. Measuring just one hundred square miles, it was far smaller than London, then the city of Shakespeare, Elizabeth, and expansionist mercantilism, or Paris, Europe's great seat of culture, immense wealth, and power. Geneva was a trading city at the tip of Lac Léman filled with goldsmiths and enamellers, a picturesque outpost of mountains and glacial lakes, hardly a bustling center of trade or industry. Its history turned on bloody conflict and owes its horological hegemony to the persecution of the French Protestants, the Huguenots, and the sober French theologian John Calvin.

Throughout the century France was riven by the struggle between its

Protestants and majority Catholics. In 1541 Calvin, the leading figure of the French Protestant Reformation, was forced to flee Paris and eventually settled in Geneva. Once there he transformed the city into a Protestant stronghold and encouraged his fellow Huguenots to join him in refuge. After the St. Bartholomew's Day Massacre in 1572, when King Charles IX ordered the slaughter of several thousand Protestants in Paris and the provinces, huge numbers flooded into Geneva, including France's most eminent artists, businessmen, scientists, craftsmen, and a disproportionate share of the country's master watch and clock makers.

Under Calvin's puritanical regime, the Genevese shunned extravagance to live a life of austerity and hard work based on Scripture. Calvin condemned the display of wealth and vanity, so jewelers were now forbidden to manufacture "crosses, chalices or other instruments of popish idolatry," but not timepieces. Ostentation was a vice, but punctuality was a virtue. With the Huguenots' knowledge of Europe's most advanced watch production and its tradition of decorative expertise, Geneva became a renowned watchmaking center. So numerous—and mindful of outside competition—were its watchmakers that, in 1601, they established the first guild, the Ordonnances et Règlements sur l'Etat des Orologiers. The craft spread north and east, and entire families worked in the watchmaking industry.

The final wave of Protestants arrived in 1685, after Louis XIV issued the Edict of Nantes, reversing nearly a century of tolerance. Two hundred thousand Protestants fled, most to Geneva, where they joined a thriving city crowded with watchmakers and their workshops, compelling many of the émigrés to decamp north to the harsh isolation of the Vallée de Joux.

Stretching from Geneva in the south to Neuchâtel on the shores of the lake and farther on to Basel in the north, the long, thin Vallée de Joux lies within an arc formed by the Jura Mountains. Concealed from France by the Grand Risoud Forest on one side, it is cut off from Lake Geneva on the other by the Mont Tendre range, leaving this rugged valley in rural seclusion. It is believed that the first people to settle in this unforgiving land were ascetic monks.

The austere land of the Jura yielded little. The snow on the ground met the snowcapped mountains, leaving the valley in a perpetual white haze, even in summer. Once the first snows fell in October, life ground to a virtual halt. The roads became impenetrable for months on end, and, naturally defended by steep passes, the inhabitants withdrew to their workshops. Working by candlelight they developed the skills of precision and repetition, cutting minuscule parts and polishing precious stones.

They became masterful builders of music boxes, which required meticulous construction. Individual families became highly skilled in crafting certain parts, say, the mechanics or the decorative boxes. Collectively, the community finished them off and during the summer brought them to Geneva for sale. This aptitude for small, exacting manufacturing translated easily to making timepieces, particularly watches.

A mechanical watch is an extremely complex object, demanding several different skills. The system that developed in the Jura Valley, of breaking down tasks into several different subtasks, put the Swiss at an advantage. At a time when watchmakers tooled their own parts, the specialization of labor meant that craftsmen who routinely endured the monotony of making and remaking the same part over and over again improved their productivity and learned to make quick adjustments.

Switzerland's horological trade blossomed. By the seventeenth century, the Genevese and the residents of the Jura who had taken to watchmaking with exceptional proficiency crafted some of the most refined, complicated timepieces in Europe. In particular, they made a great art of the fanciful *montres de fantaisie,* producing cases to look like animals, flowers, and fruits and cruciform watches of gold and enamel, decorated in precious stones with whimsical petals and wings that opened to reveal the tiniest of clocks.

While the French had introduced miniature painting on enamel around 1630, the Swiss took the art to unprecedented heights. Carefully trickling a few drops of oil or lavender into finely ground pigments, the artist applied the paint to an enamel surface, subjecting it to several firings to make the colors crystalline and permanent. The brilliance and diversity of hues that shone from the enamel made the pieces highly coveted, especially by the heads of European states. Craftsmen used a broad number of motifs; landscapes, biblical scenes, mythology, literature and poetry, and of course portraits were captured in vivid, fine detail. In France enameling had been almost exclusively a Protestant trade, now lost to the Swiss.

Switzerland's position as a trading crossroads helped spur its watchmakers to become adept at finding new opportunities for export. During the late seventeenth century, Constantinople had begun its love affair with lavishly decorated and enameled Swiss timepieces, which soon spread across the Ottoman Empire, followed by the China market. The Chinese took a particular interest in ornately decorated yin-yang pocket watches made in pairs, each a mirror image of the other, like butterfly wings.

By the nineteenth century, the workshops of Geneva and the Jura had developed the charming art of automata, little machines designed to imitate the movements of living creatures, from birds to humans. These captivating, complex little devices, particularly treasured in the East, ranged from the simple swing of an arm to full-scale concerts, automated scenes from the Bible, and the enchanting sight of a singing bird emerging out of the barrel of a jeweled pistol after its trigger was squeezed.

As the fine art of horology progressed, Europe's watchmakers relentlessly pursued new innovations. After revolutionizing the field with the addition of a pendulum to clocks in 1656, the Dutch scientist Huygens in 1675 devised the most crucial advance in watchmaking, the balance spring, transforming watches into precision instruments that no longer needed to be reset several times a day. A host of refinements and discoveries in escapements followed, all resulting in better, more reliable timekeepers.

A great period of legendary clock makers began. Thomas Tompion (1639–1713) started the line of three generations of the greatest watchmakers in British history, followed by his student George Graham, who mentored the third, Thomas Mudge. In 1704 Tompion was among the first to build watches using the balance spring. After George Graham (1673–1751) began working for Tompion, the two men became great friends and partners. Something of an open-source evangelist of his time, Graham became known for both his interest in making more innovations than money and helping his fellow horologists, chiefly John Harrison and Julien Le Roy, solve problems.

A stalwart in London's scientific and clock-making establishment, Graham perfected the cylinder escapement designed by Tompion after his mentor and friend's death. On his own, Graham racked up an impressive list of accomplishments: he built the master clock at the Royal Greenwich Observatory and is credited with designing the first stopwatch and a mercury pendulum system, which ensured better accuracy in extremely cold and hot weather. Graham's dead-beat escapements made clocks more precise overall, and the escapement that bears his name continues to be used in high-precision pendulum clocks. When Graham died, Tompion's crypt at Westminster Abbey was opened and he was buried beside him.

With timepieces' precision improved, watchmakers assiduously labored at their wooden benches to tool the most minuscule of springs, screws, and wheels for novel features in wildly curious combinations. By the mid-seventeenth century Swiss workshops began manufacturing the first complicated watches. Just as they had added a variety of astronomical indications to the great clocks, watchmakers toiled to load the greatest number of functions into the unforgiving space inside a pocket watch.

Some of the earliest efforts contained simple calendars and indicated the phases of the moon.

In 1625, merely a century after the invention of the portable watch, the French watchmaker François Pigeon devised an oval timepiece that, in addition to an hour hand, featured an alarm, the phases of the moon, and a manually triggered perpetual calendar. In 1884 the Washington Meridian Conference divided the globe into twenty-four distinct time zones, but a hundred years before it, a Genevese watchmaker named Pierre Morand invented the first World Time watch, with a twenty-four-hour dial that indicated the time in fifty-three cities. Watchmakers would spend the next five hundred years stretching the limits of their art, science, and mechanical engineering.

One of the earliest iterations of the keyless winding timepiece resulted from the ingenious watch built into a ring designed by the French watchmaker and diplomat Pierre Augustin Caron de Beaumarchais to impress Madame de Pompadour, the mistress of Louis XV. At just nine millimeters, the movement was wound by twisting a ring with a tiny peg set into the dial and turned about two-thirds around by Madame's fingernail. Caron de Beaumarchais, whose colorful repertoire included writing *The Barber of Seville* and *The Marriage of Figaro* and helping to finance the American Revolution, also equipped the ring with a manual winding option using a little key pushed into the square on the end of the cannon pinion. So delighted was the king with his mistress's ring-watch that he commissioned its maker to create one for himself.

As watches developed, so did the watchmaker, from craftsman to scientist to entrepreneur. Chief among them stood Abraham-Louis Breguet (1747–1823), a Swiss-born Parisian who set up his first workshop at 39 Quai de l'Horloge, where he devised ingenious advances in both technology and design. Crafted in what became known as "the Breguet style," his coinedge cases had a guilloche dial with intricate, finely executed patterns cut into the dial's surface and blue, apple-shaped hands.

In the 1780s Breguet's oeuvre of complications produced watches equipped with perpetual calendars, power reserve indicators, minute repeaters with *grande sonnerie,* and an independent second hand that could be halted without disturbing the movement. His own technical achievements propelled Breguet from the leading watchmaker of his day to wide recognition as the greatest watchmaker of all time. Among his astonishing range of pivotal inventions are the self-winding movement; the *parechute* suspension, which protects the balance staff against shocks; a balance spring-powered travel clock that can function during transportation; a constant-force escapement; striking mechanisms; and, perhaps his most famous and important, the tourbillon regulator. Breguet also improved the minute repeater by exchanging the bells with gongs whose steel strips, coiled inside a watchcase, took up less space and made for a cleaner sound.

Breguet's achievements brought him global renown and an enviable patronage that included two Bourbon kings, three governments of the First Republic, and Emperor Napoleon, perhaps his most loyal patron.

Beyond Breguet, the once anonymous workshops and family firms spawned in Switzerland's cradle of watchmaking became the great horological houses. Over the next two hundred years, names such as Audemars Piguet, Blancpain, Chopard, Girard-Perregaux, Le Locle, Vacheron Constantin, and a bewildering number of *maisons* became synonymous with the craft of horology. Each had forged a reputation and specific clientele, and some became known for particular specialties such as striking mechanisms or tourbillons. All vied to earn the reputation as the world's best watchmaker.

In America watchmaking followed industrialization, not gaining much traction until the mid-nineteenth century. Although the Americans lagged behind the Europeans by several hundred years, when they did enter the trade it wasn't long before they practically revolutionized the industry. Waltham in Massachusetts, Elgin in Illinois, and Waterbury in Connecticut were the main companies, but more than two dozen manufacturers soon sprouted across the country. American-made pieces initially were cheap little instruments with little artistry, but around 1850 the American firms pioneered the use of automated machines to mass-produce high-quality watches with interchangeable parts, a system that came to be known as "the American method." "Custom ordains that watches be made by hand," said a Boston jeweler, "but if silk and shoes and other things that used to be made by hand tools can now be successfully made with machinery, why on earth can we not make watches in the same up-to-date fashion?" By the mid-nineteenth century, some American manufacturers were producing fifty thousand watches a year, but the Europeans, famed for their long tradition of hand manufacturing, viewed the American industry with disdain.

In 1876 the American watchmaking industry demonstrated its ability to manufacture high-quality timepieces even compared against the best that Europe had to offer. At the first International Watch Precision Competition, held during the Philadelphia Centennial Exposition, the

Patek Philippe eighteen-carat gold ring watch, movement no. 174659, January 3, 1917.

Crafted for one of Patek Philippe's most important patrons, James Ward Packard, the watch is the only known timepiece of its kind produced during this period. Photograph by Stacy Perman. Courtesy of a private collection.

Patek Philippe cane watch, movement no. 174826, September 21, 1918.

Made especially for Ward, this polished ebony cane with a winding circular silver watch handle and a spare ivory knob is the only such walking stick clock that Patek Philippe is known to have manufactured. Photographs by Stacy Perman. Courtesy of a private collection.

Packard's walking stick fitted with its watch top (left) and ivory handle (right).

Patek Philippe eighteen-carat gold minute repeater
with musical alarm, movement no. 198014, March 8, 1927.

Ward desired a musical alarm that would play the entire *berceuse* (lullaby) from the Godard opera *Jocelyn,* his mother's favorite. The watchmaker borrowed a technique from music box manufacturers, creating this ingenious pin cylinder with steel tuning tongues (shown).   Photograph courtesy of Patek Philippe.

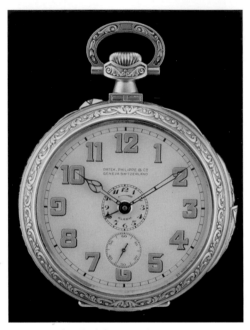

The dial features luminous
radium numerals.

The movement required
forty-eight jewels.

Patek Philippe eighteen-carat gold astronomical watch, "the Packard," movement no. 198023, April 6, 1927.

At just nineteen lignes (one and a half inches), this stunning piece has ten complications, featuring a celestial chart that mapped the night sky over Ward's beloved hometown of Warren, Ohio, with five hundred rotating stars.   Photographs courtesy of Patek Philippe.

The sunburst hand indicates
the difference between mean solar
and true solar time.

The case features Packard's
monogram in blue enamel.

Pressing down on the winding crown
reveals the sky chart.

One layer of the watch's movement.

Patek Philippe eighteen-carat gold *grande complication* with *seconde foudroyante* (flying seconds) chronometer, movement no. 174129, January 31, 1916.

Incorporating sixteen functions, the *pièce de résistance* of Ward's *grande complication* was the *foudroyante* chronometer, an intricate complication that measured time increments to a fraction of a second, dividing each second into five jump steps, each step indicating one-fifth of a second.   Photographs courtesy of Patek Philippe.

The dial displays eight hands.   The under-dial view of the movement

Patek Philippe desk clock, movement no. 197707, June 7, 1923.

Ward paid 5,275 Swiss francs for his complicated desk clock. One of the two known timepieces that Patek called *Le Presse-papiers* (the Paperweight), the second was delivered to Henry Graves, Jr., in 1927.   Photograph courtesy of Patek Philippe.

Patek Philippe eighteen-carat gold ship's bell striking, minute repeater, *grande* and *petite sonnerie,* movement no. 174876, January 31, 1920.

A skilled yachtsman, Ward commissioned what is believed to be the first instrument of its kind to feature the ancient nautical timekeeper confined inside of a pocket watch. Henry Graves, Jr., one-upped the automaker nearly five years later when he acquired his own Patek ship's bell minute repeater, featuring a double up-and-down indicator.   Photographs courtesy of Patek Philippe.

The gold dial features a thirty-hour up-and-down indicator.

The under-dial view.

Vacheron Constantin elaborately decorated twenty-carat gold pocket watch, 1918.

Emblematic of Ward's ornate aesthetic sensibility, this piece possessed a richly decorated twenty-carat gold case and an unusual dial engraved with a landscape and forged in fourteen-carat gold.   Photographs by Stacy Perman. Courtesy of a private collection.

Patek Philippe eighteen-carat gold Tonneau minute-repeating wristwatch, movement no. 97589, engraved with the Graves coat of arms, June 16, 1928.

One of Henry Graves, Jr.'s first wristwatches, it is also one of the earliest Patek Philippe minute repeaters, and one of only four made with a Tonneau-shaped case. (Henry owned three, two in platinum.) Discovered after the death of his grandson, this wristwatch fetched nearly $3 million at a Sotheby's auction held in New York on June 14, 2012. Photograph courtesy of Sotheby's, Inc. © 2012.

Patek Philippe eighteen-carat gold coin-form watch, movement no. 812471, June 29, 1928.

Henry's coin-form watch reflected both his horological and numismatic interests and was especially fitting given that the Graves family wealth rested in part on banking. Made from a 1904 U.S. $20 gold coin, it possessed a secret latch that opened to reveal the tiny watch hidden inside. Photographs courtesy of Sotheby's, Inc. © 2012.

**Patek Philippe platinum tourbillon minute repeater,
Geneva Observatory winner, movement no. 198311, July 29, 1935.**

Among Henry's many Observatory prizewinners, he possessed the only three platinum tourbillon minute repeaters ever manufactured by Patek Philippe. Awarded first prize at the 1933–34 Geneva Astronomical Observatory Timing Contest with 872 points, it went under the hammer most recently on November 14, 2011, by Christie's in Geneva, selling for $1,273,265. Photographs courtesy of Christie's. © Christie's Images Limited 2011.

Emblematic of Henry's connoisseurship, the prizewinner featured
the Graves's family crest, an engraved cuvette, and was made exclusively for him.

Patek Philippe eighteen-carat gold *grande complication,* movement no. 174961, June 1, 1926.

Henry commissioned this stunning piece in 1919. It incorporated twelve complications. After languishing in obscurity for more than seven decades, it was discovered in a shoebox stuffed under the bed of his granddaughter. First sold at Sotheby's for $640,500 in 1999, this exceptional watch returned to auction at a Christie's sale six years later, where it realized $1,980,200.   Photograph (*left*) courtesy of Christie's. © Christie's Images Limited 2005. Photograph (*right*) courtesy of Sotheby's, Inc. © 2012.

The timepiece in its fitted presentation box.

One layer of its complex movement.

Waltham Watch Company earned the gold medal for a machine-based manufacturing innovation: the first automatic screw-making machinery.

American watchmakers made their true mark for their voluminous output of cheap timepieces as well as their highly accurate ones as a result of an innovation by the railroad industry. A horrific crash on the Lake Shore and Michigan Southern Railroad between a passenger and a mail train, fifty feet east of the Kipton, Ohio, depot on April 19, 1891, occurred because the passenger train was four minutes off schedule; the conductor had consulted his watch, thinking the engineer would be on the lookout for the oncoming train, but the engineer's watch had stopped briefly, putting it off time by four minutes. Following the accident, the railway company hired the well-known Cleveland jeweler Webb C. Ball to investigate watch use among railroad employees, which he discovered did not operate on any watch standard. Ball created a new set of railroad pocket watches with white or silver dials and black Arabic numerals with each minute shown; these were accurate within thirty seconds per week. The jeweler also implemented a system with performance and inspection standards. The Ball Railroad Watch was advertised with the tagline that soon became shorthand for accuracy: "Get on the Ball."

By the time that Henry Graves, Jr., and James Ward Packard sought out Patek Philippe, the best watchmakers had long battled with one another to create new, prizewinning complications at the world expositions, a global marketplace of ideas and innovations that had begun during the second half of the nineteenth century. At these fairs, the best talents in commerce and art competed before a jury of peers, which stoked national pride and trade. At the 1878 Paris Exposition Universelle, Alexander Graham Bell displayed his telephone and Victor Hugo led the Congress for the Protection of Literary Property. Among the sensations on display was a stunning pocket watch with eleven complications made by L. Leroy & Company, quite a feat for the era.

Beginning with Charles Leroy in 1785, the Leroys of Besançon, France, were one of the great watchmaking families. Leroy had built this lavish *grande complication* that featured day, date, month, and leap year calendars, phases and ages of the moon, an independent split-second hand, an hour and minute counter, *foudroyante* (jumping seconds), an hour, quarter, and minute repeater, and the longitude of the main cities of Europe and Asia. (A replacement dial indicated the longitude of the main cities of Europe and America.) The watchmaker showed the extraordinary timekeeper in 1878 prior to delivering it to the client for whom it had been created, Nicolas Nostitz, a Russian count. Nostitz had challenged Leroy to create this *grande complication* to best the superlative timepiece with nine complications that the house had presented to the world in 1867.

After four years of laborious construction, the watchmaker Ami LeCoultre-Piguet of Le Brassus stunned the horological world at the Paris Exhibition of 1878 with his bravura piece: *La Merveilleuse*. The miraculous timepiece had between seventeen and twenty-two different functions, depending on who was describing it, giving rise to the remark that "Switzerland is unrivalled for complicated watches."

Patek & Co., like most watchmakers, ferociously guarded their technical secrets, but the World's Fairs created an irresistible opportunity to broadcast their superior craftsmanship and innovations. Patek had made its name as a house that produced watches with novel and uncommon complications such as quarter-hour repeaters, calendars, and independent second hands, and had introduced the first pocket minute repeaters in 1845. Attendance at the World's Fairs had been a decisive factor in the firm's initial success.

An exiled Polish nobleman, Count Antoine Norbert de Patek, and his partner, the highly skilled watchmaker François Czapek, had founded the firm that became Patek Philippe in 1839. At eighteen, Patek had been a cavalry officer who fought to free Poland from the Russians during the November Uprising in 1830. Wounded twice within one year, by 1833 Patek had left the service as the Russians made reprisal attacks, to work in Bavaria on behalf of refugees. He moved on to France, where he worked as a typesetter, and then to Geneva in the mid-1830s, where he began anew as an art student, painting landscapes in the studio of Alexandre Calame.

Patek's Geneva was filled with *cabinotiers,* and the watchmaking industry employed seven thousand people. The city's identity had become so inextricably tied to watchmaking that Gustave Flaubert, in his *Dictionanaire des Idées Reçues,* was moved to remark, "A watch was only worth anything if it came from Geneva."

Patek began assembling watchcases and movements and selling them to the exiled Polish aristocracy. Along with Czapek, who had also fought in the Polish uprising, he established a workshop on the Quai des Bergues, across the river from Geneva's old city.

Patek & Czapek Co., as the house was first known, produced nearly 1,120 watches over the course of six years. Their timepieces were known for their variety of movements, ruby cylinder escapements, and richly engraved or enameled cases. Catering to their Polish émigré clientele, the company created cases that displayed motifs of patriotic and religious

scenes, such as portraits of the Virgin of Czestochowa, Prince Josef Ponia-
tovski, and the revolutionary Tadeusz Kosciuszko.

Although the partners had intended to return to Poland and establish
their watchmaking business once the volatile political situation had settled
down, by 1845 about the only thing the pair agreed upon was remaining
in Geneva—although not together. Czapek launched his own watchmaking firm, taking a great deal of Patek's Polish clientele with him, leaving
Patek to start over once again.

A year earlier, the inventive French watchmaker Jean-Adrien Philippe
had come into Patek's sights during the French Industrial Exposition in
Paris, where he had earned the gold medal. An ingenious watchmaker,
Philippe had trained in London and worked in Le Havre with chronometers. His technical mastery and design were astonishing. At the Paris
fair, Philippe had displayed one of horology's holy grails: a thin keyless
stem-winding system. With this extraordinarily clever invention, pocket
watches no longer needed to be set and wound daily with a key, which
vastly improved their accuracy. Patek saw the potential in both Philippe
and his invention and extended a contract to Philippe to work for him in
Geneva. The men soon became partners in the new enterprise, Patek & Co.

By 1851 the firm was rechristened Patek Philippe & Company, the same
year that the watchmakers held court at the largest horological stall at
the World Exhibition at the Crystal Palace in Hyde Park, London. Patek's
enormous exhibition catalogue complemented the enormity of its stall and
demonstrated an astonishing breadth of charming and surprising complications: watches with insulated sea compasses, secret compartments, and
pieces known as *à triple effet,* capable of being transformed into three different shapes. Their common and repeating chronometers were tested and
awarded with official certificates from the most important astronomical
observatories, and a timepiece at 3.75 lignes (three-tenths of an inch) was
touted as the "smallest watch ever constructed."

Great Britain was the empress of the Industrial Revolution, and the
fair had been conceived to display her military and economic superiority.
Queen Victoria herself opened the Great Exhibition on May 1. During her
three visits to the fair she became enchanted with Patek Philippe, purchasing a blue enamel and rose-cut diamond pendant watch for herself (paying 612.50 Swiss francs) and a minute repeater for her beloved consort,
Prince Albert. She also appointed Patek and Philippe as horologers to Her
Majesty's Court. The queen's patronage became enormously influential for
the house, as many of Europe's crowned heads followed with their own
royal commissions.

In the Patek-Philippe partnership, Patek took on the mantle as the company's marketing genius, opening new markets, publicizing important clients, and partnering with the finest retailers. An obsessive note taker, Patek developed the company's exhaustive archives. Beginning in May 1839 with watch number 63, he recorded each watch, movement number, cost of components, purchase date, sale date, purchaser, and maintenance performed. All would be bound in thousands of tall, green leather *livres d'établissement,* which survive to this day.

Philippe was the technical mastermind who produced the innovations, patents, and technical achievements that established the house's reputation for exemplary horology. Over the years, the watchmaker would hold at least seventy patents for inventions, which gave it a spectacular array of functions that could be used in ever more audacious combinations. Together Patek and Philippe developed a house that grew to international fame.

Watchmaking was art. It was science. And it was war. Inscrutably, discreetly, yet fiercely, watchmakers competed with each other to create the most sophisticated, beautiful timepieces for the most important patrons. In turn, their competition helped fuel the race to create the *grandes complications,* watches that embodied engineering and pageantry, not always in equal measure. Creating these timepieces became a matter of pride that brought reflected glory to the country of the watchmaker, comparable to the race centuries later between nations to build a rocket ship that could travel to the moon and circle the Earth or between corporate titans to create the most technologically sophisticated operating systems, handheld devices, cellular phones, or computers.

Over time, this competition led to the creation of some legendary timepieces, the most complicated in history. Two in particular became the yardsticks by which all other *haute horlogerie* would be measured.

The first came about in 1783, when an officer of the queen's guard, Monsieur de la Croizette, hoping to impress Marie Antoinette, commissioned a watch from Abraham-Louis Breguet. The officer's expression of desire for the French queen became the highest expression of eighteenth-century horological art. Built with no consideration for expense, the watch, best known as the *Perpétuelle* (the Everlasting), more commonly referred to as "the Marie-Antoinette" (formally the Breguet no. 160), was a 63-millimeter-wide pocket watch crafted from gold and platinum, with pallets of sapphires. Contained in a rock crystal case, the piece brimmed with all of the known complications of the time, nearly a dozen in all, including a perpetual calendar, minute-repeating mechanism, thermometer, power reserve indicator, chronograph—even jumping hours. By the time the Marie-Antoinette was finally completed forty-four years later, in 1827, the watch's namesake had long lost her head to the guillotine. Breguet, who died in 1823, never saw the finished piece either.

The storied Marie-Antoinette held title as the absolute standard in complicated watchmaking for seventy-seven years, until the Leroy No. 1 was constructed for the Portuguese magnate António Augusto de Carvalho Monteiro in 1904. An eccentric, altruistic collector and bibliophile, who made a fortune in Brazil on coffee and precious stones, Carvalho Monteiro was also a horology enthusiast said to own one of the largest watch collections in existence. He had already acquired the Leroy *grande complication* originally made for Nicolas Nostitz following the count's death in 1896.

Although considered an amateur in the field, Carvalho Monteiro possessed a discriminating style and flair. Upon receiving Count Nostitz's timepiece, he was disappointed to find it incomplete. For instance, it did not have several significant complications such as a *grande sonnerie* or a function to indicate the difference between mean time and solar time at any point during the year. Carvalho Monteiro also found fault in the case design and sent it back to Paris to add further embellishments.

Each refinement led the Portuguese collector to finally commission the Leroy workshops to produce an even more impressive piece. In 1896 Louis Leroy recalled in his memoirs that Carvalho Monteiro had requested, "Create a watch that brings together in one portable timepiece all that science and mechanics can currently achieve." The final piece, equipped with twenty complications, required 975 different parts. It kept time in 125 cities in the world, the date for one hundred years, recorded the day of the week, month, year, moon phases, seasons, solstices, and equinoxes, and the time of sunrise and sunset in Lisbon. The extraordinary instrument included celestial charts of both the northern and southern hemispheres, with a mechanism that animated the sky rotated east to west. It incorporated a hair hygrometer to measure humidity, a barometer, and an altimeter up to five thousand meters. A regulator device allowed the user to adjust the timing without opening the watch.

No less attention was spared on the intricately engraved gold case, with its patron's coat of arms surrounded by a winged man representing Time and the three Fates, and the Mistresses of Our Destinies, as well as the twelve signs of the zodiac. The winding crown was ringed in pearls and other precious stones befitting the head of one of Europe's royal households, and had a compass concealed inside.

In 1904 Louis Leroy personally delivered the finished watch to King Manoel of Portugal, himself a patron, who was on a royal visit to the French capital. Upon his return to Portugal, the king summoned Carvalho Monteiro to the Lisbon Palace, where he was presented with his masterpiece, for which he was said to have paid 20,000 francs.

Such a commission would not be seen again for sixteen years, until James Ward Packard presented Patek Philippe with his wish list for his luminous *grande complication* that contained sixteen complications, delivered to him in 1916.

~

CHAPTER ELEVEN

A Gentleman's War

Perched regally on the Grand Quai, Patek Philippe's workshops overlook Lake Geneva with a spectacular view facing north. Decades of acclaim and prosperity had brought the watchmaker to the left bank of the Rhône, where it had erected an elegant, four-story building for its workshops, offices, and salons. The building, completed in 1890 (with a fifth story added in 1907), was, for the period, state of the art. Thirty-horsepower turbines in the basement furnished electricity to the entire building, supplying power for the tools and machinery on the second and third floor's engineering rooms, as well as to the fourth floor, where the escapement components were produced.

By the time James Ward Packard and Henry Graves, Jr., enlisted Patek Philippe in their collecting joust, its founders had died (Patek in 1877 and Philippe in 1894), and management of the celebrated firm had passed to Joseph Bénassy-Philippe, Philippe's son-in-law. Joseph Philippe, the youngest of Jean-Adrien's five children, became head watchmaker. In 1901 the company reorganized as a joint stock company with 1.6 million Swiss francs under the new name Ancienne Manufacture d'Horlogerie Patek Philippe et Cie S.A. But with its major stockholder, Joseph Philippe, named director of the board, the company signaled its intention to remain rooted in the founders' philosophy and stay a family firm as well. In 1913, six years after Joseph died, his son Adrien took over as director. Having established its reputation in the nineteenth century, the watchmaker focused on perfecting its celebrated craftsmanship while increasing its manufacture of complicated movements in the twentieth.

Europe had slid from its perch as the great global power since before World War I, and America had stepped into the breach. Europe's luxury watchmakers increasingly turned to America's wealthy industrialists and financiers, with their aristocratic tastes for patronage. America's flush watch collectors, Patek Philippe noted, offered its watchmakers "the

The Patek Philippe salon
and workshops, facing the
Rue du Rhône in Geneva,
have changed little
since 1907.
Photograph by Stacy Perman.

opportunity to apply the skills and techniques developed over the previous generation."

In America, Patek Philippe developed a strong network of exclusive retailers, from Tiffany & Co. in New York to Shreve & Co. in San Francisco. In Geneva, the watchmaker would usher its affluent American patrons into one of its opulent ground-floor salons, the largest of them decorated with black and gilt wood, rich Cordovan leather, and elegant bronze statuary. A large oval that framed many of the *maison*'s numerous Observatory and other important medals from over the years was prominently displayed in the salon. The genteel sales staff maintained its American Register in maroon leather, listing in florid ink the names of visiting clients.

On occasion, James Ward Packard corresponded directly with Geneva, but he appears to have preferred to conduct his business mainly through his personal emissary, F. E. Armitage. Although Henry Graves, Jr., had traveled to Geneva, he channeled most of his desires through Tiffany & Co. and, later, Patek Philippe's American headquarters in New York. Watch collecting required enormous patience; approvals and communication proceeded slowly through handwritten letters. The manufacture took years, and an instrument's delivery might take weeks, sometimes months, traveling by ship and post.

During the dozen or so years following Ward's receipt of his 1916 *grande complication* with the *foudroyante* chronograph, both he and Henry repeatedly turned to Patek Philippe to commission fantastic timepieces with multiple horological functions. From this point on, their unspoken contest thundered ahead. "First one, then the other of these two gentlemen would order timepieces," wrote Alan Banbery, the curator of Patek Philippe's private collection, many years later. He described this period as "the prime, vintage years for a number of timepieces produced by the Manufactory."

As the 1920s roared forward, the two men induced Patek Philippe to push past its own storied inventiveness and skill. They commissioned pieces with tourbillons and *grande* and *petite sonnerie,* and they combined several complications, such as perpetual calendars and phases of the moon, in novel arrangements. Every instrument was magnificent and unusual, and each became a monument to the individual gentleman's particular bravura. In something of a chess match, each watch surpassed the previous one in some manner, necessitating that the two men return once again to Geneva to best themselves, while eliciting the best from the watchmaker.

Given his engineering brilliance, Ward had an incontestable advantage in requesting his unusual approaches be translated into ticking realities. He had commissioned a unique chronograph, movement no. 157392, with a tachometer scale capable of charting speeds up to 150 miles per hour or kilometers per hour over a measured course; the spiraled calibrations on the watch's outer track converted the seconds to miles per hour. Ward's minute repeater with a split-second chronograph and hour repeater, movement no. 197505, featured two center second hands that could be halted individually and directed to fly-jump the other moving hand. It was also unusual for the period, in which an hour recorder was not only rare but highly uncommon, as on this timepiece, where both the hour and the minute occupy the same auxiliary dial. Ward desired to possess the best in a category and then rewrite all known categories.

In one highly unusual request, Ward challenged the watchmaker to add a tourbillon to a minute-repeater chronometer. His request was unprecedented. While the exquisitely intricate tourbillon was reserved for the most exceptional of timepieces, Patek Philippe had never attempted to craft such a combination of complications in order to avoid intruding on the main wheel's train adjustment, which might negatively impact the timepiece's accuracy. In the end, the Ohio engineer prevailed. He took receipt of the timekeeper with the movement no. 174720 on November 19, 1919.

Just as he had insisted on graceful lines in Packard motorcars, Ward remained committed to the singular combinations of beauty and precision in his commissioned pieces. For another piece, he requested a richly carved case of eighteen-carat gold covered by an outer casing of chiseled *repoussé* in silver, showing *L'Enlèvement d'Europe* (The Rape of Europa). Underneath the dial ticked a minute-repeater mechanism with an exceedingly rare and sophisticated carillon *grande sonnerie* on three polished spiral gongs. This particular watch, movement no. 197791, was exceptionally small, at seventeen lignes, or one and a half inches, for the degree of complications it featured.

Like Ward, Henry desired only the finest materials. In general he preferred hands made of blue steel and often requested his cases be made in high-grade gold and platinum, although Henry did not share Ward's more baroque aesthetic sense, preferring his cases polished and their designs enhanced usually with no more than his family's coat of arms.

Restlessly, Henry awaited the delivery of his *grande complication*, first commissioned in 1919. In the interim, he ordered pieces stuffed with complications: tourbillon regulators, perpetual calendars, and chronometers. As Henry came into his own as a collector, he remained fixated on Observatory prizewinning watches. One, an eighteen-carat gold pocket lever chronometer, movement no. 178448, featured thirty-six hours power reserve indication. Completed in 1921, the *Bulletin de Première Classe de l'Observatoire Astronomique de Genève* awarded the watch first prize for the timing contest in 1919. Once drawn into the remarkable discipline of timekeeping, Henry began accumulating variations, although slight, that expressed a diversity of perfection.

During their collecting years, Ward took delivery of perhaps as many as forty Patek Philippe timepieces, while Henry acquired a number closer to eighty. However, this duel was not based on the number of watches owned but on owning the watch with the most exceptional timekeeping qualities and the greatest combination of complications.

As the decade pulsed ahead, the automaker from Warren, Ohio, and the New York financial scion emerged as *haute horlogerie*'s two greatest American patrons, and their watchmakers were more than delighted to assist them. Their patronage engendered special treatment. No request, it appears, was denied. In addition to one-of-a-kind pieces with special

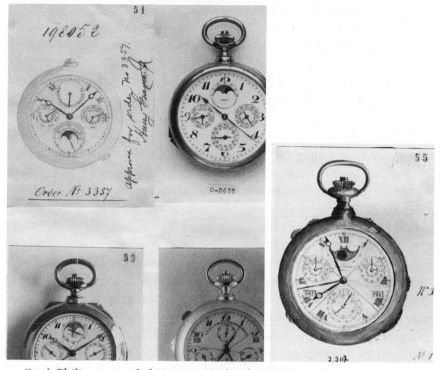

Patek Philippe provided Henry with sketches of his *grandes complications* for approval. Photographs courtesy of Patek Philippe.

features, the pair requested exclusive serial numbers and their names engraved in specific fashion; for Henry, that usually meant on the dial, and for Ward, on the movement itself.

As the master watchmakers sat at their benches delivering on each new challenge, the men at the front of Patek Philippe's house went to great lengths to keep both men enthralled. In the course of four years, the firm produced twin double-barrel desk clocks. Slanted and rectangular in shape, the pair of clocks was encased in silver with gold stylized decorations of rosettes, floral ribbons, and winged creatures; they were said to be the only two such pieces that Patek ever crafted. Hinged at the bottom, the case opened to reveal a compartment that contained the winding keys and could hold small personal items. Besides keeping time, each clock featured a complex movement with a perpetual calendar, indicated phases of the moon, and had an eight-day power reserve gauge. Patek Philippe called the stunning instrument *Le Presse-papiers* (the Paperweight). It was just the kind of extraordinary, rare piece made especially for highly regarded clients as a reward for uncommon patronage. Ward took receipt of the first, movement no. 197707, in 1923, customized with his Art Deco monogram under the dial. Four years later, Henry took delivery of the second, movement no. 198048. His clock carried the Graves eagle surmounting a crown in gold. The dial, case, movement, and cuvette were engraved *Made for Henry Graves Jr., New York.*

Patek Philippe's sundrenched workshops emitted an uninterrupted flow of craftsmanship and innovations to match the ambitions of each commission from the Americans. Each new watch was carefully considered, each a major undertaking. As one watchmaker's refrain went, "A mechanical watch is a living thing." Life began on paper. Hundreds, perhaps thousands of drawings built up the initial picture of the escapement, all demonstrating a command of applied mathematics, geometry, and engineering. Every new design was a gamble and Patek Philippe's workshops a great horological casino.

The shop ceilings featured an extensive system of pulleys, shafts, and driving belts under which the master watchmakers, draped in robes, plied their trade like surgeon-artists in a dust-free, temperature-controlled environment. Patek Philippe assiduously maintained its standing as a *fabrication complète.* The *maison* produced the majority of its own parts, watch cases, and other key elements in house, availing itself of components from the Jura Valley on those occasions when it became absolutely necessary.

The men at the bench made sheets of metal at the required thickness from which all elements were cut, hardened, tempered, filed, and turned into parts, while face-lathes and rounding tools were deployed to make the mechanisms required in each movement. At each step, every single component was washed, polished, examined, and checked once again to ensure it was delicately balanced and perfectly finished. Working with hundreds of tools packed clinically in rows of wooden drawers, the watchmakers manipulated thousands of tiny gears and bridges to shape one perfect instrument of timekeeping.

The process was painstaking and laborious. In a dictionary on the arts and sciences published in 1820, Abraham Rees asserted that a watch required no fewer than thirty-four highly trained people to manufacture, each of whom had served a lengthy apprenticeship before taking the bench. Another observer visiting Patek's workshops in 1877 noted that at least seventy-five different artists and craftsmen touched a pocket watch before it was finished. The skilled work demanded nearly four dozen processes. The burnishing of the cases, ornamentation, and decoration required at least another fourteen, and the dial another ten. Each watch was lubricated with oil, hand-assembled, and hand-finished to minimize friction and increase precision and longevity by ensuring the smoothness of the parts, the chamfered edges, and the polished screws. Before completion, every piece was heated and frozen to test against all possible temperature and climate conditions.

Furthermore, Patek Philippe strictly adhered to the twelve technical and aesthetic conditions of the Poiçon de Genève (the Geneva Seal), the strict standard of excellence dating back to an 1886 law guaranteeing the origin (mechanical movement, assembly, and regulation in the canton of Geneva), quality of workmanship, durability, and "exceptional savoir-faire." The firm was part of an exclusive circle of watchmakers allowed to use the seal, proudly stamped on the main plate and one of the bridges of each movement produced.

Before Ward's and Henry's pursuit of timepieces integrating numerous complications, *grandes complications* had generally been created by a watchmaker as a model or particular triumph. Or a purchaser would invite a watchmaker to produce an instrument that reflected his highest level of skill and knowledge of the day. In 1893, for instance, Commodore Vanderbilt had gifted his son, Cornelius Vanderbilt, Jr., an intricately engraved gold Patek Philippe minute-repeater, split-second chronograph from Tiffany for his twenty-first birthday.

Earlier, collectors had tended to belong to two camps. For one group, the watchmakers determined the design and number of complications based on their own ambition. Geneva's watchmakers lured patrons and burnished their prestige, constructing the most lavish timepieces for the world's most prestigious customers. Vacheron Constantin delighted King Fuad I of Egypt with *grandes complications* created in his honor. During his state visit to Switzerland the watchmaker presented the king with a watch of twelve complications: a split-second chronograph, *grande* and *petite sonnerie*, quarters and minutes with three-note chimes, a perpetual calendar, and moon phases. Made of eighteen-carat gold, it had eight hands and forty-six rubies.

Frequently the rich and powerful also presented those who had served them in some capacity with beautifully engraved watches, often bearing their own portraits or coat of arms. In 1963 the city of Berlin presented President John F. Kennedy with a gold Patek Philippe clock to commemorate the Geneva Conference that established the Moscow-Washington "hot line" telephone link. Decades earlier, Adolf Hitler had rewarded the meteorologist who had given him a favorable weather forecast prior to his invasion of the Low Countries and France with a gold watch engraved with his signature and the date of the invasion.

The second camp of collectors, usually immensely wealthy, took on horology as a study, bankrolling collections plucked from antiquity. Sir David Salomons, a British barrister, philanthropist, member of Parliament, scientist, and founder of the English Society of Engineers, also held the distinction of being the greatest historian and collector of Breguet watches. Much like Ward, Salomons's deep passion for horology stemmed from his great interest in engineering. Born in 1851, he had taken out patents for electric lamps and had developed one of the first electric cooking devices. His home north of Tunbridge Wells, Broomhill, was one of the first in Britain to be lit with electricity. But it was for his watch collection that he would forever be known.

During his lifetime, Salomons assembled the largest, most varied Breguet collection in the world, some 124 examples of the watchmaker's finest work. He wrote, "To carry a fine Breguet watch is to feel that you have the brains of a genius in your pocket." In 1921 he published the first complete Breguet biography, noting, "My object is not to advertise my Collection . . . for I dislike advertisement."

On May 3, 1917, while making his way home to Grosvenor Street in a downpour, Salomons happened to come upon Breguet's *chef d'oeuvre*, the Marie-Antoinette, sitting in the window of a jeweler's near Regent Street, and after careful contemplation, he purchased the remarkable piece. "Evening after evening, I studied this watch," wrote Salomons, "which is most complex and interesting, with the result that I formed the opinion that no other maker of watches could approach such work, and I have had considerable experience of the productions of other makers."*

*Following Salomons's death in 1925, his family divided the collection among his wife and children. Christie's sold nearly seventy pieces in three sales between 1964 and 1965. Another fifty-seven, including the Marie-Antoinette, were donated to the Museum of Islamic Art in Jerusalem, founded by Salomons's daughter Vera. It is there that the tragic queen took another fateful turn. A career criminal named Na'aman Lidor (née Diller) broke into the museum on April 15, 1983, and stole it, along with nearly one hundred other clocks and watches worth hundreds of millions of dollars, leaving behind no clues other than a half-eaten ham sandwich, empty Coca-Cola bottles, cables, wires, and a dirty mattress. Lidor was no common thief but a man obsessed with Breguet. He never attempted to sell the watches; rather he took them apart and stashed them in safety deposit boxes across Europe with detailed instructions on how to put them back together again. His odd scheme came to light only in 2008, four years after he died. A lawyer contacted a Tel Aviv antiques dealer on behalf of an anonymous client who said she had inherited the watches. In turn, the dealer called the museum. The client was Lidor's wife, and the crime unraveled as Lidor had confessed to her all his crimes before his death and suggested she sell the watches. The Marie-Antoinette, along with forty-two other stolen pieces, were secretly returned to the museum.

In America the princes of commerce assembled collections with great historical provenance. In 1910 J. P. Morgan purchased the celebrated watch collection of the German publisher Carl Marfels, eighty seventeenth- and eighteenth-century timekeeping masterpieces (including an egg watch in Limoges enamel, considered the most expensive in the world at the time) for $360,000 (the equivalent of $8.3 million today). The coal-mining baron and former senator Clarence W. Watson amassed a collection of 314 historical watches, much like Henry Huntington collected Gainsboroughs, some dating back to the seventeenth century.

Men like Willard H. Wheeler, the son of the prominent New York jeweler Hayden Wheeler, collected timepieces in homage to human ingenuity. Over twenty years, beginning at the start of the twentieth century, Wheeler collected precisely one hundred rare antique timepieces representing the evolution of the watch, with the notable exception of the wristwatch. He maintained this number by adding new, superior pieces and removing lesser ones, including examples made by British masters such as Thomas Tompion and Thomas Mudge for their royal patrons. Among his prizes, Wheeler possessed "the Fountain," a clock elaborately enameled and inlaid with pearls made for Kea-king, the fifth Chinese emperor of the Mantchow Dynasty, and a large coach watch that spoke the time when a string was pulled, made for British Vice Admiral Lord Horatio Nelson.

Once Wheeler had set his heart on a watch, he wouldn't stop until he possessed it. It might take him years to persuade an owner to part with a prized watch, but he always managed to procure the pieces he fancied.

James Ward Packard and Henry Graves, Jr., changed the game. Until these two connoisseurs made their desires known, few men had created entire custom-designed collections based on their whims and fancies, specifically for their own usage.

CHAPTER TWELVE

~

Time Stop

As Henry and Ward commissioned their fancies, lingering over drawings and agonizing over details—the choice of color for the hands or dials, the preferred striking mechanism—the American economy was swept up in a tide of speculative mania and incredible excess. Buying on margin had become popular as a growing number of middle-class citizens invested in the stock market. Luxury buildings pierced the skies and prestige automobiles populated highways. In the decade following World War I, the nation's wealth soared, manufacturing jumped, and Americans rushed headlong into a fast-paced world of luxury and glamour not thought possible only a few short years earlier.

As American society became increasingly concerned with material goods, the nation's industrial output required new markets, which led to the growth of a new business of persuasion: advertising. Dedicated to creating and stoking an endless desire for new products, advertising did not simply announce a product but encouraged people to believe that they needed to own it. "Your teeth need a vacation," coaxed one ad for a twenty-five-cent tube of Colgate toothpaste.

The decade's motto could be summed up in one word: *more*. Mass production, new technology, and lower prices on consumer goods made more items available to more people. Between 1921 and 1925 Ford slashed automobile prices six times, to $290 (around $3,700 today)—less than three months' wages for the average worker.

Radio and aviation, the telephone and the power grid made it possible for millions of people to improve their lives. Before World War I, roughly 42 percent of all Americans lived on a farm. By the end of the 1920s that number had fallen to close to 25 percent. With the greater amount of goods available to a greater number of ordinary Americans came a society of greater standardization. For the first time, Americans wore ready-to-wear clothing. They ate processed foods from cans and out of frozen

Ward tested out his Packards on numerous road trips. Here he drives past Grant's Tomb in New York City.
Courtesy of Betsy Solis.

cartons. New fads, from the Charleston to pulp fiction, swept up the entire nation at once. Chain stores like Woolworth's and the A&P grocery store proliferated across the country.

Consumerism became the new secular religion. Throughout the 1920s American families spent less of their income on necessities—food, clothing, and utilities—and more on appliances, entertainment, and consumer goods. Cigarettes, cosmetics, and electric vacuums became staples. New consumer categories emerged to service entirely new products and services. Brooks Brothers, once the makers of bespoke clothing for boys and men, now manufactured twenty-two different types of "motor attire," from raccoon-fur motor coats to fleece-lined automobile boots.

Not everyone who heard the roar of the Twenties appreciated it. The writer Willa Cather found the headlong hurtle toward modernity troublesome. In a speech she gave in Omaha in 1921, she cautioned, "We now have music by machines, we travel by machines—soon we will be having machines to do our thinking."

Where thrift and savings were the hallmarks of Victorian society, spending and borrowing marked America's new consumer society. In

1899 the American economist Thorstein Veblen coined the term *conspicuous consumption*. Two decades later his words had sprung to life. "To gain
and hold the esteem of men it is not sufficient merely to hold wealth and
power," he wrote in *The Theory of the Leisure Class*. "The wealth and power
must be put into evidence."

A broad swath of the population who had not been born into money
now had purchasing power, thanks to credit. For the first time, manufacturers allowed consumers to buy everything from cars to appliances on
the installment plan. Seventy-five percent of all radios were purchased this
way. Banks offered home mortgages. In 1923, when four million new cars
were sold, there were 635 companies specializing in installment financing.

The public's attitude toward millionaires had drastically changed, from
cynicism and resentment at the turn of the century to admiration. "When
the making of millionaires is accompanied by such an increase of general
prosperity," wrote the editor of the *Pittsburgh Gazette Times*, "the country
may well pray for more of them."

Wealth spread in America like warm butter on toast. At the turn of
the century, bank deposits in New York totaled $1 billion (equal to over
$13 billion today); by 1926 the top four banks (National City, Chase,
Guaranty Trust, and National Bank of Commerce) held more than $2.5
billion (equal to some $32.5 billion today) in combined deposits. The
rest of the country fared similarly: sixty-three banks and trust companies held deposits of more than $75 million (about $976 million today),
each.

Finance thrived. Wall Street's windfall spilled over to Main Street. Or
so it seemed. With margin buying both cheap and easy, by some estimates upward of 25 million Americans had jumped into the stock market, borrowing 50 percent of the money from their brokers to buy stocks
on alarmingly low 3.5 percent interest rates. They then turned around and
used the stock as collateral for those loans. When the stock spiked, investors collected their dividends; when it dropped, they simply raised some
cash to cover the loss and waited for the market to rise once again.

The prosperity of the times brought new opportunities for women.
They could vote, drive cars, and carry on with men as they pleased. They
wore rouge, smoked cigarettes, and drank furtively from silver flasks hid-
den within the folds of their skirts. In 1921 Margaret Sanger founded the
American Birth Control League. Alice Paul drafted the first version of the
Equal Rights Amendment introduced to Congress in 1923, the same year
that Edna St. Vincent Millay became the first woman to be awarded the
Pulitzer Prize for poetry.

This fast-living behavior among the fairer sex found a variety of new expressions, chief among them divorce. During the 1920s America's divorce rate surged like the stock market, particularly among the upper crust. The women who once had crossed to Europe hoping to land a prince or baron were now heading to Paris to divorce their husbands in high fashion.

The rising tide did not lift all boats. By some estimates, 60 percent of the population still lived below the poverty line. The richest 1 percent of Americans owned approximately 40 percent of the country's wealth. The tremendous profits realized from mass production went to investors, not workers, while some 80 percent of Americans had no savings (much like the American economy that cratered more than eight decades later). Farm prices between 1920 and 1921 fell at calamitous rates. Wheat prices bottomed out, dropping by nearly half, while cotton, the *elán vital* of the South, collapsed by three-quarters. Farm foreclosures and rural banks failed with startling frequency, as farmers could not make payments on their loans. Grinding poverty and the threat of racial violence forced a migration of southern African Americans to the North. In 1919 riots engulfed three American cities: Chicago, Washington, and Omaha. The Ku Klux Klan, founded in 1866 by former Confederate soldiers opposed to Reconstruction, was reinvigorated during the 1920s, targeting blacks, Catholics, foreigners, and Jews.

This was a suspicious boom, but few cared to look beyond the caviar and illicit flowing champagne when a man could buy stock like a Rockefeller on a fishmonger's salary and a young lady could party like Mrs. Astor in a dress bought on the installment plan. For those who cared to notice, however, the days of the American oligarchs were numbered.

For the Graves family, these heady days seemed to have just gotten started. At fifty-four, Henry was more or less retired. In 1920 the "family" bank, Liberty National, was in merger talks with Bankers Trust. Both concerns had been prominent in financing the Great War. Following the armistice, Liberty National had focused much of its attention on extending mercantile and banking relations across Europe. With resources well over $500 million (roughly $5.7 billion today), this new entity was hailed as "one of the most powerful and aggressive banking institutions in the world." The *New York Times* reported that only National City Bank and the Guaranty Trust Company exceeded its resources.

Henry's two eldest sons had joined the family business at 30 Broad Street. Harry took up his legacy with a position at Banker's Trust, while his brother Duncan traveled across Mexico to Veracruz, Guadalajara, and

Mexico City on survey trips, exploring potential property and mining interests. George Coe II, the youngest of the Graves boys and the family's reigning intellectual, entered Yale, graduating from the Sheffield Scientific School, where he belonged to the St. Anthony's secret society, before setting off to explore the world.

At nineteen, Gwendolen assumed her role as a debutante and member of the Junior League. She had graduated from the Masters School in Dobbs Ferry, New York, a finishing school with a sound liberal arts program that included Latin, mathematics, and astronomy. All that was left for the petite young heiress was to make herself as attractive as possible and to marry well. To that end, a three-month trip abroad filled with couture, culture, and exotic ports was planned for her in the spring of 1923. Accompanied by Florence, Gwendolen planned to sail from New York for a spin through Spain, Gibraltar, Algeria, Tunisia, Italy, Greece, Constantinople, Syria, Palestine, Egypt, Monaco, France, and the British Isles.

The top of the decade should have prophesied brilliance for Henry. The stock market boomed, his fortunes swelled, and in a very short period his watch connoisseurship had put him in elevated company. Having commissioned a trio of rare *grandes complications,* he had joined the ranks of James Ward Packard. But forces beyond Henry's control would remind him that neither his great wealth nor his privileged pedigree could insulate him from the greater world.

On the evening of February 28, 1922, Harry and his wife, Margaret, decided to dine quietly at home. Incandescent creatures, the couple were part of Ardsley's fast set, a circle of young elites who could often be found dining at the Ardsley Club or in the city at the St. Regis or any number of soirées suffused with cigarette smoke and jolly fun. Before dinner, Margaret removed her rings and collected several pieces of jewelry from her dressing table and put them in a strongbox. Without a thought, she left it unlocked, placed it in the drawer of her bureau, and walked downstairs, where she met her husband in their dining room at around 7:30.

As they ate, a man scrambled onto the property, making his way behind the gates and hiding in the trees. Gently he placed a ladder against the wall at the rear of the house leading up to a window in the master bedroom, smashed the glass, releasing the window lock, and crawled through. Once inside the darkened master suite, the intruder made his way to the bureau and began rummaging through. Just as he reached Margaret's strongbox, one of the couple's maids, Mary O'Connell, who had gone up to the second floor to check on the children sleeping in the nursery, spied the man dressed in a fedora, light suit, an overcoat and gloves standing at the bureau in the dark. Smiling, he looked directly at the maid and spit out a laugh. For a split second she thought he was the plumber. Thinking it odd that the workman was still in the house, Mary exclaimed, "What on earth are you doing here at this time of the night?," before noticing that the man held one of Mrs. Graves's jewelry boxes. Quickly realizing he was no plumber, she cried out, "It's a burglar!"

Hearing her screams, Harry called the police and then, grabbing his revolver, gave chase. In the dense night air Harry could barely make out a figure in the shadows. The intruder had jumped from the window and run off. Harry began indiscriminately shooting off several rounds. Alarmed, many of his wealthy neighbors ran to the house. By the time the police had arrived, the burglar was long gone. At the bottom of the ladder they found several pieces of jewelry scattered on the lawn. But the thief had made off with $25,000 worth of Margaret's valuables (a sum equal to about $343,000 today), including several solid gold bracelets—two set with diamonds and another with square sapphires—a four-carat square-cut diamond ring, a three-carat diamond ring, and her engagement ring.

The Graves robbery was the third such theft in as many weeks in the area and quickly set off a chain of events that would end terribly. Incensed, the very next day Harry and several of his neighbors formed a vigilante group, the Gold Badge Squad, empowered by the deputy sheriff to arrest criminals, and started patrolling the streets.

On March 21, just past midnight, Harry, his brother Duncan, and Henry Wilson, a neighbor and fellow vigilante, chewed over the situation, their anger rising with each retelling, amplified with drink. Soon their argument moved outside, where they jumped into a car with Duncan at the wheel and took off into the night at top speed. Careening at sixty miles an hour along Saw Mill River Road in Westchester County, Duncan rounded an S-curve at Woodlands Lake when another vehicle with "blinding headlights" approached in the opposite direction. Swerving to avoid the car, Duncan lost control and ran into a ditch. The car skidded and plunged forward, crashing through a stone wall and then into a large tree, before overturning. Harry was thrown through the windshield forty feet and landed on a pile of stones, fracturing his skull. He was killed instantly. Wilson was trapped in the wreckage and also killed. Thrown from the car, Duncan landed opposite the wall, where he lay unconscious. When sheriffs arrived at the scene, they found Harry's gold badge pinned to his vest and a pistol affixed to his hip.

At twenty-five years old Henry Graves 3rd left behind his twenty-three-year-old widow, their three children, and a piercing sorrow that the Graves family had not known before. Although his own personal achievements were slim, news coverage of the tragedy portrayed the young vigilante as the bravest and finest of men who had lost his life in pursuit of scoundrels in the night. "He represented the highest type of America's young manhood," declared *The Historical Register: A Biographical Record of the Men of Our Time Who Have Contributed to the Making of America.* "He was favored by birth and breeding, travel and culture and by personality and charm for a brilliant career."

His widow had a different perspective; she would bitterly recount to his children and grandchildren decades later that the heroic Harry and Duncan Graves had gotten drunk and agitated and lit out into the night on a drunken joyride.

Earlier than he had planned, Henry built the family crypt in the storied Sleepy Hollow Cemetery in Tarrytown, New York, just steps from William Rockefeller's towering mausoleum carved from a thirty-two-ton

After Harry died, Florence placed his photo and a lock of her eldest son's hair in a gold locket. Photograph by Stacy Perman. Courtesy of Gwendolen Graves Shupe.

block of granite. Two large elm trees framed the Graveses' white marble vault with a motif of grape vines and leaves along the frieze and archway over double-glass doors. As was the fashion among affluent families, a five-panel Tiffany stained-glass window, this one designed with a scene of Jesus tending his sheep, illuminated the dark crypt from the rear. Elegant and commanding without being showy, it had all of the hallmarks of Henry's connoisseurship and taste, and it occupied one of the choicest pieces of real estate.

Florence Graves had her own private mausoleum for her eldest son, which she kept next to her heart, an oval locket of solid gold in one half of which she placed a photograph of Harry standing in the woodland at Shadowbrook, his camera in his hand. In the other half she placed a lock of his hair tied with silk.

In the wake of the tragedy, Florence and Gwendolen pressed on with their travel plans. The three months away allowed Gwendolen to emerge into society with the news of the accident and any whiff of scandal a faded memory. For Florence, the Mediterranean idyll among the ruins and teahouses of ancient lands allowed her to put an entire continent between her and her grief without spectacle.

To outsiders, Henry's bearing, with his placid brown eyes, betrayed little of his inner turmoil, but Harry's death proved a sorrowful break from Henry's collecting. During the nearly three years that followed the accident, until 1925, Henry virtually withdrew from his collecting duel. Among the known Tiffany watch ledgers and the Patek Philippe archives, except for one piece that he obtained in 1924, none are inscribed with a new purchase or commission from him.

CHAPTER THIRTEEN

~

The Final Windup

In the fall of 1925, a sense of urgency took hold of Ward. His doctors had discovered a tumor. Although surgeons soon removed the malignancy, initially giving him a positive prognosis, the specter of his mortality had revealed itself. Scarcely two years earlier, on November 11, 1923, his brother, Will, had finally succumbed to his chronic litany of health woes, including paralysis and blindness, and died at his home on North Mahoning Avenue. The engineer, now sixty-two, made certain to put his affairs in order.

At the top of his list, after taking care of his wife, Elizabeth, were plans for the dispersal of his watches. While few would accuse Ward of vanity, he recognized the important legacy that his remarkable assemblage represented. In drawing up his will on November 12, 1925, he made the decision to preserve the core of the collection built up over decades, bequeathing his most complicated and important timepieces to the Cleveland Museum of Art.

Ward was never ostentatious about his timepieces. For him the joy was in the challenge, to push horology beyond its current boundaries. He and Elizabeth had remained childless (he did make ample provisions for his brother's son, Warren Packard, with whom he remained quite close), and it seemed only logical for a man so devoted to the public's understanding of technology and art to offer his watches to the kind of forum that would allow the greatest number of people to appreciate them. When the Cleveland Museum opened its doors nine years earlier on the edge of Wade Park, it had declared its mission: "For the benefit of all people, forever."

Keeping his watches intact was an unusual step. Throughout the ages few historically significant collections had survived more than a generation unbroken; most were carved up among descendants and passed quietly among dealers and collectors, eventually making their way under the hammer. Even the great Breguets that Sir David Salomons had assembled so lovingly were eventually dispersed. To ensure that his watches remained together, Ward generously stipulated the sum of $2,000 (about $26,000 today) to "provide for the proper housing and care of such a collection."

Ward's decision had broad implications. For starters, he didn't explicitly list the number of watches he wished to bestow on the Museum. At the time he signed his last will and testament, he had commissioned several new pieces, with a number already in the process of manufacture for delivery upon their completion, including his astronomical watch, the Packard. Within the next two years, he would receive at least a handful more, among which were some of the most important and inventive of his entire oeuvre.

For a collection already brimming with incomparable mechanical specimens, Ward reached higher still. With his favorite complication, the chiming mechanism, he had arrived at a rather sentimental desire. In addition to a minute repeater, he commissioned Patek Philippe, possibly after he drew up his will, to create a musical alarm that would play the lullaby from the Benjamin Godard opera *Jocelyn*, a favorite of his mother's. The watchmaker had long become accustomed to the automaker's requests for specific functions, but this latest one posed an exceedingly difficult technical dilemma. At the time, an alarm watch was a rarity, and one playing a melodic alarm almost unheard of. In addition, this particular lullaby was quite long, and Ward insisted that it be played in its entirety.

Ordinarily, creating such a mechanism would have necessitated a thick pin barrel that required an even larger watch. Patek Philippe struck upon an inspired solution, which borrowed a technique used by nineteenth-century Swiss music box makers. As the barrel turned, with the first half tripping the steel prongs of the musical comb, the watchmakers fashioned a second set of pins (separated by less than half a millimeter) that could take over the melody when the cylinder paused halfway after completing a full rotation, creating a slight shift. By pressing down on the gold pushpin on the dial's edge, a series of wheels were activated that turned the alarm hand in the small center sub-dial, releasing the beautiful "berceuse" that his mother had played for him as a boy.

Despite his illness, or in spite of it, as Ward watched his last minutes and hours of life slip by, he continued to commission superb instruments to chronicle the passage of time.

And it was the passage of time that led Henry Graves, Jr., back into the game, or at least revived his competitive spirit.

• • •

On June 5, 1926, before the Graves family left for the season at Eagle Island, Henry had the happiness of escorting his twenty-three-year-old daughter, Gwendolen, down the aisle at St. Thomas's Protestant Episcopal Church. Thirty years earlier he had married Florence in this very church. Clutching a bouquet of white orchids and lilies of the valley, Gwendolen wore a gown of white satin and a veil of rose-point lace covered with seed pearls, which her mother had worn on her wedding day, to which she fastened orange blossoms.

Since the couple's engagement four months earlier, the society columns churned out a reliable stream of copy concerning the union between the wealthy financier's only daughter and Reginald Humphrey Fullerton, a thirty-six-year-old Bankers Trust vice president, reporting every detail, from the couple's prewedding luncheon at Pierre's to the final lineup of Gwendolen's ten bridesmaids. The *New York Times* printed a large engagement portrait of the bride-to-be staring coolly at the camera, draped in pearls. Following a grand reception at the Park Lane Hotel, Mr. and Mrs. Fullerton embarked on a six-month European honeymoon.

Reginald Humphrey Fullerton and Gwendolen Graves Fullerton at the time of their wedding in 1926. Photograph courtesy of Sotheby's, Inc. © 2012.

Not long after the newlyweds returned to New York, Henry gifted his daughter a twelve-room duplex encompassing the seventh and eighth floors in the new, lavishly appointed building at 1030 Fifth Avenue, across from Central Park and the Metropolitan Museum of Art. The thirteen-story neo-Italianate building contained yet more luxury apartments to stake their claim over the prime stretch of real estate from which Old New York's graceful mansions had long stood guard over society. Henry had purchased the apartment from L. Gordon Hamersley, the tobacco heir and President's Cup yachtsman. Two years earlier Hamersley had demolished his family's five-story French Renaissance mansion that had occupied the spot since 1899 and built the luxury co-op, taking up residence in its twenty-three-room penthouse.

In the lingering shadow of sadness that followed Harry's death, Henry and Florence immersed themselves in the familiar rhythms of public life. Henry kept up his calendar of social engagements. Florence resumed her charity work, organizing a benefit for St. Faith's House, an Episcopalian home for unwed mothers, among other good works. They followed the rituals of society, attending their country clubs and the opera. They moved with the seasons: summer on the water at Eagle Island, the fall on the horse trails at the Homestead in Virginia, winter at the Park Avenue apartment.

In February 1925 Harry's widow, Margaret, married a man Henry viewed as a gold digger after her considerable assets, Dexter Wright Hewitt, a thirty-five-year-old widower who worked in advertising and had two children. None of Henry's criticisms of Hewitt's intentions or character concerned his former daughter-in-law, who moved her new husband into the Ardsley mansion, the one she had shared with Harry, and resumed her old life without missing a beat. Margaret and Dexter golfed, danced at dinner parties, and spent the season in Palm Beach. In Ardsley they raised champion German shepherds at Mardex Kennels, which Margaret established after her husband died.*

For Henry, Irvington had become unbearable. Not long after his son's death, he and Florence left the Millionaire's Colony for New York and did not return. In May, just three months after Margaret wed Hewitt, Henry sold Shadowbrook to Dr. Joseph A. Blake, a noted surgeon. Henry severed his relationship with Margaret and cut off his grandchildren from

---

*The Hewitts earned great acclaim as breeders, and their shepherds took Best in Show at the world's most important dog shows. In 1929 their Utz von Haus Schutting won the Grand Championship of his breed at the German Shepherd Dog Association in Berlin. Some two hundred pedigreed German shepherds would trace their lineage to Utz von Haus Schutting.

their share of the trust set up by his father in 1906. In time, both Henry and Margaret would come to regret their decisions, but for entirely different reasons.

Gwendolen's marriage was a welcome tonic. Henry had a special affection for her and was extremely fond of Reginald Fullerton, a Yale graduate who traced his lineage to the American Revolution. His father, William Dixon Fullerton, an Ohioan whose birth was noted in *Americans of Gentle Birth and Their Ancestors,* had listed his profession as simply "capitalist." Gwendolen and Reginald would have two children, Reginald "Pete" Humphrey, Jr., and Nan Trimble.

As a token of the fondness and warmth he felt for his new son-in-law, on the day of their nuptials Henry presented Reginald with a Patek Philippe pocket watch. Engraved with the Fullerton coat of arms and the Latin motto *Lux in Tenebris* (Light in Darkness), the eighteen-carat gold trip minute repeater offered Reginald a rather intimate link to his new father-in-law's most private and exclusive world.

Three years earlier, on June 26, 1923, Gwendolen's brother Duncan had married Helen Johnson, a pretty society girl who made her debut at the Colony Club, Manhattan's most prestigious private women's social club. After a honeymoon spent briefly at Eagle Island before cutting across eighteen countries in Asia, Europe, the Middle East, and North Africa, the couple moved into an apartment at the exclusive Ambassador Hotel on Park Avenue at Fifty-first Street. The Ambassador billed itself as "New York's Most Aristocratic Hotel," with a five-room apartment renting for upward of $30,000 (about $390,000 today) a year. As the *New York Tribune* reported, its residents paid "some of the highest [rents] demanded by any hotel in the country." In due course, Helen and Duncan held sway over their own Park Avenue duplex and a country mansion, first on Long Island and then in Connecticut, with their four children: Henry, Duncan Jr., and twin girls, Marilyn Preston and Helen Mitchell.

Only the Graveses' youngest, George Coe, a dashing adventurer with movie star good looks, who was finishing his studies at Yale, had yet to take a bride. George Coe had long demonstrated his preference for hunting caribou in Alaska, surveying the South African bush, and chasing adventure in his seaplane to the social chase.

On May 7, 1926, just weeks before Gwendolen's nuptials, Henry's long-awaited package arrived from Tiffany's. It held the Patek Philippe pocket watch that he had commissioned seven years earlier. A singular piece of timekeeping, the eighteen-carat gold open-face watch with an enamel dial, Roman numerals, and blue-steel moon-style hands signaled Henry's determination to place his name on history's most exclusive horological lists. The beautiful watch featured twelve complications, including two-train minute repeating, *grande* and *petite sonnerie,* a perpetual calendar, split-second chronograph, and ages and phases of the moon. The gold cuvette, engraved per Henry's instruction, read, *No. 174961 Made For Henry Graves Jr. New York 1926 by Patek Philippe & Co. Geneva, Switzerland.* He paid the monumental sum of $2,650.50 for the piece (equal to $34,498 today).

Not only did this particular watch elevate Henry into the realm of storied *grandes complications*—James Ward Packard territory—but it marked his return to the game. One of horology's most voracious consumers was back on the hunt. Indeed between the fall of 1925 and the spring of 1927, Henry acquired at least thirty-eight watches, the majority purchased through Tiffany & Co. During this period of prodigious acquisitiveness, his commissions displayed his intense desire to own prizewinning *grandes complications* for which he paid handsomely. One exceptional pocket watch, with the movement no. 197506, cost the financier $3,875.50 ($51,312 today).

Henry's Patek Philippe watches were exquisite examples of his sophistication as a collector of complicated timepieces. As he chased down Ward's venerable expertise, his main point of distinction was that nearly all of his commissions displayed his intense obsession with proof of world-class merit: a *Bulletin de Première Classe de l'Observatoire Astronomique de Genève.* One, with the movement no. 198050-1913, was an exceptional piece of engineering and featured a thirty-two-hour power reserve, up-and-down indicator, and a platinum dial. Its movement had won first prize at the Geneva Observatory timing contests in 1925–1926. The movement of his eighteen-carat gold pocket chronometer, with blue steel hands and subsidiary seconds dial, no. 170358, had captured a first prize for "best average running" at Geneva in 1925. Henry commissioned a third Patek Philippe, an eighteen-carat gold open-face keyless winding watch, with the movement no. 198052, incorporating a minute repeater on three gongs, *grande* and *petite sonnerie,* perpetual retrograde calendar with phases and ages of the moon, and power reserve indications; in all, there were eleven complications. In the quiet whiplash of commissions that marked this gentlemen's contest, the men often went head-to-head, as this particular piece was quite similar to another *grande complication,* an eighteen-carat gold, open-face pocket watch that featured a perpetual calendar, moon phases, *grande* and *petite sonnerie,* minute repeating the Westminster chime on four gongs, and up-and-down indications (no. 174749), first delivered to James Ward Packard.

In 1925, about five years after Ward received his rare Patek Philippe ship's bell pocket watch (no. 174876) with the eighteen-carat gold dial, the *maison* delivered a near identical piece to Henry. His no. 198061 contained an open-face enamel dial minute repeater, *grande sonnerie,* and two up-and-down indicators, one for the winding train and one for the striking train. Engraved with the family crest, Henry's eighteen-carat gold pocket watch triumphed over the automaker's, as Ward's timepiece featured but a single up-and-down indicator.

Neither of these two men could have known what they started when they challenged Patek Philippe to craft the world's most complicated timepieces. Henry and Ward dictated the terms of their timekeeping arms race, a contest that would end only when one conceded defeat, withdrew, died, or simply tired of the game. Even then, some other contender could be out in the horological universe. This gentlemen's contest did not play out publicly. The watchmakers, bound by discretion, walked a tightrope of delicate diplomacy. Outside of the World's Fairs, Patek Philippe kept its technological achievements close, protecting them almost as zealously as they guarded their patrons. Yet rumors were rife, and news of an incredibly inventive feature continuously burnished the *maison*'s reputation among collectors and horological aficionados, as did word of a marquee client professing fealty to the firm. After all, when Queen Victoria purchased a pair of lovely watches for herself and Prince Albert at the Crystal Palace in London in 1851, Europe's crown heads soon beat a path to Patek Philippe. A collector's greatest weakness was his or her extraordinary will to possess, which was the watchmaker's most tantalizing weapon.

Henry could be certain that, of all the wondrous ticking universes that Patek Philippe had created, none could claim to be the most complicated watch ever made. His gold pocket watch with twelve complications, while unique and significant, had four fewer complications than Ward's magnificent *grande complication.* Perhaps he recognized that in terms of mechanical intellect and creative technology he would always have to concede to the engineering genius of Ward. But Henry, a practiced connoisseur, was not one to hold a candle to anybody else's flame.

Already as he walked Gwendolen down the aisle at St. Thomas's and even as he received delivery of his *grande complication* Henry had leaped to his next decisive move. In 1925, as Ward laid out plans for his watches following his death, Henry made the decision that would alter the balance between the exclusive circle of collectors throughout history, finishing the game of one-upmanship.

• • •

Henry approached Patek Philippe apparently through Tiffany's for a meeting to be held in "strictest secrecy." Not particularly comfortable practicing the art of small talk, he had one epic request and got to the point abruptly. He desired another *grande complication*, although not just another complex timekeeper. His marching orders were simple: he wanted "the most complicated watch," one that was "impossibly elaborate" and contained "the maximum possible number of complications." Putting a fine point on his very explicit instructions, he added, "And, in any case, certainly more complicated than that of Mr. Packard!"

With that, Henry put into motion a nearly eight-year odyssey that sent Patek Philippe's craftsmen, horologists, scientists, jewelers, artists, and engineers to their sunlit ateliers to create what would become known as the Graves Supercomplication. According to Patek Philippe, Henry's request "had not been heard in the watchmaking industry for a generation."

Over the next three years, the watchmaker undertook in-depth studies in astronomy, applied mathematics, and precision mechanisms. The mechanicians drew up a list of specifications of existing functions and those not yet tested, categorizing all possible functions. In some ways Patek Philippe was competing with itself, just as it was contending with the glory of the past. The magnificent timepieces it had crafted for James Ward Packard and other luminaries gave it a measure of experience—but only to a degree, for the firm set out to push past all that had been created before that time.

Gradually the design took shape: an open-face pocket watch, requiring two main dials to contain the various displays in an unadorned eighteen-carat gold case. There would be twenty-four complications in all, to include a full Westminster carillon on five gongs with a *grande* and *petite sonnerie* in passing, a minute repeater, and an alarm. This number alone would make it the most complicated watch ever crafted. The watch also contained two power reserve indicators for the movement and chime. Several of the complications, such as sidereal time, the equation of time as indicated by a sundial and mean time (the average of solar time), were hardly necessary to understand the passage of hours in the twentieth century. This pursuit of obscure complications not even remotely essential to the movement typified Henry's determined connoisseurship. In spectacular fashion, just as the Packard watch had, the Graves watch would feature a celestial chart representing the night sky over Henry's beloved New York City. In its storied history, this was only Patek Philippe's second attempt to produce such a magnificent sky map (the Packard was its first). The Graves Supercomplication enabled Patek Philippe to demonstrate its watchmaking genius in a single instrument.

Remarkably, given the five centuries that preceded this undertaking, outside of patent designs, the construction of complicated movements had never been enshrined in a manual. Traditionally, watchmakers worked from memory, passing on their craft and wisdom to their apprentices. The small horological army involved in the Supercomplication marshaled their encyclopedic knowledge, having built complicated timepieces over eight decades. During the course of five years, Patek's master watchmakers engaged in the meticulous continuous process of trial and error until the mechanism took shape, producing ratchet wheels, pinions, bridges, barrels, balance springs, jewels, and discs. As always, pieces were hand-finished to minimize friction and increase their precision and longevity. They worked out the three-dimensional jigsaw puzzle of building one layer on top of another, with layers of complications composed of microscopic parts.

Patek Philippe assembled the finest watchmakers in Switzerland to produce the Graves Supercomplication, nearly all members of the country's most important horological families. The endeavor recalled a time when Europe's kings called to their court expert scientists and specialists to craft instruments that defined an era. Le Fils de Victorin Piguet in Le Sentier, founded in the late nineteenth century, specializing in the production of complicated watches, blank movements, and dial trains, headed up a great deal of the craftsmanship and manufacture. Michel Piguet of Le Brassus designed the *grande sonnerie* mechanism, and Henri Daniel Piguet built the hand-setting mechanism. The case, built with a depth to accommodate the highly complex movement, was the work of Luc Rochat of L'Abbeye.

Jean Piguet, chief technician on the project, was kept awake at night attempting to solve the extreme difficulty posed by the setting mechanism. Three pairs of hands—for mean time, sidereal time, and the alarm—all shared the same winding-crown. When, during one of these sleepless nights, Piguet hit upon the solution, he jumped from his bed and immediately wrote it down, terrified that he might forget the answer by morning. His solution: a double winding method—pushing the crown forward wound up the striking method, and turning it backward wound up the watch. Pushing on the left of the pendant set the alarm. Pulling the crown out set the mean and sidereal time functions.

Once the watch had taken shape on paper and the drawings had been checked and rechecked, Patek Philippe sent sketches to Henry for his final endorsement. Detailed designs, two sheaves of paper showing the two dials, a side view, and the engraving above the sky chart, were returned with the signature *Design Approved Henry Graves Jr.*

Yet neither Henry nor Ward knew that, as they stayed in the game, time itself had become an even more powerful player in their drama.

CHAPTER FOURTEEN

~

Game Over

In November 1926, a year after his first surgery to remove the tumor, Ward's cancer returned and he checked himself into the renowned Cleveland Clinic. For weeks, Elizabeth made the sixty-mile trip from Warren, sleeping at the nearby Bolton Square Hotel, returning home occasionally, but only for half days. While the doctors cared for Ward's body, Elizabeth tended to her husband's spirits, making sure that he received his correspondence, supplying him with news from home, conferring over his ongoing treatment, and reading to him. Mostly, however, she just kept him company with her steadying, pleasant presence.

Ward had not chosen the Cleveland Clinic randomly. Some years earlier he had become interested in the Clinic's novel multidisciplinary medical practice, dedicated to education and research. Its founders, led by Dr. George Washington Crile, established the hospital in 1921, after serving in the Lakeside Unit during World War I. Deployed in the spring of 1917, the Unit tended to more than eighty thousand Allied troops near Rouen, France, over twenty months. Intensely ambitious and curious, Crile, the son of Ohio farmers, had already earned international standing as a physician when he opened the Clinic. One of the first surgeons in the United States to use blood transfusions, in 1903 he also designed a pneumatic rubber suit to control blood pressure and prevent patients from going into shock during surgery. (The device was later used during World War II to prevent pilots subjected to high gravity forces from blacking out.) On the battlefields of Europe, he introduced new methods of preventing infections. Upon returning home, Crile and his battle-hardened fellow medics, Drs. William Lower and Frank E. Bunts, along with a local internist, John Philips, decided to build a practice resembling their combat experience, one that included every branch of medicine while integrating research and direct patient treatment; they founded the nonprofit Cleveland Clinic Foundation. The endeavor struck a deep chord with Ward, and he became one of its benefactors, donating $200,000 (roughly $2.6 million today) to the Clinic.

From the outset, the Cleveland Clinic pioneered a number of new fields, such as x-ray therapy, endocrinology, and orthopedic and neurological surgery. When Ward arrived, surgery continued to be the foremost treatment for cancer. Radiation as a medical treatment remained in its infancy. With great foresight, the Clinic's founders used a portion of their initial building funds to purchase a gram of radium and installed a radium emanation plant that produced radon seeds, the first such plant in the Midwest. Still, in these early experimental days, the ability to accurately measure dosages often made the cure worse than the disease. With radiation therapy reserved for inoperable or recurrent tumors, its use earned grim notoriety as the "last hope." As Ward entered this bleak stretch of the disease, he still believed that science and technology had the answers to improve everyday life and health.

At the Clinic, the engineer's own body became a laboratory. While the Clinic's physicists and radiologists developed the first dosimeter to accurately measure the amount of radiation administered to patients,* Ward submitted to a series of treatments in which small glass tubes called capillaries injected radium emanation directly into his tumors. A course of deep therapy X-ray waves followed, intended to destroy any remaining cancerous cells.

Far from home, Ward spent his days surrounded by scientists, the smell of disinfectant, and the sound of metal wheels against cement floors. As the weeks passed into months, Elizabeth helped to ease the tedium by reading aloud to her husband. She tore through Thackeray's *Vanity Fair* and Dickens's *Barnaby Rudge*. As she finished P. C. Wren's adventure novel *Beau Geste,* the blockbuster silent film *Resurrection* starring Dolores Del Rio had come and gone at Cleveland's Allen Theatre Movie Palace, and construction on the fifty-two-story skyscraper, Terminal Tower, at the Public Square downtown was well under way. By the time the building was completed in 1928, it was the world's second largest skyscraper; the first "talkie," *The Jazz Singer,* had premiered; and Ward was still in the hospital.

Before arriving in Cleveland, Ward had received an attractive twenty-nine-page pamphlet from Lehigh University. His alma mater was in great

---

*In 1928 the Clinic's Otto Glasser, Valentine Seitz, and U. V. Portmann developed the first condenser dosimeter that accurately measured x-rays and radiation. This device was so highly regarded that the U.S. military later used it to test nuclear weapons.

need of funds to build suitable laboratories for its growing electrical and mechanical engineering departments, which had long outgrown their facilities. The carefully prepared pamphlet presented a description of the proposed building. Walter Okeson, secretary and treasurer of Lehigh's board of trustees and secretary of its alumni association, asked, "Who will build it?" For Ward, lying in his hospital bed, Okeson's appeal became something of a paper *memento mori.*

Ward turned the question over in his head. Given his tremendous success, he believed that he owed a debt of gratitude to Lehigh and that any gift he might make would be a "partial payment." He moved quickly and, on November 20, 1926, put pen to paper and addressed a series of blunt questions to Okeson, starting with "Will you please give me a little more dope regarding the proposed new building?"

Over the next two months a series of letters ricocheted between Ward, his personal secretary, his attorney, and the eager university. On December 8, after examining his financial arrangements, Ward agreed to donate $1 million (some $13 million today) in Packard Motor Company stock to construct the building in his name. Ward's "partial payment" turned out to be the entire building. In any era, a bequest of this size would be considered a colossal act of philanthropy, and Ward's gift quickly catapulted Lehigh into one of the richest universities in the country at the time. News of the endowment made national headlines. "Million to Lehigh Is Packard's Gift," shouted the *New York World.* The *Philadelphia Record* described it simply as "Lehigh University's Windfall." At the time Ward made his generous bequest, he had not visited the university since graduating forty-two years earlier.

Ward spent the Christmas of 1927 and rang in the New Year from his private room at the Cleveland Clinic. Elizabeth, exhausted and battling the flu, watched as her husband lay in bed on the hospital's fifth floor, growing ashen and frail. The top physicians had exerted their best efforts, but the cancer had wound through his body. In a veiled acknowledgment that Ward's condition had reached the point of no return, Dr. William Lower, a stout man with a patrician bearing, suggested on April 20 that it might make better sense for him to return home to Warren, where he would be more comfortable. Ward refused to consider the idea. The following day the doctor once again urged Ward to remove himself to the more familiar surroundings of his Oak Knoll mansion, and once again Ward refused.

As Ward's universe had been reduced to a hospital room, the world outside was expanding. On the morning of May 20, 1927, Charles Lindbergh had bounced down a muddy Roosevelt Field on Long Island in a wobbly silver monoplane and, equipped with 451 gallons of gas, four sandwiches, and two canteens of water, lifted off the ground, clipped a tangle of telephone wires, and then glided east across the sky. An entire ocean and thirty-three and a half hours later, Lindbergh landed his *Spirit of St. Louis* in Le Bourget Field near Paris.

In these postwar years, the world had unbuttoned itself and the contours of the American landscape shifted in darkness and excess. Al Capone moved his headquarters to Chicago's Metropole Hotel, where he presided over an empire of speakeasies, distilleries, gambling houses, horse tracks, whorehouses, and more. The Yankee slugger Babe Ruth was blasting his way toward a sixty-home-run streak. The Harlem Renaissance was in full swing. The play *Porgy* would soon debut on Broadway. Although the British novelist and futurist H. G. Wells had predicted the "speedy decline of radio," by 1927 nearly every American household owned one, and families eagerly gathered in their living rooms to listen to sermons, sports, dramas, and news of the "Red Menace."

After bobbing up and down, the stock market surged ahead, transforming the way investors looked at the market. As the Dow Jones Industrial Average hit 200, it was no longer a place to park long-term investments; the stock market had become a wheelhouse where anybody who entered would leave rich. Attempting to quash critical murmurs of trouble ahead, on January 12, 1928, E. H. Simmons, president of the New York Stock Exchange, proclaimed, "I cannot help but raise a dissenting voice to statements that we are living in a fool's paradise, and that prosperity in this country must necessarily diminish and recede in the near future."

As the decade approached its end, the automobile had firmly established itself as the most popular method of American transportation, and the business of automobiles was fast becoming *the* American industry. Ninety-five percent of the world's cars were now manufactured in the United States. After raw cotton and oil, autos were the third largest American export. More than four million people earned their livelihood from automobiles, and the industry consumed 18 percent of American steel production, 85 percent of rubber, 74 percent of plate glass, and 27 percent of aluminum. It was the third largest consumer of railroad equipment. When Ford retired the Model T in 1927, having sold 15 million of them, America registered one car for every 5.3 people.

As the number of cars mushroomed, the number of auto manufacturers plummeted. From the sidelines, Ward watched as the American automobile industry, once a model of competitive ingenuity, was radically

altered. Economies of scale and mass production had led to consolidation, bankruptcies, and a tremendous shakeout. Between 1921 and 1927
the number of major auto manufacturers in the country dropped from
eighty-eight to forty-four. Ford's emergence as an auto giant, followed by
the massive success of General Motors and the late entry of Chrysler at
the tail end of the 1920s, put the smaller makers in a headlock. Packard
occupied a shrinking pool of major independent car makers, once the
lifeblood of the industry. Names like Jordan, Kissle, Hudson, Essex, and
de Soto soon became footnotes in the industry's history. In the spring of
1928 Chrysler Corporation announced its merger with the Dodge Brothers, signaling an entirely new automobile world, in which three companies
now produced 80 percent of all cars. As the *New York Times* of May 31,
1928, summarized, "New Union Creates 'Big Three' in Autos."

Although Ward had removed himself from active involvement in the
motorcar company he launched, his stock kept his fortunes growing.
Rather than go mass-market, as many American manufacturers had done,
Packard leveraged its luxury position. In a letter to shareholders Packard's
president Alvan Macauley, the man *Time* magazine described as "cool,
self-possessed, quiet, sure of his facts & figures," emphasized the company's commitment to prestige motoring: "We know that the single standard
of high quality will produce better motor cars than were we to attempt to
secure the business of the world by building to all the pocket books in it."

The company continued to cater to the special requests of its devoted
clientele. King Alexander I of Yugoslavia, the Swiss-educated monarch,
was in the habit of ordering six custom automobiles at a time, with hood
ornaments that resembled his crown. Packard built gorgeous inlaid cases
made from rare woods into the front and back seats of his limousine
where the king kept his royal sword, maps, a picnic service of gold and silver, and his monogrammed silver cigarette case. His Highness the Maharajah of Gwalior in India, who had earlier replaced his pack of two dozen
elephants with ten Packards, ordered one for his royal consort, with a special request. He asked that the upholstery match the texture and delicate
pink of a dress slipper he sent to the Detroit factory. The Packard factory
not only set up special looms to weave the cloth but also developed special
dyes to perfect the match.

The last name in American luxury had a global spin. Packard Motors
Export Corporation, headquartered in New York, managed 450 international sales outlets, where its associates demonstrated the company's dedication to craftsmanship, detail, and engineering. When a mining engineer
in Colombia ordered a Packard expressly for joyriding over a thirty-mile

road that circled the top of the 6,000-foot mountain where he lived, the company went to great lengths to deliver it. At the mountain's base, the car was disassembled and its parts loaded onto pack mules and carried up a long and winding trail, with the chassis, body, and engine fastened to parallel poles.

Even as the company expanded abroad it remained deeply committed to Ward's founding principles. Packard had earned the moniker "America's master builder." The company erected a million-dollar proving grounds on nearly six hundred magnificently manicured acres in Shelby Township, twenty-two miles north of downtown Detroit. Packard's famed chief engineer, Jesse Vincent, built a state-of-the-art version of the crude test track that the Packard brothers had built more than two decades earlier. A four-lane, 2.5-mile paved oval would soon claim title as the world's fastest closed track. Packard boasted that the track was "so beautifully banked" that you could "take a Packard safely around the turns at one hundred miles an hour without even having your hands on the wheel!"

By the end of 1927, with the Dow breaking records, the company would ship 32,122 cars and record a net income of $11 million (nearly $146 million today). Its board of directors would increase the regular dividend on the common stock from twenty to twenty-five cents per month per share. Macauley told the *New York Times*, "Prospects never looked better for the Packard Company. I believe there is a pent-up demand for automobiles and for the products of general manufacturing that will be realized. There is no reason, I believe, for a closing down in business."

On July 20, 1927, a week after Babe Ruth had swatted his thirtieth home run, Ward marked his eighth month of hospitalization and Elizabeth traveled once again to Cleveland. By now the trip had become more or less a ritual, and she brought with her more books, correspondence, and news of home. On this trip, however, she also brought something that was sure to lift her husband's spirits: three pocket watches, a Vacheron Constantin and two Patek Philippes. Ward had commissioned the luminous pieces before he fell sick, and the two Pateks had actually arrived in the spring, a month apart, on March 8 and April 6.

Although quite weak, Ward undoubtedly brightened at the appearance of this trio of timepieces. The Vacheron Constantin was something of a departure from the many complex pocket watches that he had acquired, a highly unusual, ultrathin skeleton pocket watch that boasted a transparent rock crystal case, with the edge set in sapphires and the bow, crown, and wheel made of platinum. The transparent case had no metal around the

Ward received this ultrathin Vacheron Constantin skeleton watch while at the Cleveland Clinic in 1927. Encased in rock crystal and rimmed in sapphires, it was a stunning timepiece and something of a departure for the automaker.

movement, allowing Ward to view its elaborate construction. The movement, a work of art in itself, was made from fourteen-carat gold, beautifully damascened and polished; even the mainspring coils were visible, like a dainty princess sitting in a glass carriage. The watch, no thicker than a small child's finger, resembled none of his other pocket watches. Its bridges were made of solid gold, and the minuscule amount of metal surrounding the crystal was hand-engraved and ornamented. The dial's numerals were painted in black enamel on the inside of the crystal. Its unusual hands incorporated Ward's initials into its design.

Picking back a tiny latch on another of the wooden boxes, Ward opened the lid and gingerly lifted a minute repeater from its cradle. The Patek Philippe pocket watch, movement no. 198014, was eighteen-carat gold with richly carved rims and edges. The dial was made of gilded silver, and large and luminous radium numerals appeared just as Ward had imagined. It was perhaps his most sentimental commission. Delicate skeletonized hands, like the numerals, were filled with radium-saturated yellow wax, allowing it to glow in the dark. The same radium that the Clinic's doctors used to destroy Ward's cancer, the substance would later be out-lawed from commercial use because it contained dangerously high levels of carcinogens. At twenty-nine lignes (two and a half inches), it was a large timepiece, weighing nearly two pounds, necessary to accommodate the watch's *raison d'être*: the customized musical alarm that played Godard's *Jocelyn* lullaby.

The third watch, however, most enthralled Ward. It was the piece that he had contemplated on that cold, rainy night five years earlier, as he entertained his father-in-law and Dr. John Kingsley of the Essex Institute. This was the watch whose commission had caught the attention of Henry Graves, Jr., and awakened his own enthusiasms. It had taken five years to produce and set the horological world on fire. Unquestionably, the watch, with movement no. 198023, known as the Packard, was the masterpiece of his entire collection and the apotheosis of his relationship with Patek Philippe.

The dial with Arabic numerals housed four sub-dials. The eighteen-carat gold case shimmered in the harsh hospital light; it was elaborately engraved with Ward's initials in blue enamel encircled by a burst of rays. Just nineteen lignes in diameter (one and two-third inches), the watch contained the greatest number of complications in the smallest amount of space.

In all, there were ten complications, each customized to Ward's fancy and calibrated precisely to his hometown. Manifesting nearly nine decades of experience and knowledge, Patek Philippe had mobilized its best engineers, craftsman, and mathematicians from Geneva to the Jura Valley to produce this masterpiece. The watchmakers had walked a tight-rope between technique and art. Dozens of escapements were drawn. Teeth were cut on one wheel and then the next. The tiniest of pieces were built, polished, tested, and tested again. Layer by layer the movement took shape. The craftsmen sat pressed at the edge of the bench, loupe to eye with forefinger and thumb cradling the slimmest instruments and work-ing the lathe; every move made a difference.

On the silver dial, the minute and hour hands of blue steel marked the time, but that was almost an afterthought. A sunburst minute hand displayed the difference between mean solar time (the time we go by) and true solar time (the time told by a sundial). The moon phase aper-ture, located at 12 o'clock, included a moving moon disc made of blue enamel. The numbers surrounding the disk were calibrated to show the moon's phase, and a pointer indicated the month. At 6 o'clock a small dial with three hands running on concentric tracks indicated the day of the week, day of the month, and seconds. The perpetual calendar automatically adjusted to compensate for uneven months and leap years until the year 2100. A dial at 9 o'clock pointed to the time of sunrise, and one at 3 o'clock to the hour of sunset, both calibrated specifically for the latitude and longitude of Warren, Ohio. Ward had a soft spot for minute repeaters, and this watch was no exception: the mechanism rang on three gongs set to indicate the hours, quarters, and exact minute.

Pressing down on the watch's winding crown, Ward popped open the back case to reveal an inside cover aperture and its true charm, a blue enamel celestial chart of five hundred stars, each represented by a gold point in six sizes according to their precise magnitude, and designed to match the night sky above Warren, at exactly 41 degrees 20 minutes. This miniature nocturnal scene rotated daily, following the same heavenly path that appeared outside Ward's mansion bedroom. Looking at the stars in the palm of his hand, Ward could imagine that he was back in Warren gazing into the night sky.

Blissfully unaware of the Graves Supercomplication under production in Geneva, Ward believed his place in horology was secure. After thirty years, he had outdone himself. The Packard stood alongside the two other most celebrated *grandes complications* in history, the Marie-Antoinette and the Leroy No. 1. Although it possessed ten complications (six fewer than his earlier *grande complication*), it represented a tremendously complex world contained inside the space of just one and two-third inches, and for decades horologists would debate its supremacy as *the* ultra complication of all time. Fittingly, the debate was based on a technicality. In the annals of watchmaking, it was only the second watch produced in modern times that was equipped with a celestial chart. Ward emerged on top. For those keeping score, the auto pioneer had won the war of complications, and the Packard had won the battle.

Henry Graves, Jr., wasn't the only aficionado who took an intense interest in Ward's watches. Over the years Paul Moore had followed news of Ward's acquisitions with a high level of curiosity. As head of the research division at the National Research Council in Washington D.C., Moore spearheaded government surveys for the National Academy of Sciences, but it was in his position as secretary of the Horological Institute of America that he had become acquainted with Ward's watches. Like Ward, Moore admired mechanical perfection, specifically complicated watch mechanisms. But more than just an admirer, Moore believed that timepieces and watchmaking represented civilization, and he worked tirelessly "to elevate and dignify the art, science and practice of horology." For Moore, the Packard collection did more than just demonstrate timekeeping; it revealed "the highest development of human ingenuity as applied to horological mechanisms."

Moore believed innovation and watchmaking were inextricably linked. Since the end of World War I, he had feared that America's ability to invent and manufacture was on the decline, in the process dragging down the nation's capacity for mechanical watchmaking. Under the aegis of the Research Council, he worked to develop standards for horological schools, certifying requirements for watchmakers in this country. His great desire was to raise American workmanship and horological science on par with Europe. "While the factory system and mass production have done much for America," he wrote, "they have wrought evils also. Many watchmakers all over this country are not equal to the demands of this age for repair work. On the other hand the public should not expect too much of a cheap watch."

Moore's greatest ambition was to establish a time museum in Washington under the auspices of the Horological Institute of America, dedicated to the science of time and timekeeping. As he explained, "Time is as important as music or art or many other 'causes,' for which money is forthcoming to erect buildings." His push to build such a dedicated monument in the United States gained little momentum, but Ward's standout collection fit in perfectly with Moore's desires. He believed that, if he could acquire the Packard watches, they would attract interest on a large scale and would broadcast the significance and importance of his planned museum. Furthermore the museum would serve as a magnet to attract other individual pieces and collections over time.

Aware that Ward's collection was set to be handed over to the Cleveland Museum of Art, Moore traveled from Washington to the hospital where Ward lay gravely ill and laid out his case. After describing his vision for a time museum, he explained that, if Ward's intention in giving his watches to the Cleveland Museum of Art would be to give the pieces the widest possible audience, they would in actuality be better placed in the care of the Horological Institute. There, his magnificent watches would be used "for educational work in developing for America the finest mechanicians."

Ward and Elizabeth apparently agreed, and the Cleveland Museum of Art graciously relinquished its rights to the collection, transferring them to the Horological Institute, along with the $2,000 for the collection's future care.

• • •

In 1928 Ward spent his last New Year's Eve at the Cleveland Clinic. In the fiercely cold January, he anxiously passed the time awaiting word on the progress of his namesake engineering building. Elizabeth's brother and Ward's personal attorney, Judge R. I. Gillmer, traveled to Lehigh shortly after the New Year to check on the building's development. The meeting was cordial but troubling, as the project managers informed Gillmer that it now appeared that Ward's gift would not cover the undertaking. Walter Okeson showed Gillmer drawings of the proposed building. The architects had designed a massive four-story, Gothic-style structure of cut limestone. Equipped with the latest in heating and electricity, it was planned with an eye toward the most modern boilers, generators, and measuring devices. The designs called for two main labs for electrical engineering and mechanical engineering, housing every phase of study, as well as several specialized labs for the study of internal combustion engines, wired and wireless telegraphy, radio, high voltages, and refrigeration. Provisions were made for mechanics' shops, drafting rooms, offices, instrument rooms, classrooms, and an extensive library. Extremely well thought out, the design left nothing unconsidered, including plans for live steam, exhaust steam, and concealed wiring.

But the incoming bids exceeded the million-dollar pledge, and the university was considering revising its plans in order to make the building smaller or use cheaper construction materials. Gillmer disagreed. Although Ward put no conditions on his endowment (except that it bear his name and be the "finest plant of its kind in existence"), he had been apprised of its preliminary design and wholeheartedly supported it. Before leaving Bethlehem, Gillmer told the university representatives that he was "quite sure that Mr. Packard would not wish to have the building other than [the university] desired it and he [Gillmer] was confident that if necessary Mr. Packard would provide additional funds to enable the building to be completed as planned."

On February 1, 1928, Okeson received a letter from Ward's secretary and a check for $200,000 (about $2.69 million today). "Mr. Packard realizing the importance of the undertaking and the efforts being set forth by everyone concerned," the enclosed correspondence explained, "does not wish to leave a single item omitted in making this gift of his to Lehigh the finest of its kind."

On March 18, 1928, Ward's health took another turn for the worse. Drifting in and out of sleep, he momentarily broke free from his lethargy to plead with his wife not to marry after he had gone. "Seems so trivial at this time," she wrote in her diary later in the day.

The following morning Elizabeth found her husband's condition "alarming." At nine in the morning on March 20, she arrived once again at the hospital and, to her horror, her husband lay weak and incoherent, unable to recognize her. She sat beside him as his breath grew increasingly shallow for two and a half hours before stopping. Sixteen months after he first checked into the Cleveland Clinic, Ward died, "peacefully and without realizing that the end had come," as Elizabeth recorded in her diary. Ward had believed in deploying science and technology to improve the human condition, but science could not save him.

At his death, James Ward Packard was sixty-four years old. One of America's 15,000 millionaires, he left an estate valued at $7 million ($94.3 million today). At the time of his death, net annual profits of the Packard Motor Car Company had reached $10 million. Packard Electric had become the leading manufacturer of automotive, appliance, and aircraft wiring, employing more than six thousand people. Four years later General Motors would buy the company.

Newspapers and magazines celebrated the remarkable inventor who had quietly helped transform America. "With his passing," reported the

Ward was buried at the Packard family gravesite at the Oakwood Cemetery in Warren, Ohio. Photograph by Stacy Perman.

New York Times, "the little circle of men who built the first 'horseless car-riages' in America near the close of the last century and were the forerun-ners of the mammoth automobile industry of today loses one of its most prominent members."

The Packard News Service, the company's internal marketing arm, wrote a lengthy obituary that it sent to the national media. Newspapers and magazines across the country reprinted chunks of the lavish praise. Henry R. Luce, the managing editor of *Time* and a proud Packard motor-car owner, personally sent a copy of the magazine's planned obituary with a telegram to Packard Motor asking its marketing department to kindly read it and report back "if you detect any error of fact." The final read in part, "Yet few of the men who built the first automobiles are still alive; Maxwell, Haynes, the Dodge Brothers—these were among the most important and all of them are dead. Last week Death, in his quick chariot, overtook one more. This was James Ward Packard, famed maker of Packard cars."

At 2:30 p.m. on March 26, friends, relatives, and many of Ward's busi-ness associates from across the country filed into the Oak Knoll Drive mansion, where the funeral took place. For a full hour before the Rever-end R. E. Schulz, rector of Christ Episcopal Church, began the service, its bells chimed in requiem.

On June 8, Ward's nephew Warren traveled to Lehigh University in Beth-lehem, Pennsylvania, where more than one thousand alumni watched as he laid the cornerstone for the James Ward Packard Laboratory of Elec-trical and Mechanical Engineering. Within months, arrangements began for the transfer of Ward's watches to the Horological Institute of America.

Ward's death marked the beginning of the end of the golden age of mechanical watchmaking. The pocket watch that had captivated Ward and Henry Graves, Jr., was a rich man's dinosaur on the brink of extinction.

~

CHAPTER FIFTEEN

Across the Sea

In 1928, as spring turned to summer, Florence and Henry Graves, Jr., sailed to Europe aboard the RMS *Olympic,* the grandest ocean liner of the White Star Line. The sister ship of the ill-fated *Titanic,* it weighed nearly 47,000 tons and its hull, at just over 882 feet long, ensured that the *Olympic* was the largest vessel at sea. It was also the most glamorous. Henry had booked passage on the ocean liner simply because it was the best. The Graveses had chosen to forgo Eagle Island altogether and instead spend the season among the glitterati on the Continent, with no agenda other than pleasure-seeking—although, for her part, Florence planned to remain in front of New York society's fashions. Her husband, it appeared, intended to stay one step ahead of James Ward Packard.

Over six days plying the Atlantic, Mr. and Mrs. Graves, as did their ilk, passed the time exploring the *Olympic*'s more sumptuous quarters, ignoring the ship's exotic novelties such as the mechanical camel in the fully equipped gym and indulging in more familiar diversions. Under the molded ceiling of the first-class drawing room, Florence immersed herself in the delicate pursuits of a proper gentlewoman, writing postcards, doing needlepoint, and reading. Henry availed himself of the ship's male haunts, such as the opulent Turkish bath. Enjoying the proximity to the sea, together they sauntered along the promenade, breathing the salty air. Even engaged in such informal activity, Henry commanded deference among his peers. His obvious polished bearing suggested a gentleman of substance who lived extremely well—or at least one who had mastered presentation as a vocation. Dressed in one of his bespoke three-button, four-pocket jackets, razor-edged trousers the color of cream, his straw boater, and a tie carefully knotted around his collar, he blithely strolled the decks. With a jaunty hand on his hip, he casually nodded to the other passersby with Florence by his side; emancipated from the corsets she had

Florence Graves relaxing with
Henry and one of her favorite
pastimes, needlepoint.
Courtesy of Cheryl Graves.

worn during the early years of her marriage, she now donned a fashionably low-waisted dress, t-strap shoes, and a sun hat.

At sunset, the pair changed into their evening finery. Henry dressed in
tails and top hat, while Florence wore an au courant French frock. Stepping onto the upper landing leading to the first-class reception room, they
were bathed in the purplish nocturnal light streaming through the massive dome of wrought-iron and glass. Although making an entrance never
appealed to Henry, taking one of the three elevators hidden behind the
stairway meant missing another opportunity to appreciate the immense
grand staircase that fanned out at each landing and to walk past the intricately carved clock made of oak that depicted *Honor and Glory Crowning
Time* by Charles Wilson.

The couple's evenings on board followed a ritualized set of manners.
After dinner and perhaps a dance, Florence accompanied the women for
small talk and coffee in one of the ornate drawing rooms, while her husband repaired with the men to the first-class smoking lounge, where they
relaxed, unshackled from the bonds of Prohibition back home. At the end

of the evening the couple retired to their private suite, where Louis XIV-style décor was bolted to the carpeted floor, a subtle reminder that this grand hotel was in fact a massive ship gliding across the ocean.

Arriving in Cherbourg, Henry and Florence boarded a first-class train and headed south for an extended European stay through the fall. The couple planned to spend much of their time in Paris based at the Hôtel de Crillon. There, Henry, officially "retired," could still conduct some private business at the branch of Bankers Trust Company on the Place Vendôme, Florence could visit the couturiers, and both could sip aperitifs side by side at the Ritz Hotel bar on the Rue Cambon. The pair also cut a decadent swath across the most exclusive resorts along the French Riviera and the Swiss Alps, with a side trip to Geneva, where Henry had some personal business to attend to.

The crescent-shaped Lac Léman appeared sapphire blue against the snowy peaks of the Alps when Henry and Florence arrived in the dry June heat. The belle époque paddle steamers cut across the lake that Mary Shelley, her husband, Percy Bysshe Shelley, and Lord Byron had immortalized in prose and poetry following their scandalous summer in 1816. At the water's edge, the Jet d'Eau spewed a stream of water nearly one hundred feet into the air. It wasn't Henry's first visit, but much had changed.

A drowsy village masquerading as an international city, Geneva gently unfurled at the foot of the Alps, a postcard of French style and German efficiency. Rows of little shops, ateliers, luxury jewelers, chocolatiers, stone churches, cobbled alleyways, and medieval ramparts gave way to grand châteaus. A watchmaking capital and the cradle of global private banking, Geneva had Protestant ethics and Calvinist leanings that made it a city of money and piety, with aristocrats, craftsmen, and a contained elegance. By the time Henry and Florence strolled the Rive Gauche, Geneva had also become a safe harbor for the world's moguls and their growing fortunes.

During the Great War, Switzerland's stubborn neutrality, outsized army, and mountain fortress had left Geneva largely unscathed. In the war's aftermath, the city lost none of its calibrated restraint. With the establishment of the League of Nations, Geneva was made the seat of international diplomacy. The city of loupe-eyed watchmakers and bowler-hatted bankers was now a global village of pinstriped functionaries.

Henry made his way to the five-story building overlooking the lake at the Quai Général Guisan housing Patek Philippe's headquarters and manufacturing workshops. With the type barely set on the headlines announcing James Ward Packard's death, the crank on horology's war of complications was to turn yet again, even though it was nearly impossible to escape Packard's legacy. Among the national flood of obituaries describing the automobile pioneer's life, many were filled with breathless accounts of his remarkable watch collection. As the *New York Times* noted, "Mr. Packard's hobby was collecting clocks and watches . . . some of which were not duplicated anywhere in the world." Word of his final commission, the exceptional astronomical watch, the Packard, had begun to spread even before its deathbed transfer to the Horological Institute of America. In death he had been anointed America's preeminent horological connoisseur.

In Ward's passing Henry could see more than just a moment of reflection; this was an irresistible opening. Approaching sixty, Henry had lost none of his competitive nature. Even as the newspapers crowed over Ward's watches, Henry's greatest timepiece, the Supercomplication, was undergoing the laborious process of turning dream into reality.

On March 5, before he departed for Europe, Henry had made another trip to Tiffany & Co. to collect his latest complicated pocket watch, which he had commissioned some three years earlier, at the height of the Graves-Packard steeplechase. Patek Philippe had delivered the eighteen-carat gold open-face keyless winding watch, with movement no. 198052, to the jeweler. For this highly complex minute repeater, the one with eleven complications, Henry paid $2,700 (equal to $36,375 today). Admiring the exquisitely complex craftsmanship, Henry was not yet satisfied.

In Geneva, entering the doors at 41 Rue du Rhône, Henry would be ushered through the Salon Napoleon III. The ground floor *salon d'exposition* was the size of a small Fifth Avenue ballroom and smelled faintly of rose oil and leather. Hanging prominently against the sculpted paneling and above one of the main glass cases was the famous large gilt-edged oval displaying the many medals of honor Patek Philippe had won over the years at the various World's Fairs and timing competitions.

Along the back wall stood the elegant, enormous, former Tiffany vault, with its painting of a bald eagle grasping a pair of American flags encircled in stars. The vault was a vestige of Tiffany & Co.'s short-lived adventure in Geneva and its later joining of forces with Patek Philippe. In 1872 Tiffany had established its own factory at the Place Cornavin, said to be the largest of its kind, as part of the jeweler's grand experiment to marry American mass production with the Swiss tradition. Four years later the factory was shuttered and sold to Patek Philippe, and the two parties entered into a five-year contract. Tiffany agreed to amplify its representation of Patek in

America and service its watches. For its part, Patek assumed control over the Cornavin factory and agreed to service Tiffany watches in Europe.*

In the workshops above, some of the watchmakers sat in front of their face-lathes boring holes in bridge plates; others polished and oiled the tiniest of parts, all of them robed and silent, going about their work as if dedicated monks in the service of a greater power. Given Henry's elevated status as one of the house's most important clients, he was likely whisked upstairs to one of the private salons with its spectacular view of the lake. Small pleasantries were exchanged. Hands were shaken. Henry sank back into a regal wood and velvet chair. His visit coincided with the end of the three-year period during which Patek Philippe's watchmakers had considered his watch's intricate design and the bench had just turned to its actual production. As with his previous complex pieces, Henry approved drawings that led to the next steps, a prototype, and manufacture.

Henry had at least three complicated pocket watches in various stages of manufacture, including two rare platinum tourbillon minute repeaters. While at the *maison*, however, he appears to have turned his attention from pocket watches to the increasingly popular wristwatch.

Patek Philippe had entered into one of its most vibrant periods, taking the lead in miniaturizing popular complications for the wristwatch. In 1922 the watchmaker introduced the first single-button split-second chronograph, followed three years later with yet another first: the perpetual calendar wristwatch. By 1927 Patek had begun producing the first wristwatch chronographs with and without split seconds. In Geneva, Henry acquired some of the first of what would be a tremendous trove of wristwatches. On this trip they included two Tonneau-shaped minute repeaters, one encased in eighteen-carat gold and the other, which he took receipt of in August, in platinum, both engraved with the Graves family crest. The earliest minute repeaters of Henry's collection, these timepieces were also some of the earliest that Patek Philippe had manufactured.

In a nod to one of his other passions, Henry had also arranged to purchase a coin-form watch. It too was one of the first of its kind that Patek Philippe had ever crafted. Made from an eighteen-carat gold U.S. $20 coin minted in 1904, the watch, with the movement no. 812471, had a secret latch embedded in the coin's edge that triggered a minuscule spring to reveal a gold watch dial with blue steel hands and Breguet numerals in black enamel.

---

*Patek Philippe eventually sold the Place Cornavin factory, which was later converted into a hotel.

• • •

Just days before Amelia Earhart took off on her transatlantic flight, Henry picked up a copy of the most recent *Observatoire de Genève* and, with his wife, boarded a train and crossed into southern France. On June 20, 1928, they arrived at Nice's grand Hotel Ruhl. Rising like a frothy white wedding cake on the Côte d'Azur, the hotel was a sun-splashed carnival for the wealthy on the Riviera, long the fashionable playground for a set of global habitués. Here Coco Chanel had introduced the scandalous idea that a suntan was a mark of wealth and health and not the province of low-class laborers. Before the Bolshevik Revolution of 1917, the Ruhl's casino had drawn a regular parade of Russia's grand dukes and duchesses. Under the Beaux Arts coffered ceiling, they had piled stacks of gold francs on the roulette tables, tossing a few to the lucky croupier. Dripping in jewels and poured into the finest couture, the oligarchs had thrown dinner parties and spent with abandon. By the time Henry and Florence descended the Ruhl's curling stairs in their Parisian couture, the Russians had been supplanted by wealthy American plutocrats and their magnificently turned-out wives, who now mingled with Europe's aristocracy, ordering exotic flowers and choice wines and throwing a few gold coins at the croupier for a bit of luck.

Each morning Henry and Florence rose to a *petit déjeuner* of coffee and croissants on delicate china spread over crisp linens. Their days fit inside a postcard. The couple relaxed under the Mediterranean sun along the palm-fringed Promenade des Anglais or sailing on the sea. They exchanged pleasantries and afternoon tea with their circle. These gentle amusements apparently did little to soothe Henry's spirits, as he had left Geneva somewhat unsettled and fixated on another timepiece. While in Geneva, he had read through the *Report of the Observatoire* for 1927–28 and learned of a tourbillon chronometer awarded the "First First Prize" in the timing competition at the Geneva Observatory. It was made by Vacheron Constantin, and Henry could not get it out of his head.

Pulling a sheet of hotel stationery from the bureau, Henry sent off a handwritten note to Messrs. Vacheron and Constantin in Geneva, inquiring about the availability of the piece. The oldest watch manufacturer in the world, in addition to royals and other luminaries it had crafted pieces for the industrialist John D. Rockefeller, men of letters Henry James and his brother William, and of course James Ward Packard. Yet Henry was not among Vacheron's long list of wealthy and powerful patrons and ostensibly remained unaware of Ward's relationship with the house. To date, Henry had remained largely faithful to Patek Philippe, but as summer turned to fall he engaged in an intense correspondence with Vacheron Constantin. Like Ward before him, it appeared that Henry had found a new object at which to direct his ardor.

Arriving at the Hôtel de Crillon in Paris on the evening of June 23, Henry received word from the watchmaker that the piece was indeed available. The very next day he replied, "I will purchase the chronometer." He also inquired about when he might receive the piece, noting its price and taxes. Over the coming weeks, he remained utterly consumed with the goal of possessing the prizewinner. As he and Florence flitted from one glamorous resort to the next, he continued to correspond with Vacheron Constantin. Each letter demonstrated his growing enchantment and impatience. He worried over the watch's "safe delivery" and obsessed over its "Official Bulletin Rating," insisting that he receive not only a copy of the bulletin but the diploma showing the award of the "First First Prize." He also requested certain changes to be made to the watch. He wanted a gold dial rather than enamel, and his preferred blue-steel hands. He dictated precisely how the inside cover of the case, the cuvette, should be engraved, and he asked that the movement number be displayed on the new dial. Finally he wrote, "Please have the watch cleaned and freshly oiled before delivery." Anxious to possess the watch, he did not agonize over its price, $1,000 (about $13,400 today), or any additional charges his requests might add to the final invoice. With each dispatch, Henry eagerly invited Vacheron to respond with its acknowledgment "at your earliest convenience."

With such a potentially important client, Vacheron Constantin remained solicitous, assenting to each new request and desire. Also befitting a house of such stature, it remained discreet. Increasingly restless, Henry instructed the firm to send the watch via special messenger to the Hôtel de Crillon in Paris, agreeing to pay the firm $50 to do so. This way, he explained, "I would be assured of the safe delivery to me of the watch without unnecessary jarring."

On Friday, July 13, 1928, Vacheron Constantin's emissary arrived. Charles Constantin, the great-grandson of François Constantin, the watchmaker's cofounder, delivered the watch personally. The two men would have repaired to one of the public salons among seventeenth- and eighteenth-century tapestries, arranging themselves on the gilt and brocade furniture, not far from where Queen Marie Antoinette had taken piano lessons a century earlier. Constantin produced a fitted wooden box, and Henry reached for his spectacles. Barely keeping his impatience under wraps, he sprung open the box and examined his shiny new toy.

At sixty millimeters in diameter, sixteen millimeters wide, and 117.25 grams, by pocket watch standards the watch was rather large. The case was made of solid eighteen-carat gold and in the Louis XV style with a gold empire dial. Henry pressed his thumb gently against the watch's side, unlocking the case to reveal the cuvette. The engraving was exactly as he had requested: *Awarded First First Prize (866 Points) Geneva Astronomical Observatory Timing Contest 1927–28, No. 401562, Henry Graves Jr., New York, by Vacheron & Constantin, Geneva, Switzerland.*

Henry was pleased. And it appears that Monsieur Constantin was intrigued. His great-grandfather had adopted the motto "Do better if possible, which always is possible." This meeting offered him an opening. Henry Graves, Jr., had two tantalizing qualities: he was a man of precise tastes obsessed with watches (preferably those with diplomas from the Geneva Observatory) for whom price was merely a detail. Constantin gently coaxed this initial flirtation further. The great watchmaking houses had built their legacies over the centuries fawning over and winning favor with important clients from popes to kings and now the modern royals, America's industrial princes. Constantin mentioned that the atelier had produced a unique skeletal watch, exceptionally thin, with a dial of rock crystal, its movement forged out of fourteen-carat gold. Crafted entirely by hand, it was quite simply a tour de force of artistic brio and technical workmanship.

In a letter dated July 18, 1928, Henry followed up with Vacheron Constantin. Delighted with both the chronometer and the way the watchmaker had carried out his wishes, he reminded the *maison* of his expressed interest in prizewinning timepieces. "Also do not forget," he wrote, "to advise me if you ever produce a watch that secures from the Geneva Observatory a higher record than the chronometer I have just purchased." As his return plans to America were still in flux, he asked that Vacheron send him a leather case and spare mainspring for his prizewinner, in care of the Banker's Trust Company in Paris. The skeletal watch that Charles Constantin had so delicately mentioned intrigued him. Vacheron had made only two. Henry asked the house to send him further details.

By the last week of July, Henry and Florence had pushed on to St. Moritz, settling into the exclusive Hotel Suvretta House, nestled in the alpine woods with glorious views of the Champfèr and Silvaplana Lakes on the Chasellas Plateau. There he received not only a letter from Vacheron Constantin and the information he had requested but also a series of photographs that showed the skeletal watch's front, back, and side. Although exceptional, the piece did not strike his attention, as he had hoped.

As for the other Observatory pieces Vacheron mentioned in its letter, he replied curtly, "[They] do not interest me at all, but when you produce one that secures from the Observatory a better record than the one you have just sold to me I would then be interested to hear from you at my New York address."

Henry dismissed the skeleton watch, despite its intricate beauty and complexity. He was in this for sport, and the skeleton, though beautiful, possessed no record. But Henry had missed one fine detail. In the photo series that he received, the hour hands were designed to form the initials of its owner, the man who had first commissioned the piece. In blue steel the letters were *J* and *P*.

On October 24, 1928, Henry and Florence boarded the RMS *Olympic* at Cherbourg for the return trip home. They had been away for nearly six months. Sailing through rough waters, the couple watched the Normandy Coast recede into the sunlight. As they crossed the Atlantic, the Woolworth Building was about to end its reign as the world's tallest building, and Henry was about to begin his as the king of *haute horlogerie*. They stood on the first-class promenade blissfully unaware that their shimmering world was winding down its final days.

CHAPTER SIXTEEN

~

The James W. Packard Collection of Unusual and Complicated Watches

The dispersal of Ward's watch collection was swift. He had named his nephew Warren Packard, to whom he'd always been close, one of his estate executors. At thirty-four, Warren had grown up at his uncle's knee and had moved up the Packard ladder. After working at the Packard branch in Cleveland, he was named president of Packard Engineering. Most recently he had become an advertising executive with Packard Motor Company in Detroit. He lived in a red-brick Federal-style house in Grosse Point with his wife, Dorothy (née Braden), and their two children, a son, Warren Jr., age three, and a two-year-old daughter, Rosalie.

In April, just a month after Ward's death, Warren accompanied Elizabeth Packard to Cleveland for an appointment at the Union Trust Company. She took thirty of her husband's watches for an appraisal by the Cleveland Museum of Art. "Feel my loss more as I begin to dispose of things," she wrote in her diary after depositing the watches.

The Cleveland Art Museum, assisted by the watch expert George E. Lee, made an inventory of each piece, grouping the watches in order of importance by manufacture, with a reference number and a brief description. There were thirteen Patek Philippes, constituting nearly all of his *grandes complications,* among them the Packard, the *Jocelyn* musical alarm, the tourbillon minute repeater, and the ship's bell, as well as the first Patek he had acquired, in 1905, the chronograph with a minute repeater, perpetual calendar, and *grande* and *petite sonnerie.* The grouping also included the Vacheron Constantin skeleton pocket watch and his custom-designed set of British *grandes complications* from J. W. Benson, Smith & Sons, and E. Dent Company. Lee catalogued nine pieces as "miscellaneous" and informed Elizabeth that these were not of museum quality. The Horological Institute of America would later note of these timepieces, "They do not enter into the same realm of appeal as the watches which Mr. Packard acquired through personal effort."

Two weeks later, on May 2, Elizabeth received the final appraisal, typed out neatly on two pieces of onionskin paper. The entire collection was valued at just $15,240 (about $205,320 today), a rather low sum considering that 1928 was still a boom year. Furthermore some of the individual prices were well below what Ward had paid for the watches; the Packard alone had cost $16,000. Although the low appraisal was likely for estate purposes, Paul Moore, writing in the *Jewelry Circular* a year after Ward's death, declared the collection's value at $80,000, the equivalent of $1,077,796 today.

The watches represented a fraction of Ward's total collection. Not included were the walking stick, gold ring-watch, wristwatch, the Paperweight desk clock, and other Patek Philippe and Vacheron Constantin pocket watches that Elizabeth retained, as well as several pocket watches from Agassiz, Waltham, Wittnauer, and Jules Jürgensen; numerous eight-day instruments; and ship clocks and timekeepers of French and American lineage. In the coming months Elizabeth took several of them to the Ball Company, the famous jeweler in Cleveland, and to F. E. Armitage, Ward's longtime broker and friend, apparently for appraisals, servicing, and perhaps dispersal. She also gave a number of her husband's timepieces to family members, and she allowed one of her nephews to choose one as a Christmas present.

On July 26, 1928, almost two weeks after the estate inventory was completed, Elizabeth invited Warren Packard to make his own selection. He picked several mechanical pieces, such as barometers and two of Ward's complicated Swiss pocket watches, a Patek Philippe and a Vacheron Constantin. Ward had commissioned the pair during his most fruitful period of partnership with both watchmakers, between 1918 and 1920, when he acquired perhaps dozens of timepieces. The Patek Philippe, with movement no. 174907, incorporated an extremely unique combination of complications: a minute repeater with a power reserve framed in a one-of-a-kind eighteen-carat gold Murat-style case. The Vacheron Constantin, movement no. 375551, was the open-face twenty-carat gold chronograph that featured a rare set of complications: a trip minute repeater, *grande* and *petite sonnerie,* and half-quarter repeating functions.

Before finally relinquishing the collection, the Cleveland Museum of Art exhibited Ward's watches for a brief period. Elizabeth, along with a representative from the Union Trust Company, traveled to see them on December 18, 1928. "Much pleased with the display of watches," she wrote in her diary. Following the final distribution of Ward's timepieces, she traveled to the couple's Lakewood mansion, their summer estate, which would now be her primary residence. She turned over the Oak Knoll mansion, only finished in 1924, to her niece Katherine Summers, allowing the family to live there as long as they wished. In Lakewood, Elizabeth got around in her chauffeured Packard limousine. She enjoyed her vast gardens and resumed many of the philanthropic activities she had shared with Ward, especially concerning children, making her one of the most beloved and generous benefactors Lakewood had ever known. Never remarrying, Elizabeth remained there until her death in 1960.

In January 1929 "The James W. Packard Collection of Unusual and Complicated Watches" was formally transferred to the custodianship of the Horological Institute of America. News of the transfer surfaced in papers across the country, from the *Miami News* to the *Hartford Courant* to the *New York Times*. The astronomical watch, the Packard, quickly assumed stature as "the gem of the collection."

As Paul Moore moved to make his dream of a time museum a reality, he used Ward's collection in the Horological Institute's educational programs, displaying them at events primed to attract enthusiastic parties that might share Moore's desire. The Institute sent Ward's watches on tour. In June 1929 it lent the pieces to the Smithsonian Institution in Washington, which exhibited them in the Arts and Industries Building with the understanding that they would remain on display until the Institute had its own museum.

Periodically the Institute recalled the watches for its own purposes. Between August 4 and 9, it presented a complete exhibit of the collection at the Fourth Annual Jewelry Show in Chicago. In the catalogue accompanying the exhibit, Moore wrote, "No man who has any appreciation of fine work in watchmaking can look at these marvels of mechanical ingenuity and perfection without a desire to improve his own ability." The following year the Institute put the Packard watches back on display during the Fifty-second Annual Banquet of the Chicago Jewelers Association, and once again in 1939 for a private occasion in San Francisco, before returning them to the Smithsonian.

Ten years on, Moore's great hope that the Packard collection would ignite a movement to establish an American monument to time had not been realized. The Institute still had no facility to properly exhibit the watches. It was a short road for the Packard collection from celebration to relative obscurity.

An aviation enthusiast, Warren Packard had organized the Detroit Air Yacht Club. On August 26, 1929, he took to the skies above Grosse Pointe, Michigan, with his flying instructor, Talbott Barnard. The warm, clear summer evening provided the perfect excuse for a flight lesson, but the two-seater light sport plane had barely cleared one hundred feet when it began to tailspin, dropping from the sky and crashing into the Detroit River near the railroad bridge on the west side of Grosse Ile Township. When rescuers arrived at the scene, the plane was submerged in the water. Barnard was severely injured, with a broken jaw, concussion, and fractured skull. Warren had died instantly, leaving behind a widow and two young children.

It had been just fourteen months since he had buried his Uncle Ward. The beautiful pair of pocket watches that Warren had received, the Patek Philippe and the Vacheron Constantin, were still in their original fitted boxes. They were placed in a bank vault and in the chaos following the tragedy, seemingly forgotten.

CHAPTER SEVENTEEN

~

A Supercomplication

In the spring of 1929, Henry bought into a new Fifth Avenue co-op under development. Thrumming with victory, the financier had all but snared the crown in the war of complications. With Patek Philippe's craftsmen painstakingly working on his *pièce de résistance,* all he had to do was await final delivery of his prize. Still, Henry lost little of his enthusiasm for watches. With his quiet intensity he continued to monitor the results of the Geneva Observatory timing competitions, and before the year was over he snapped up a handful of new first prize winners, as well as a pair of superlative minute-repeater wristwatches.

These were good times in America, in particular for American capitalism. In his final State of the Union Address, President Calvin Coolidge buoyed an already optimistic nation, telling Congress, "In the domestic field there is tranquility and contentment . . . and the highest record of prosperity." With little skepticism, news of the exploding stock market regularly made the front page. It was not unheard of for an investor to realize gains of ten to twenty points in a single day. Hot stocks like RCA tripled, even quadrupled in value.

In 1927 brokers borrowed $4 billion (about $52.9 billion today), lending the money to stock buyers. By the end of 1928, brokers' loans had vaulted to $6.4 billion ($84.7 billion today). More led to more, which led to more still, and by 1929 almost four dollars out of every ten that banks loaned was earmarked to buy stocks. The extended boom allowed the middle class to skip up the rungs on the nation's wealth ladder.

And yet there were troubling signs: steel production slipped; housing construction slowed; car sales dipped; the number of slums, racial unrest, poverty, and debt all continued to grow. But most chose not to see them. The chorus of naysayers was small, while the circle of professional utopians was large and loud. In June the financier and presidential advisor Bernard Baruch pronounced, "There is no reason why both prosperity and the market should not continue for years at this high level or even higher." The future, for those who peered into it and did not look too closely, appeared quite rosy.

Between June and August the market's bull run coursed ahead like a runaway circus train and stock prices reached their highest levels to date. On September 3 the Dow peaked at 381.17. The country's mood was euphoric.

At sixty-one, Henry had decided to spend this latest chapter in life with Florence in a palatial new Rosario Candela–designed apartment hotel. This was the golden age of the grand luxury apartment in New York City, and no architect built grander apartments than Candela, the Sicilian-born son of plasterers, who set the standard for upper-class urban living. The fourteen-story building at 834 Fifth Avenue was his most massive and elegant to date, encompassing 13,000 square feet on the site formerly occupied by the homes of three of the city's most influential families—the Guggenheims, the Goulds, and the Lewisohns—as well as the private Bovee School.

Candela buildings were known for their immense scale, open views, and amenities such as private laundries. At 834, Candela designed a series of simplex, duplex, and maisonette apartments, and no two dwellings were alike. On the east side of the building the architect built a private 2,000-square-foot garden. The elegantly understated exterior with its limestone façade and elegant Art Deco–style cartouches would soon be Manhattan's most majestic and exclusive address, housing some of the city's most prominent residents.

As the developer Anthony Campagna secured an initial $7 million (about $94.3 million today) in financing, Henry and Florence were one of the first ten to buy in on the building, billed as the city's first 100 percent cooperative. They joined Carl J. Schmidlapp, vice president of Chase National Bank; Mrs. Elden C. DeWitt, whose husband had made a fortune manufacturing proprietary medicines such as the One Minute Cough Cure; Ezra D. Bushnell, the director of the Hamilton Trust Company; and Mrs. John E. Berwind, a widow whose husband had been one of the largest coal operators in the country and the brother of Edward Julius Berwind, a man with whom Henry and his father had tangled over paintings at auction some four decades earlier. Hugh Baker, president of National City Bank, the largest investment bank of its day (decades later renamed Citibank), took the 8,000-square-foot three-story penthouse with a small conservatory built inside the living room.*

*In 2004 the media mogul Rupert Murdoch purchased the triplex penthouse for $44 million, at the time the largest amount ever paid for a New York City apartment.

• • •

After peaking in September, the market's mighty surge began to wobble. With prices fluctuating wildly, on October 17, 1929, the Yale economics professor Irving Fisher optimistically proclaimed, "Stock prices have reached what looks like a permanently high plateau." Not a week later, on October 24, "Black Thursday," the market plummeted. Stocks continued to drop over the next six days, losing nearly a third of their value; $25 billion (equal to about $336.8 billion today) in savings was obliterated. With stunned crowds flocking to the New York Stock Exchange, stories rippled through the throngs on Wall Street that people were committing suicide.

The bottom continued to fall. By the end of November investors had lost some $100 billion (about $1.3 trillion today) in assets. The avalanche of economic misery wreaked havoc over the widest possible area; families lost their life savings, and companies were wiped out. Attempts to rally the market were brief. Rumors ran rampant that the bankers were selling off, and investors panicked. The former General Motors president William Durant, along with the Rockefellers and other big investors, bought large blocks of stock to shore up the market. This proved detrimental for Durant, who found himself bankrupt within seven years.

As the country's economy all but collapsed, not only did construction continue apace at 834 Fifth Avenue, but the building expanded. In the summer of 1930, its developers purchased the mansion owned by the family of the late multimillionaire James B. Haggin on the adjacent corner at Sixty-fourth Street. As it turned out, 834 Fifth Avenue was the last great apartment house to rise in the city before the Depression halted such projects. Seven decades later it would be called "the most pedigreed building on the snobbiest street in the country's most real estate–obsessed city."

On October 1, 1931, as the country spiraled into mass unemployment and bankruptcy, Henry and Florence moved into their fifteen-room duplex overlooking the Central Park Zoo. A private elevator took the couple to a grand entry gallery, where a curving staircase stood, connecting the ninth and tenth floors. The place had the ambience of a country house, with plaster moldings, working fireplaces, mahogany doors, and dark wood paneling. Its massive size and darkness entranced all, except for Henry's grandchildren, some of whom found it unsettling during formal visits.

Henry installed a burglarproof vault, where he kept his watches. By the time he put his feet up on his Chippendale desk, he had several new Patek Philippe watches to store inside it.

The Graveses appeared to glide through the economic morass. Although the idle rich who primarily lived off the interest and dividends from their securities were devastated following the Crash, the economic cataclysm hardly seemed to scratch the surface of Henry's upper-class privileges. Many in their milieu, terrified as they watched their fortunes disappear while their debts mushroomed, had reduced some of the expenses of their cozy lifestyle: dismissing staff, curbing travel, and dropping club member-ships. Others in the smart set made a great show of frugality. In 1935 Doris Duke, the wealthy tobacco heiress, courted public empathy when she told a phalanx of reporters on the Hawaiian leg of her year-long, around-the-world honeymoon with husband James Cromwell that her bathing suit was three years old. The "richest girl in the world" failed to mention that she had just broken ground on a $1.4 million ($23.5 million today) estate in Honolulu. The eclectic mansion called Shangri La was the most expen-sive home built in Hawaii at the time.

At the Fifth Avenue apartment, Henry retained his four maids, five fewer than the couple maintained at Shadowbrook, but then again that was a twenty-seven-room house on ten acres that necessitated a small army of gardeners, groomsmen, and chauffeurs to tend to the grounds as well. One of the benefits of apartment dwelling was that it required fewer servants.

The Graveses lived as if someone had stopped the family's clocks some-time in 1927. In addition to spending summers at Eagle Island, they began dividing the season, spending part of it at the exclusive cottage colony called the Homestead in the Virginia Hot Springs, often accompanied by Gwendolen, Reginald, and their children. An accomplished rider, Henry rode the resort's famous trails with his daughter and granddaughter. At Christmas, Henry and Florence followed the same ritual, spending the morning at their daughter's Fifth Avenue apartment before driving off to Duncan's house on Long Island. In the evening they returned to their own duplex, where all of their children and grandchildren assembled for Christmas dinner.

Although Henry had long retired, he continued to list his title as vice president of the Atlas Portland Cement Company, in which he retained a considerable financial interest. The company certainly rode the boom of prosperity. Between 1906 and 1910 it paid out 8 percent dividends, stop-ping only to expand into new plants and enlarge existing ones. In 1929 Atlas Portland operated seven American-based factories that churned out more than 36 million barrels of cement a year, making it the larg-est cement concern in the world. Two months after the stock market imploded, the U.S. Steel Corporation acquired Atlas Portland Cement

Company in a stock swap worth $31,320,000 (about $432 million today). The transaction involved 180,000 shares of U.S. Steel common stock in return for Atlas Portland's outstanding stock and assets, equal to one share of U.S. Steel for every five shares of Atlas. The de Navarros, Maxwells, and Graveses had retained large stakes in the company. With some luck, Henry got out just in time, and the deal created something of a windfall, for the market would bottom out in 1931. Henry had preserved the core of his family's wealth. Many decades later, his grandson Reginald Fullerton, Jr., would claim that the family's fortunes had remained untouched by the stock market crash.

The Graves family was rocked in other ways, however. On June 2, 1930, Henry's older brother, Edward Hale, had died at sixty-five. (His wife, Jean, had died three years earlier.) Four months later, Duncan Graves's first-born, Henry, named after his grandfather, died just before reaching his sixth birthday. Equally startling, without warning on December 10, 1932, Henry's fifty-nine year-old younger brother, George Coe, suddenly and mysteriously died. He had been traveling aboard the SS *President Coolidge* across the Pacific Ocean, making his way from San Francisco to the South Sea Islands.

Unmarried, George spent his days visiting many exotic ports, such as Cuba and throughout Southeast Asia. Outside of the family trusts, he had an estate worth more than $1 million ($16.8 million today), excluding his real estate holdings. A trustee at the Metropolitan Museum of Art, he left one-quarter of his estate to the Museum, along with forty-one etchings and paintings. To Henry and their sister, Daisy, he gave his Orange, New Jersey, home on the Graves homestead, and he left Gwendolen his Cape Cod estate, Sylmaris. He divided the residue of his securities and cash among his nieces and nephews.

The terms of the Graves family trust divided the core among the remaining descendants of Henry Graves, Sr. With the deaths of Henry's two brothers (the family matriarch, Isabella Hale Graves, had died on February 23, 1926), he had gone to court to contest his father's trust to reapportion the remaining shares. With each death of a Graves family member, the amount received by the immediate succeeding descendants of Henry Sr. grew. In a morbid twist, as Henry's family got smaller and the wealth of his friends and contemporaries shrank, his fortune continued to swell.

If anything exemplified Henry's financial health, perhaps it was his ability to continue underwriting the decidedly glamorous life of his youngest son, George Coe II. An intrepid explorer, George spent much of his time in distant lands, following uncharted paths, bringing back mementos (black bear skins, stuffed heads of wildlife, and aboriginal figurines inlaid with mother-of-pearl) from the most isolated corners of the world. At Yale he had become great friends with the similarly inclined Bruce Thorne, the wealthy grandson of a founder of the Montgomery Ward department store, and together they hunted brown bears in Hokkaido, Japan, prospected for gold in isolated pockets of the Yukon, and trekked across Kenya and Rhodesia in search of big game and adventures. An enormous pair of curved elephant tusks flanked the foot of Henry and Florence's spiral staircase, a souvenir from their son, as was the copper-lined baby elephant foot used as an umbrella stand. (A second was kept at Eagle Island, where George's numerous trophy heads hung.)

In 1929, following George's graduation from Yale, Henry financed the Thorne-Graves Arctic Expedition to the Polar ice caps on behalf of the Field Museum in Chicago to capture the Pacific walrus. George spent three months among the ice floes at the edge of the Arctic on the *Dorothy*, an old 105-foot halibut schooner equipped with a 270-horsepower diesel engine and an iron bark hull. Anchored near Koliuchin Island, the team brought back seven specimens, later stuffed and displayed in

The great adventurer George Coe Graves II with his kill in Hokkaido, Japan.
Courtesy of Elizabeth Pyott.

a diorama at the museum's Marine Mammals Hall. Writing in the *Field Museum News,* Wilfred Osgood, the curator of the Department of Zoology, said the museum was "indebted" to Henry Graves, Jr., for his "substantial contribution." In addition to underwriting the expedition itself, Henry bankrolled the installation, an elaborate *in situ* re-creation of the walruses sitting on the ice pack under the midnight sun.

With Henry shunning publicity for his financial support, all of the glory belonged to George. He was nominated into the exclusive, secret society of the Explorers Club in New York, whose membership boasted the likes of Theodore Roosevelt, Charles Lindbergh, and Admiral Richard Byrd. (Women were not admitted until 1981.) On the heels of his Arctic exploits, on November 1, 1929, George traveled to Khabarovsk in the Soviet Far East to head up the Morden-Graves North Asiatic Expedition under the auspices of the American Natural History Museum in New York. George's traveling party included the famous explorer William J. Morden, director of the Explorers Club, and George Goodwin, the museum's assistant curator in mammals. (It was Goodwin who introduced the use of beetles to clean the carcasses of small animals.) Laden down with supplies, the trio left on the Trans-Siberian Railway to stop in Amur, from which they would make the rest of the arduous journey on horseback and dog sleds. The group headed toward the windswept steppes of eastern Siberia, with stops in Samarkand and Bukhara, to find the elusive long-haired Siberian tiger and Saig antelope. In the end, George, with the aid of local trappers, secured three of the tigers, two male and one female, using improvised trap guns placed along the spoors in the snow, as well as six antelope.

Six months later, as George trekked across Africa, a tuxedo-clad Henry attended the dedication of the American Museum of Natural History's new South Asiatic Hall, where the Siberian tiger and Saig antelope would eventually be displayed.

On January 19, 1933, Henry received his prize: the Graves Supercomplication. The full thrust of winter had yet to arrive, but a distinct chill hung in the air. Three months earlier, an angry, anxious nation tossed out Herbert Hoover and elected the Democratic candidate Franklin Delano Roosevelt as America slipped further into the depths of the Great Depression.

The golden pocket watch was breathtaking. The presentation box appeared fit for royalty. Made of pink-yellow tulipwood inlaid with ebony and a mother-of-pearl panel, it was engraved with the Graves family coat of arms. The interior of the silk-lined lid concealed the Patek Philippe certificate. The pocket watch was technological perfection and aesthetically flawless. On the front mean-time dial, a "IIII" was used instead of the more common Roman numeral "IV" for a better visual balance of the numbers around the dial.

Patek Philippe had sent the watch from Geneva on December 16, and it took nearly a month to cross the Atlantic. In order to ensure that the pocket watch was not disturbed en route to New York, the watchmakers had stopped it from running at 12 o'clock. Patek Philippe sent a special letter to the U.S. Customs Bureau, requesting that its officers not attempt to wind the watch in order to avoid any improper handling that might disrupt the mechanism. A slim envelope containing three single-typed sheets with detailed instructions in English accompanied the piece, under the title "Complication for Mr. Graves' Supercomplicated Watch #198'385." (A duplicate was sent to Patek Philippe's New York offices.) The instructions explained how to set every recondite measure of time, along with advice about how to handle such sensitive features as the double winding system that activated both the striking mechanism (by turning the crown forward) and the winding of the watch (by turning the crown backward), which had kept chief technician Jean Piguet awake a night. For good measure, Patek Philippe explicitly warned that, before starting the process, the watch must be placed in a flat position with the enamel dial facing up, the hands set forward, and *never* while the chimes were ringing.

The timepiece's stunning twenty-four complications were everything that Henry had desired when he first approached Patek Philippe nearly eight years earlier. The zenith of artistry and technology, it was *the* most complicated watch in the history of time. It surpassed the Marie-Antoinette, the Leroy No. 1, Ami-LeCoultre Piguet's *La Merveilleuse*, and all of James Ward Packard's *grandes complications*. It was also the measure of a period that was slowly crumbling.

In Geneva the economic turbulence had dissolved many companies that had resided in the golden web of Patek Philippe's elite dealerships and retailers built up more than a century earlier. Business with America's glitterati declined severely. Henry's payment of $15,000 for the Supercomplication was more than the price tag of a very expensive vanity project for the firm; it was financial lifeblood.

Internally the *maison* underwent a series of wide-ranging shifts as a result of the Great Depression. By 1932 severe financial realities had forced the founders' successors to sell their majority in the firm. Adrien Philippe, who took on a directorship in 1913, was the last blood relative of the founders to run the company. An offer from Jacques David LeCoultre from Jaeger LeCoultre in Le Sentier was passed over in favor of the one presented by Charles and Jean Stern of Fabrique de Cadrans Stern Frères. The Stern brothers had enjoyed a very long relationship with Patek Philippe, having manufactured dials for them over many years.

Under the Sterns, Patek Philippe remained committed to preserving its nineteenth-century watchmaking traditions of superior craftsmanship and technological innovation, while maintaining the house as a family enterprise. In 1934 Charles Stern sent his son Henri, born in 1911 and trained as an engraver, to temporarily run the company's American operations at 607 Fifth Avenue in New York. Even while America foundered, its market for luxury watches remained much brighter than Europe's. Alfred Stein, who had opened the firm's first distributorship at 68 Nassau Street just off Maiden Lane and who, over the years, had assiduously nurtured Patek Philippe's relationship with retailers, found his relationship with Geneva foundering. When the Sterns took over in 1932, he sold his shares to the brothers, who in turn became majority stakeholders, and left both the company and its board of directors. Two years later Stein died.

In 1935 Henri Stern moved Patek Philippe's offices, opening the firm's new American headquarters, the eponymous Henri Stern Watch Agency, in the International Building at Rockefeller Center in Manhattan. His wife and his son Philippe, born in 1938 in Switzerland, soon joined him there. Stern, who did not intend to remain stateside indefinitely, traveled to Geneva only at the end of World War II, in 1945. After a brief trip, he once again returned to New York, where he ran the firm's corporate offices until 1958. In New York he worked to cultivate the firm's wealthy American clientele, preserving its relations with its best patrons and collectors. He also further cemented the watchmaker's reputation among collectors for complicated movements, proving its inventive aptitude, miniaturizing the most sophisticated complications: split-second chronographs, minute repeaters, and perpetual calendars inside of the exceptionally confined space of the now popular wristwatch.

Henri also groomed his son Philippe, who went on to study computer science in Germany and business administration at the University in Geneva, to eventually take over Patek Philippe. From a very young age Henri instilled in his son two essential principles: the first was that Patek Philippe must stay small and independent; the second was that the *maison* remain dedicated to crafting the world's highest quality, most beautiful timepieces.

As well as being a shimmering trophy, the Supercomplication appeared in Henry's life as an ominous talisman, its presence coinciding with a series of devastating events.

On the evening of July 17, seven months after its arrival, Henry's great friend Edwin Gould complained of feeling unwell after dinner. He retired to his bedroom at his mansion in Oyster Bay, New York. Just after midnight Gould, age sixty-seven, suffered a heart attack and died.

Two months later, Henry's former daughter-in-law, Margaret, returned from a trip around the world with her three children aboard the SS *President Pierce*. Since her wedding to Dexter Hewitt, she and her children had remained largely estranged from the Graveses, although, given their social circle, from time to time they drifted into each other's lives. On September 9, 1933, Margaret and her brood—Henry Dickson, Florence Barbara, and Mary Dickson—sailed into busy New York Harbor, past the gleaming skyscrapers that lined the harbor. The day the family arrived the sky heaved with moisture, the residue from a major Atlantic storm that had blanketed the coast the previous two days. The children, dressed in their formal travel clothes, gloves, and collars, disembarked with Margaret and stood on the platform as porters carried the family's trunks and packages. Craning their necks above the scurrying passengers, food carts, and newspaper boys, they waited for their car and driver to whisk them back home to Ardsley. As the swarm thinned and then disappeared, Margaret's driver was nowhere to be found.

Frantic, Margaret tried to find her husband, unaware that, four days earlier, the *New York Times* had reported that he had died of a sudden heart attack. The truth was far murkier. While Margaret had been visiting the souks of Asia and the boulevards of Paris, her husband had charted a course through his wife's fortune and lost nearly all of it. Just days before she returned, Hewitt, her family later maintained, had committed suicide, making Margaret a widow for the second time in a decade. This time she was also nearly penniless. Jettisoning the Hewitt name, Margaret resumed calling herself Mrs. Henry Graves 3rd and soon sold the Ardsley mansion. Embittered and seemingly abandoned by the smart set, Margaret told her children, "They come to the party when you've got it all. Where are they now?"

Feeling herself backed into a corner, Margaret took her former father-in-law to court in order to reinstate the shares from the Graves trust to provide for her children. After negotiations the parties came to an agreement to once again start payments to the grandchildren. The children resumed a formal acquaintance with their grandparents, but their life of privilege had taken a swan dive.

A year later Henry and Florence found themselves once more in mourning. After financing for a six-month Pacaraima-Venezuela Expedition fell apart due to the economy, their son George Coe II hung up his guns and moved to Los Angeles to become a partner in the newly formed brokerage firm Simmons & Peckham, with offices in the Pacific Mutual Building. He took an apartment at the swank twelve-story Arcady Apartment Hotel Building on Wilshire Boulevard. George transferred his taste for adventure and the adrenaline charges he got piloting his seaplane across the Yukon and big game hunting to hurtling along the streets of Los Angeles in one of his big, souped-up cars. On November 3, 1934, Henry received a devastating call: while speeding down Valley Boulevard in Pasadena, George had been killed in an automobile accident. His body was returned to New York for the funeral at St. Thomas's Episcopal Church, and he was buried alongside his older brother in the Graves mausoleum in Sleepy Hollow. Unmarried and childless, George had an estate valued at nearly $1 million (mainly in stocks, bonds, his Bellanca airplane, and hunting trophies), which was left to his parents. As she had done nearly ten years earlier, Florence memorialized her youngest son, pressing his photograph inside a golden locket.

At each end of this topsy-turvy decade, Henry had buried a son, each a bookend to an era that began flush with optimism and prosperity and

For the second time in a decade, Florence Graves memorialized a son, her youngest, George Coe II (left), in a gold locket, alongside a photo of her husband and their children taken on Eagle Island in 1909 (right).
Photograph by Stacy Perman. Courtesy of Gwendolen Graves Shupe.

ended in gloom and panic with the fear that the ebbing of economic malaise remained in some unknowable distance.

The arrival of the Supercomplication signaled a period of great personal grief and high public anxiety. Rather than a magnificent private trophy, the watch became something of a horological celebrity, as had the *grandes complications* of earlier periods. Its fame put Henry in the white-hot glare of the kind of publicity he found most distasteful. For the first time, the exceptionally reclusive financier found himself the center of unwanted attention.

He was said to have asked Patek Philippe to refrain from mentioning his pieces in any sort of marketing, but photographs of "the Graves," described as owned by a "private collector," showed up in magazines in both France and the United States. "Time Is Only One of Many Things This Two-faced Watch Tells," claimed the caption in one without revealing the owner's identity, while a French publication, headlined "Geneva Watchmaking, Racy Watches," displayed images of both dials and the movement. Despite Henry's intense desire, publicity over the world's most complicated timepiece emerged repeatedly in the years after he received it. Among collectors and enthusiasts, "the private collector" was known to be Henry Graves, Jr.

C. B. Driscoll's society column, "New York Day by Day," chronicled tableaux of life in Gotham that ran in newspapers from Florida to Montana to California. On June 21, 1938, Driscoll opened his column with the words, "Henry Graves, Jr., has one of the most expensive hobbies in New York. He is a watch collector. Mr. Graves pays $2,500 or $3,000 for most of his specimens, and he has to be on the ground to get them for that." Readers learned that Henry owned "more than fifty Geneva observatory winners"—"none of them," Driscoll noted, "a trick watch."

The publicity more than grated on Henry. This was a time when the nation's wealthy went to great lengths to mask their fortunes in the midst of massive unemployment and economic upheaval. Stories of burgled mansions littered the papers. Bankers had become Public Enemy Number 1, with Congress regularly railing against the villains.

Earlier, the kidnapping of Charles Lindbergh's twenty-month-old son from his nursery had left Henry deeply rattled. The fraught saga began with a $50,000 ransom note on March 1, 1932, and ended with the infant's remains found just four miles from his home. J. Edgar Hoover directed the FBI to undertake an investigation that stretched over two years before Bruno Richard Hauptmann, a German immigrant with a criminal record, was arrested, convicted, and finally executed on April 3, 1936. The kidnapping horrified and riveted the entire nation. The newspaperman H. L. Mencken called the kidnapping and trial "the biggest story since the Resurrection."

For Henry, already shattered by the deaths of his two sons, the Supercomplication seemed to bring the worst kind of attention. He apparently explored the idea of selling the pocket watch, and a potential buyer was said to have been identified. But just as suddenly as this avenue was taken up, Henry appeared to drop it without further discussion.

While at Eagle Island one afternoon, Henry took Gwendolen out on the lake in one of his motorboats. Remarkably he carried the Supercomplication with him. As they moved out into the deep cerulean waters, Henry became agitated. Pulling the watch from his pocket, he began to share his misgivings with his daughter. Owning such objects "brought nothing but trouble," he said. "Notoriety brings bad fortune." At one point he cut the engine. "What is the point of being wealthy and owning such objects if something like this could happen?" he asked. With that he pulled his arm back to toss the Supercomplication into the lake when Gwendolen cried out, "No, Daddy!" Stopping him just in time, she reasoned, "Let me hold on to that. Someday I might want that."

From this point forward, Gwendolen kept her father's masterpiece for safekeeping. Eight years in the making, the Supercomplication had been in Henry's possession for scarcely three. And for the second time in his

Henry cruising Upper Saranac Lake in his favorite speedster, *The Eagle*.
Courtesy of Cheryl Graves.

long collecting career, his acquisitions virtually ceased. It would be nearly four years before Henry obtained another watch.

As the Great Depression entered its seventh year, America was suffering great anxiety and pessimism. President Roosevelt won a landslide reelection with many of his New Deal policies, such as Social Security and unemployment benefits, popular with the public.

The financial forces that had rocked the world finally made their way to Henry's doorstep. The owner-shareholders of 834 Fifth Avenue went into foreclosure, encumbered by two mortgages totaling more than $3 million ($49.7 million today). Hemmed into ninety-nine-year leases with terms that made it nearly impossible to cover the mortgage through sales or rentals, the original resident-stockholders, including Henry, took over the developer's shares and formed a new building corporation, and soon after the Metropolitan Life Insurance Company took it over at auction. While the building's solvency and legal status gyrated through the courts, Henry and Florence held onto their apartment. In 1946 the building was reorganized under new owners. Laurance S. Rockefeller, the venture capitalist, philanthropist, conservationist, and grandson of John D. Rockefeller, purchased the building from the Metropolitan Life Insurance Company for $1.55 million ($18.3 million today) and took over the triplex penthouse.

When he needed money, Henry did what the wealthy had done for centuries: dipped into the family treasures and sold them off, in pieces. Three years earlier, in the fall of 1933, he had led executors in selling the Graves homestead in Orange, New Jersey, to the real estate developer Colyer Homes, Inc. The property, site of his childhood home, some of the Oranges' most famous balls, and his father's orchid conservatories, was turned into a plot of tract homes.

In April 1936, Henry arranged to sell 115 of his finest etchings and engravings, including the works of Dürer, Frank Weston Benson, Rembrandt, Whistler, and James McBey, through the American Art Association Anderson Galleries, Inc. It was among the art world's highlights of the year. The bound catalogue for the sale exclaimed, "No other collection so rich in beauty, so carefully chosen, and in such splendid condition has ever been offered at public sale in this country." At the time, Old Master print sales provoked the same kind of unrestrained excitement that Picassos and Van Goghs would five decades later.

The auction took place at East Fifty-seventh Street on the evening of

April 3, with the bidders arriving in their dinner jackets for the 8:15 sale. Henry's Dürer masterpiece, *Adam and Eve*, went for $10,000 ($165,740 today), the biggest sale of the night, while Rembrandt's *Hundred Guilder Print* brought $7,300 ($120,990 today). The firm of Charles Sessler, the prominent Philadelphia dealer in rare books and prints, bought both. Despite the economic free fall, serious collectors continued to shatter prices for important art (much like the explosive sales at auction for art, jewels, and wine that would smash records during the Great Recession of the early twenty-first century). The entire sale realized $79,635 (about $1.32 million today), which would certainly allow Henry to maintain his grand standard of living for some time.

The astonishing prices aside, the evening was notable for another reason; until the spectacle of Henry's single-sale auction, the public at large had little knowledge of the scope of the financier's great art assemblage.

In November the Italian dictator Benito Mussolini described his country's alliance with Germany as an "axis." A month later King Edward VIII abdicated the British throne to marry the American divorcee Wallis Simpson. The Chinese leader Chiang Kai-shek declared war on Japan. Germany continued its goosestep toward war. In America, unemployment hovered at nearly 17 percent and FDR raised the income tax on America's wealthiest to 79 percent.

The heavy tax levy, along with increasing property taxes, made maintaining the rustic palaces of the Great Camps a burden. Many tycoons attempted to offload their camps; some simply abandoned them to the elements. Henry decided to sell Eagle Island. Following the death of their son George Coe, he and Florence had begun spending more of the season in Virginia at the Homestead, and in 1936 they skipped Upper Saranac altogether. A four-page brochure advertising the island noted, "Adirondack Lodge will make its strongest appeal to someone in quest of a summer home which combines to an unusual degree the seclusion of forest and lake with the comforts of a town house." At the time the camp was valued at upwards of $1 million ($16.4 million today). It languished on the market for a year.

After an acquaintance of Mrs. Graves, who sat on the Girl Scouts Council of South Orange and Maplewood, New Jersey, mentioned that the Scouts were looking for a new camp, the couple offered to sell Eagle Island to them for $20,000 ($331,480 today). But in the middle of negotiations, without explanation, the couple decided to gift the entire island and its contents to the Girl Scouts in memory of their two sons. The cache included all of Graves's legendary boats, the trophy heads, and a silver and china set for fifty. The transfer was the one and only significant donation Henry made during his lifetime.

Before the handover, Henry told his caretaker, William Meagher, that he could take anything he liked. Meagher took the monogrammed silver and a number of trinkets, then boxed up the china and dumped it in the lake, where, according to locals, a windfall of numerous treasures found their final resting place.

Despite having pruned his considerable art collection, Henry still owned more than a hundred exceptionally rare English, French, Old New York, and historical American naval prints, one of the most remarkable and largest collections of its kind. And the auction had represented but a small sample of his etchings and drawings. His discerning and eclectic tastes had led him to acquire exhaustively—his collection of Chinese porcelains numbered in the thousands and covered the most important dynasties, some pieces in the rarest glazes—and intensely: his collection of French paperweights were all produced during the classic period. His profoundly large coin collection remained untouched, as did his prizewinning, custom-designed mechanical watches—except for the Supercomplication, now safely in the care of his daughter, Gwendolen.

In 1940 the Supercomplication surfaced once again in the pages of *Life* magazine as part of the feature "Watches: These Are the Best in the World." Under four images depicting each dial and two of the corresponding layers of the intricate movement was the caption "World's most complicated watch." The magazine noted that it was made for a "private collector." Henry's name, engraved on the enamel dial, had been deleted in the photograph. In the same article, James Ward Packard was described as "America's greatest watch fancier and collector," and photographs of the Packard astronomical watch and his beautiful musical alarm watch that played the *Jocelyn* lullaby accompanied it. The article was something of an epitaph for the golden age of pocket watches and the grand collectors who so passionately acquired them. Nowhere did the name Henry Graves, Jr., appear. The man behind the most complicated watch in the world became a shadow victor in a phantom race.

CHAPTER EIGHTEEN

~

Age of Quartz

After more than two decades of American isolationism and official neutrality, on December 7, 1941, Japan pushed the United States into World War II with its surprise attack on Pearl Harbor. Four days later, Germany and its Axis partners declared war on the United States.

"The James W. Packard Collection of Unusual and Complicated Watches" had toured the country, showing at jewelry conventions and private exhibitions, where Paul Moore continued to advocate for an American time museum to educate and develop the nation's "finest type of mechanicians." The war interrupted his ambitious plans. Concerned over the wartime conditions in Washington and the possibility that the capital might come under attack, the Horological Institute of America and the Smithsonian Institution, where the watches were based, decided to move the collection for safekeeping. Each pocket watch was individually swathed in a protective covering, boxed, and sent to the Museum of Science and Industry in Chicago. Almost fifty years earlier, Ward had sent his older brother, Will, to that very spot. The museum occupied the Palace of the Fine Arts in Jackson Park, originally constructed for the World's Columbian Exposition in 1893, where Packard Electric had its own exhibit and Will examined the Daimler, the first gasoline-powered automobile to touch American soil.

A visitor roaming through the Packard watch exhibit would see a distinctly unimpressive display. On three wooden tiers against a wall, sixteen of the watches sat under protective glass, while the remaining pieces lay flat and exposed on the bottom row. The collection remained in Chicago throughout the war and several years afterward. With each passing year, Moore's great hope of constructing his museum of timekeeping faded.

At seventy-two, Henry Graves, Jr., emerged from his self-imposed collecting exile. During the war years, his interest in pocket watches had largely faded, along with their place in popular fashion. After learning of a Jules Jürgensen chronometer that took the prize at the Neuchâtel Observatory, however, Henry ordered the movement, no. 15954, in a platinum open-face pocket watch, which arrived via steamship on February 28, 1940. Having recovered from his personal and public *Sturm und Drang* and following an almost four-year absence, he also resumed his patronage of Patek Philippe. While he acquired a small number of the *maison*'s pocket watches, for the most part he moved on to a new sensation: the wristwatch.

Once Henry had acquired the Supercomplication and after James Ward Packard's death, he appeared to have lost some of his competitive urgency, and his acquisitions shifted greatly. Continuing to enjoy a privileged relationship with Patek Philippe in the postwar years, he still purchased his watches from Tiffany while conducting his business through Patek's American headquarters, the Henri Stern Watch Agency in New York. When Henry visited the firm's corporate offices, the attentive staff pulled timepieces in anticipation of his arrival. Einar Buhl, the sales manager, instructed his salesmen to "always listen carefully to Mr. Graves." Werner Sonn, who began working for Patek Philippe at the Henri Stern Watch Agency in 1939, always remembered Henry's visits. "We understood that Mr. Graves wanted special watches. Sometimes his wife came with him. He was a very kind and pleasant man. He looked at watches to get ideas. He asked us if we do this or that." As Sonn recalled, "He came in and he wanted things, and we did what he wanted."

In 1928, while visiting Geneva, Henry had apparently made arrangements to obtain his gold and platinum, Tonneau-shaped minute-repeater wristwatches. Two years later he would acquire yet another, also in platinum. During the early part of the twentieth century, Patek Philippe is known to have produced only about a dozen minute-repeating wristwatches, four with Tonneau cases. Of this rare group Henry owned three.

In a short time, Henry came to own more than two dozen Patek Philippe wristwatches, usually buying one or two a year, each encased in gold or platinum. Among them, he acquired an eighteen-carat gold perpetual calendar chronograph wristwatch with moon phases, register, and tachometer in a circular case (reference no. 1518). Manufactured in 1947, this was the first perpetual calendar chronograph model that Patek Philippe made available to the public, and over the course of thirteen years the watchmaker produced less than three hundred examples. The same year Henry also took delivery of a pink gold rectangular wristwatch with unusual lugs (reference no. 2425). As he did with his pocket watches, with a few select exceptions Henry had the backs of his cases engraved with the Graves coat of arms.

The same year, 1947, Henry took receipt of a rather unusual timekeeper. The platinum wristwatch featured a movement (no. 922902) manufactured in 1942 and a massive, curved rectangular dial with raised alternating Roman, double bullet, and faceted square indexes, for which he paid $1,500 (about $15,500 today). Although this was a time-only piece, the price equaled that of Patek Philippe's most complicated wristwatch of the period, the perpetual calendar chronograph, reference no. 1518 (which Henry also owned). The astonishing price tag came about because the discriminating Henry insisted upon the platinum case. But this was wartime, and platinum was declared a strategic metal, prohibited for nonmilitary use. Yet, determined to please one of their most important clients, not only did Patek Philippe craft the enormous platinum watch, but according to records of the Henri Stern Watch Agency, the watchmaker actually took a $525 loss on the piece.

Interestingly, the stunning watch did not contain the Graves family crest, nor did the case possess, as was customary, its own designated number. Devising such an instrument with the precious metal would likely have drawn attention from customs authorities as the piece crossed into the United States and back into Europe for engraving, then back again across

Although platinum was declared
a strategic metal during World War II,
Patek Philippe manufactured this massive
wristwatch at Henry's request.
Photograph courtesy of Sotheby's, Inc. © 2012.

the Atlantic to its final destination. On behalf of Tiffany & Co., the Henri
Stern Watch Agency ordered the large and heavy platinum case especially
for Henry from the famous case maker Villaret-Dauvergne of France. One
theory posited for this unusual situation was that the French firm was
chosen to manage the costs of such an undertaking. There was another
unconfirmed but dramatic possibility as to why the case was absent such
markings: the platinum was smuggled in to avoid arousing suspicions.

Four months before Henry's eighty-second birthday, in 1949, he
received one of his last acquisitions, a highly complex, eighteen-carat gold
split-second chronograph wristwatch with Breguet numerals in gold, his
favored blue steel hands, and a tachometer scale (reference no. 1436), first
produced in 1938. In a span of thirty-three years, Patek Philippe would
produce only about 140 of this particular reference.

As he approached his ninth decade, Henry had pushed complicated
movements and precision timekeeping to their technological limits. With
nothing further to prove, his collecting ambitions simply stopped.

By 1952 Henry's health was in decline. He organized his business, made
arrangements for his finances, and received family that he hadn't seen in
years. In the nearly three decades since Margaret Dickson Graves Hewitt
had taken Henry to court to provide for her three children from the Graves
family trust, he'd seldom seen her children, who were now grown, married, and with children of their own. During one visit, his granddaughter Florence Barbara Graves Jenks arrived from Rhode Island, with her
eight-year-old daughter, Sharon, for tea. Dressed formally in Mary Janes
and a coat, the girl was told to be on her best behavior. It was the first
time she remembered meeting her great-grandparents. Finding the apartment enormous and dark, she was shocked to see the pair of elephant
tusks at the foot of the staircase. "They were taller than any of us," she later
recalled. "I was told to keep my mouth shut." Just as unsettling was her
grandmother's lock-jawed elocution. "Do you like *daahgs*?" she asked her
great-granddaughter and presented her with a small stuffed dog.

On March 21, 1953, Henry died quietly in his apartment. He was eighty-
five. True to form, his funeral was private. Unlike other men of his stature,
there was no long and florid obituary outside of a perfunctory four-line
notice published in the *New York Times*. He was interred with his two sons
in the Graves family mausoleum at the Sleepy Hollow Cemetery.

Not long after Henry died, one of his great-grandsons, Airell Jenks,
visited Gwendolen and Reginald in New York with his mother. She had

remained in touch, "friendly distant," as he described it, through the years, sending Christmas cards. Airell recalled that as they sat on plush couches in sumptuous surroundings high above Fifth Avenue, Reginald suddenly turned to his wife and said, "I think we should show it to him," and left the room. When he returned, Reginald placed the Supercomplication in Airell's hands. "It was this fabulous watch and it filled the palm of your hand like a big bar of precious soap," he remembered. "It was mammoth. They opened it up and there was an exquisite, beautiful moon and sun and stars. I held that watch in my hands but I didn't know what to make of it."

Appraisers and insurance agents combed through Graves's nearly $8 million estate (worth more than $69 million today). Henry's forty-six-page will arranged for nearly half of his estate to go his wife, including the shares in their duplex at 834 Fifth Avenue. A year earlier, Laurance Rockefeller had reorganized the complex into a co-op, and Henry and all of the remaining tenants bought back their apartments. Henry owned 1,300 shares worth $35,000, equal to nearly 2 percent of the building, then valued at $1.95 million. He divided the rest of his assets among his children, providing an annuity for each of his grandchildren and great-grandchildren, including Margaret and Harry's children—with the exception of Duncan's son, Duncan Graves, Jr., who after some unforgiveable transgression had been disinherited. Henry had established a multimillion-dollar trust for his descendants. Gwendolen received Henry's French paperweights. His spectacular coin collection had been originally earmarked for his daughter as well, but not long before he died he added a codicil to his will requesting that the entire collection, which had been moved to a safety deposit box at the Bankers Safe Deposit Company, be sold and the proceeds split between Gwendolen and Duncan. The collection with the rare Double Eagles fetched $67,600 (about $583,000 today).

Henry's affection for Reginald Fullerton had grown deeper over the years, and he named him as one of the executors as well as leaving him a slice of his estate, "in appreciation of the advice and counsel of my son-in-law." Underscoring the great esteem and warmth with which Henry held him, he arranged to have the capital stock of his apartment at 834 Fifth Avenue transferred to Reginald upon the death of Florence.

Surrounded by his life's accumulation of art and priceless treasures, when he died Henry had a mere $76.88 in cash in his Fifth Avenue apartment.

Six years after his death, in 1959, the Graves family sold off a portion of Henry's art. In May the Kennedy Galleries held an auction of "Notable American Prints: The Collection of Henry Graves, Jr.," accounting for 166

The Graves family mausoleum, Sleepy Hollow Cemetery, Westchester, New York. Photograph by Stacy Perman.

of his pieces, which it hailed as "one of the outstanding private collections in this country." Describing the portraits of Manhattan that dated between 1764 and 1869, an art critic for the *New York Times* observed, "Many of the buildings represented in these etchings and aquatints were beautiful in design and worthy of being preserved. But they are long since gone." As was the world Henry had inhabited for nearly nine decades.

In the fall of 1959, the family dispersed another round of Henry's art. In September a great deal of his French and English antique furniture and silver went under the hammer during a two-day auction held at the Parke-Bernet Galleries. In October a selection of his Chinese porcelains, jades, and sculptures were also sold off; the sale included his rare set of eight Buddhist temple emblems carved in polished white jade.

Like James Ward Packard, Henry had not formally catalogued his watches. During his lifetime a conservatively rendered estimate would have placed the total number somewhere close to one hundred. After his death most of his watches were quietly dispersed among his descendants. The Supercomplication had already gone to Gwendolen, but by the time Henry died, only twenty-eight timepieces were accounted for in his estate appraisal. (Of these, fifteen were wristwatches: fourteen Patek Philippe and one Jules Jürgensen.) These represented only a small portion of the pieces he had collected over his lifetime, and the estate appraisal did not document the great number of his custom-designed Patek pocket watches (including his stunning ship's bell) or make note of his Vacheron Constantin. They too would soon slip into obscurity.

On the thick, gummy afternoon of August 11, 1953, Elizabeth and James Ward Packard's niece Katherine Summers and her husband, Myron, traveled to Chicago to view the Packard watches at the Museum of Science and Industry. Finding the pieces poorly displayed and needing care, they left exceptionally disappointed. In an era of television, the Corvette's first all-fiberglass body, and passenger jets, these watches elicited little public excitement. The collection would spend the next years shuttling between Chicago and the Smithsonian, eventually residing in a bank vault in Ohio.

More than one era had ended.

The Packard Motor Car Company, once *the* name in aristocratic, luxury automobiles, had survived the Depression only to lose its way in the postwar years. In 1934, when the economy imploded and the last of the stalwart independents—Peerless, Pierce Arrow, Locomobile, Franklin, and Duesenberg—went under, Packard managed to continue producing its classic, exquisitely handcrafted Twelves. A longtime model of corporate stability, the company deftly navigated the sour climate, introducing new vehicles, targeting the midpriced consumer, briefly casting off its prestige image to thrive during desperate times.

But the strategy that served the company so well during the global bust failed it during the postwar boom. Consumers found themselves unable to distinguish between expensive Packards and its lower-end models. Its once untouchable image as a luxury marque blurred further as management pushed harder into volume, selling affordable cars. The Packard Four Hundred dwindled. Having cannibalized its brand, the company lost scores of status buyers but never picked up enough lower-tier buyers to replace the losses. Before long, Packard lost the luxury market altogether, to Cadillac. In the early days of automobiles, America had counted some 2,700 car companies. After the war, Packard found itself too small and without enough cash to compete with the firepower of the Big Three.

The remaining independents, fearing they could no longer survive alone, took cover in a series of mergers. In 1953 Kaiser and Willys united to become Kaiser-Willys, while Nash and Hudson became American Motors. A year later, Packard's president James Nance merged the company with the Studebaker Corporation of South Bend, Indiana. The union ended badly. Studebaker was the larger partner, but Packard had the stronger management and bigger coffers. The new entity bled dry Packard's once healthy balance sheets. Bankers left the company for dead, and the government refused to throw a lifeline to the ailing manufacturer (in stark contrast to the $1.5 billion loan guarantee that the government gave Chrysler in 1979 and again thirty years later, when it propped up Chrysler and General Motors with an $80 billion bailout).

In 1959 Studebaker removed the Packard name from the company. After sixty years, Packard was history. Its elegant East Grand factory, covering 3.5 million square feet of forty-seven interconnected buildings, would be abandoned, left to become a desolate stretch of broken windows, twisted metal, creeping vines, and brick piles. Packard automobiles, once the car of kings, became expensive collectibles, lovingly maintained and restored around the world in various museums and among ferocious top-tier auto enthusiasts. James Ward Packard, the mechanical engineer and great collector, had produced for a select group of individuals the ultimate object of desire.

Postwar America was the world's economic and military power. The country was in the throes of the golden age of television. Like the radio before it, television captivated the public but with an even greater influence.

A time of unmistakable upward mobility, these years witnessed a return to the kind of unhinged lust for consumerism that had marked the 1920s. As the historian David Halberstam observed, "In the years following the traumatic experiences of the Depression and World War II, the American Dream was to exercise its personal freedom not in social and political terms but in economic ones."

The days of WASP splendor drew to a close, and with it, its narrowly calculated social hierarchies. The center of wealth's gravity not only shifted but also spread out from New York's financial orbit. Oil money poured out from the South, while in the West self-made entrepreneurs of various stripes emerged, ushering in new professional ranks of the rich, giving what was left of the Old Money classes quite a shove.

The G.I. Bill, signed in 1944 to aid veterans in their transition from battlefield combat to stateside citizen, did much to expand the country's growing mass affluence. Offering veterans such generous benefits as college tuition, job training, and, perhaps most significant, low-interest, no-money-down loans for homes and businesses, the Bill was the "magic carpet to the middle class." As the nation's middle class grew, the chasm between the upper and middle classes shrank substantially.

Social standing, once definitively measured by one's house, car, pocket watch, or family name, had become trickier to discern. The once rigid status hierarchy had all but dissolved. Escalating incomes, dropping prices, and easily accessible credit put the totems of wealth in reach of a great deal of the country. A middle-class family might have a color television in the living room and a swimming pool in the yard. With such broad access to high-end products, the bottom fell out of the meaning of status.

Along with the explosion of mass affluence, the mass production of technology transformed the country. The transistor became a part of industry and daily life. Refrigerators, electric mixers, and washing machines saved housewives hours and labor. The decade of the '50s saw the introduction of everything from Superglue to nonstick Teflon pans to the first oral contraceptive pills. The Soviets launched *Sputnik* into space. Back on Earth industrial-strength automated machines increasingly ran lives. Fueled by rampant consumerism, the economy was also now based in part on planned obsolescence.

In this new universe, a handcrafted, intricately made, expensive mechanical watch that needed manual winding was not simply outdated but about as desirable as a corset.

By the late 1950s, a wondrous new era in science and technology, watchmakers discovered that quartz crystals, a combination of silicon and oxygen, could power a watch with incredible accuracy and efficiency—at an absurdly low price. Quartz oscillators vibrate in the presence of an electric field, and the high frequency of the vibrations creates exceptionally precise timekeeping: within one minute a year. (Mechanical watches are accurate to within one minute a day.) Although the properties of quartz had been discovered a hundred years earlier, it took that long before it was possible to produce integrated electrical circuits small enough to work inside of the confined space of a wristwatch. The quartz age had begun.

In 1962 the Swiss watch industry formed the Centre Electronique Horologer (CEH) in Neuchâtel to develop electronic timepieces and see just where this new technology might take horology. Within five years, its sober staff of twenty-five invented two quartz-based watch prototypes. When the CEH entered these revolutionary timekeepers at the Neuchâtel Observatory's time trials in 1967, they earned ten first-place prizes for wristwatch accuracy.

Though the CEH researchers offered this exciting new technology to the Swiss industry, the watchmakers saw very little advantage to adopting it as their own. They viewed the ignoble quartz watch as beneath them, preferring instead to concentrate their energies on improving traditional mechanical technology. It was as if IBM found the microprocessor chip déclassé and, rather than embrace the development of the personal computer, simply turned up their noses and remained focused on the typewriter.

Others saw things differently. The Japanese raced to bring the first quartz watch to market. In 1959 the Japanese firm Seiko assembled a dedicated team to reduce the size of conventional watches and sell them at a much lower price, attainable through high-volume production. One of the team's first triumphs, the Seiko Crystal Chronometer QC-951, a portable quartz clock, was used as a backup timer during the 1964 Tokyo Summer Olympics. Five years later, on Christmas Day, Seiko introduced the 35SQ Astron, the first analog quartz watch.

The Seiko Astron delivered a quantum leap in accuracy. It also brought down Switzerland's hegemony. Within eight years of its introduction, quartz watches surpassed mechanical watches in global popularity. The ranks of the venerable watchmaking houses located in the densely forested Vallée de Joux soon thinned. Between 1970 and 1984 the number of Swiss watchmakers dropped from 1,600 to 600.

Quartz also gave America's watchmakers a great shot in the arm. The heir to the old Waterbury Clock Company and later the United States Time Corporation shifted gears to produce cheap, durable, unjeweled watches with a pin-lever escapement, called Timex. (The x was added to time to denote technological innovation.) The company, led by a Norwegian refugee named Joakim Lehmkuhl, completely changed the game. Priced between $6.95 and $7.95, Timex watches were sold at decidedly down-market venues: drugstores, supermarkets, and hardware shops. Lehmkuhl turned the key in the opposite direction. Subordinating luxury for cheap utility, he began selling affordable time, and consumers bought Timex watches in droves. Timex advertisements featured the watches undergoing "torture tests" (spinning on a ship's propeller, attached to the hooves of running horses), with one of the most famous ad slogans of all time: "It takes a licking and keeps on ticking." By 1962 one out of every three watches sold in the United States carried the Timex name.

The slow disappearance of the mechanical watch became a harbinger of the times. If watches were symbols of their era and their owners, then the Age of Quartz was emblematic of America's postwar prosperity and the explosion of consumerism, upward mobility, and the growth of the middle class.

• • •

After more than five hundred glorious years of spectacular innovation and artistry, the manufacture of mechanical watches appeared to be under the executioner's blade. Pressure to consolidate was brought to bear on Switzerland's venerable watchmaking community, from Geneva to all across the Jura Valley. At the time, Gérard Bauer, president of the Fédération Horlogère, was convinced that Switzerland needed only one watchmaking company, and he worked hard to make it the country's new reality.

Henri Stern believed otherwise. As early as 1948, Patek Philippe launched an electronics department. Four years later, the firm introduced its first quartz clock, with a luminous dial made of circular dots that indicated the passage of the hours and minutes, but without any mechanically moving parts. Not long afterward, the firm began manufacturing public clocks in railway stations, airports, and factories. Working with physicists, Patek Philippe's engineers experimented further with quartz-based clocks, producing marine chronometers, radio-guided clocks, and centralized quartz clocks.

But Patek Philippe also continued to cater to the small circle of influential connoisseurs who desired mechanical timepieces. The firm relied on Henri Stern's guiding principle of sticking to what the house did best: producing the most beautiful, technologically superior watches. As the 1970s wrought devastation across the Swiss watchmaking industry, many watchmakers went under, were sold off, or became part of larger conglomerates. "I took the decision to keep our tooling," Philippe Stern once explained. "If one Swiss brand were going to stay in the mechanical field, it would be Patek Philippe."

With an emphasis on tradition, Patek Philippe exhibited its historically important watches around the world throughout the years. At a retrospective held at Tiffany's in 1967, the firm displayed a watch made for Admiral Richard E. Byrd during one of his Antarctic expeditions to operate with Observatory accuracy at 104 degrees below zero, as well as the world's first stem-winding timepiece, introduced by the company in 1841. But of all the watches Patek Philippe had crafted through the years, the Supercomplication spoke the loudest to the firm's tradition and ingenuity. "In a business like ours, there are two elements," as one insider acknowledged, "the technical side and the commercial side. One helps the other." The Graves was important too. "It was good business to talk about having created something unique."

Indeed in 1949 Patek Philippe showcased its ingenuity at an exhibition of its watches held at the Cartier boutique in New York. Among the many historical pieces and horological wonders, the show featured a double-faced pocket watch with a celestial chart of visible constellations at every hour and at sunrise and sunset. The *New York Times* reported the timepiece as having taken five years to construct and placed its value at $15,000. Described as the "most complicated watch on display is a modern one," this unnamed marvel boasted particulars that, somewhat oddly, lined up with those of the Graves Supercomplication. There was only one Graves Supercomplication, however, and it was unlikely that this particular piece was the Graves. Certainly, the *Times* article did not identify the watch as such, or note Henry Graves, Jr.'s name that only appeared on the face of his Supercomplication's dial.

A similar demonstration occurred about twenty years later, during one of Patek Philippe's exhibitions at the jeweler C. D. Peacock's in Chicago, where the firm also displayed a highly complicated double-dial pocket watch with a list of its many complications, and again touted it as the most complicated and unique timepiece ever crafted. One gentleman who came to see the pieces became confused. Looking at the watch from every angle, he called over one of the Patek Philippe executives and began asking a series of questions about the watch. According to an individual familiar with the episode, the executive took him through the twenty-four different complications. Agitated, the gentleman erupted, "That's not the Henry Graves Supercomplication!" The man revealed himself to be Henry's grandson, Reginald Fullerton, Jr., and the staff explained that the watch was actually a mock-up of the Supercomplication used at exhibitions and did not contain the actual movement.

Apparently the Chicago event also happened to be the last in which the mock Supercomplication made a public appearance. And following the event, Fullerton and the Patek staffers had a good laugh over it.

CHAPTER NINETEEN

❧

Collecting Time

In 1970 the industrialist Seth G. Atwood opened the Time Museum, housing his spectacular collection of clocks and watches, including the envy of the watch collecting world, the Graves Supercomplication. Ninety miles from Chicago, the museum became an unlikely mecca. During its opening year, three thousand visitors wandered through the museum, and eventually experts and scholars would come from all over the world.

While quartz now dominated horology, there remained a clutch of men who continued to find mechanical timepieces endlessly captivating. For them, watches represented more than priceless antiques. They offered something else: the enduring measure of mankind. An obsessive lot, often magnificently eccentric, they were not necessarily patrons in the mold of men like Henry Graves, Jr., or James Ward Packard, but they were immensely wealthy and became the self-appointed caretakers of the world's timepieces. These men canvassed the globe, poking into drafty attics and forgotten closets, shaking loose museum inventories, and sizing up known collections. A few, like Atwood, still commissioned pieces from the best watchmakers. They often found themselves backslapping or back-stabbing each other in their pursuit of the world's timekeeping patrimony.

Winthrop Kellogg Edey was one such creature. Since childhood, he'd been spellbound by the transparent precision of antique mechanical clocks and claimed to have acquired his first at age six. Born in 1937 to a New York society family that could trace its origins to the eighteenth-century colonization of Barbados, Edey's grandfather, the engineer Morris W. Kellogg, struck it rich building and designing American oil refineries and wartime atomic bomb plants. A bon vivant, Edey lived off his family's fortune while pursuing four categories of personal interest: ancient Egypt, photography, diary writing, and collecting European clocks made between 1500 and 1830.

A man of extreme habits, Edey lived in an unrestored nineteenth-century town house on Manhattan's Upper West Side, refusing to refurbish the Victorian décor or upgrade its nineteenth-century plumbing and heating system. Known to amble through Central Park wearing plaid Bermuda shorts, black socks, and brown wingtips, he slept all day and tinkered all night, enjoying the dead silence to obsessively work on his clocks and pen exhaustive notes about them in his diaries. Through the years his collection grew to some sixty pieces, which he kept in his bedroom, lined up on the fireplace mantel and along an old wobbly bookcase. Edey became something of an éminence grise in the watch world, consulting with Christie's in New York and the J. Paul Getty Museum in Los Angeles and ultimately publishing two volumes on French clocks in 1967, the first on this subject in English. While the public at large had lost interest in mechanical timekeeping, Edey made it his life's work to keep its history and scholarship alive.

Although Edey spent a fortune on acquiring his clocks, said to be worth several million dollars, he wore a Timex on his wrist. When he died in 1999, he bequeathed thirty-nine of his clocks to the Frick Museum in New York.

George Daniels was another seminal keeper of the horological flame. A carpenter's son from Edware, London, he single-mindedly took up the trade, becoming one of the few individuals left on the planet to craft watches completely by hand. Born in 1926, he became intrigued with horology at age five after taking apart a cheap watch he bought on the street. In the 1960s Daniels had established a clock and watch repair business in London where he found himself restoring so many Breguet timepieces that he became an incomparable expert on the famed French horologist.

A passionate collector in his own right, by 1968 Daniels went from restorer to master watchmaker, building his first timepiece, a gold and silver one-minute pivoted-detent chronometer tourbillon that captured the eye of a wealthy collector. Soon a clutch of affluent connoisseurs, including Seth Atwood, began commissioning Daniels for their own specialized pocket watches. A great believer in the survival of the mechanical watch, he dedicated himself to redesigning an escapement to compete with the quartz watch, and in 1976 Daniels unveiled the coaxial escapement. Acclaimed as nothing short of a revolution, Daniel's mechanism was a modification of the lever escapement. It reduced sliding friction and was impervious to corrosion of the lubricant, long an enemy of accuracy. Considered the greatest horological advancement in 250 years, Daniels was hailed as the greatest watchmaker since Abraham-Louis Breguet. In 2010, the year before he died, Queen Elizabeth named Daniels a Commander of the British Empire, the only watchmaker to receive the honor for services to horology.

Among collectors, Seth Atwood wielded his enthusiasm like a blunt instrument. The son of a prominent Rockford, Illinois, family, in 1909 his father, Seth B. Atwood, and his uncle James T. Atwood founded the Atwood Vacuum Machine Company. Six years later, Atwood senior came up with a small adjustable rubber shock absorber to stop the shaking and rattling noise coming from his car door. His little invention pushed the family business from vacuums to car door hardware. Eventually the Atwood Vacuum Machine Company became the umbrella for a full line of auto body hardware, expanding to five manufacturing plants in the United States and Canada as well as a number of other family-owned businesses involving banking, venture capital, and real estate.

Both wealthy and entrepreneurial, the younger Atwood appeared the perfect amalgam of Henry Graves, Jr., and James Ward Packard. Born on June 2, 1917, he was a tinkerer, always fixing and building things, trying to find out how they worked, attempting to improve their efficiency. (In later years he designed yachts and developed a three-wheel energy-efficient car.) As a result people mistook him for an engineer. Tall and thin with granite-cut cheekbones, Atwood had been captivated by the mysteries of timekeeping since childhood. While attending Carleton College, he recalled, he would sit on the grass looking at the stars overhead, filled with questions. "Why is there a universe? How did all of this all start?" Later he decided that "nobody can define time," finding the word "an arbitrary human construct for a phenomenon, a dimension, about whose usefulness there is no doubt."

After graduating Phi Beta Kappa with a BA from Stanford in 1938 and an MBA from Harvard in 1940, Atwood served as an officer in the U.S. Navy from 1942 to 1946, earning the rank of lieutenant commander. Returning to Rockford, he joined the family business. Although his enormous fortune afforded him luxury and choice, he chose to remain in his small hometown.

Atwood's entry into the realm of serious collector occurred in the 1960s, when mechanical timepieces had fallen out of fashion and he was already in his fifties. "I decided to try to collect a few items, artifacts that showed the development of time-finding and timekeeping devices as we normally use them to order our lives," was how he described his motivations.

Hardly impetuous and certainly not a vanity connoisseur, Atwood spent four years immersing himself in exhaustive study, traveling the world visiting museums, talking to dealers, and consulting experts, before he opened his wallet. Doggedly he read G. H. Baillie's landmark book *Watchmakers and Clockmakers of the World,* first published in 1929, from which he drew up a wish list of one hundred must-have pieces. His philosophy was singular. "Collecting clocks and watches is not like collecting art," he once said. "You can't just walk into a gallery, see something you like, and buy it. When you are looking at a timepiece you have to consider its use, its function and its place in history. You have to see whether it's mechanically important and 'right.' You have to be knowledgeable about what's inside."

On a visit to London in 1968 Atwood met with a group of dealers and experts, including Robert Foulkes, a preeminent British appraiser who was also the secretary of the Antiquarian Horological Society. Atwood showed them his wish list. It included the name of every master watchmaker in history and every significant piece that added to technology or horology through the centuries. Foulkes sneered, "There isn't a chance in the world that you will ever get a fraction of this lot."

Few grasped the full contours of Atwood's ambitions. In less polite company, they found him to be a crazy amateur hobbyist, completely out of his depth. "Many thought he was nuts," recalled William Andrewes, a British-born clockmaker and restorer working at the Old Royal Observatory, whom Atwood later recruited to curate the Time Museum. "It was as if someone came into a gallery and said they were starting an art collection and they wanted a Michelangelo, a Cézanne, and a Rembrandt." Atwood, however, had two weapons in his arsenal: unbridled determination and unlimited funds.

Seven years after Atwood first stunned London watch dealers, he had amassed 1,200 pieces, all of great historical importance and rarity. Eventually this number would expand to more than 3,500 timepieces.

The essential strategy that Atwood would use over the years when he desired an object for his collection took shape during his very first acquisition. In 1968, the same year that he presented his list in London, he learned that one of the few remaining Thomas Tompion quarter-repeating gold pocket watches in existence was coming up for auction, previously owned by the Irish American mining magnate Sir A. Chester Beatty, more familiarly known as the King of Copper. Made by England's most famous clockmaker in 1697, the watch had an elaborately engraved movement with scrolling foliage, signed and marked "144." Intriguingly, the movement was secured to the case with secret latches to prevent other watchmakers from copying the notoriously secretive Tompion's movement.

Foulkes bid on behalf of Atwood at the auction held at Sotheby's in London, with instructions to "pay whatever was necessary." The sale quickly heated up, and the price soared skyward like a missile. When it was over, the price, £4,400 (the equivalent of $92,013 today), broke all previous records for an English watch. Shocked, not to mention embarrassed, Foulkes delayed calling Atwood back in Illinois to inform him of the results. While Foulkes hesitated, the underbidder, the formidable American collector Sam Bloomfield, contacted Atwood, inquiring whether he might turn around and sell the clock to him, offering him twice the amount that he had paid at Sotheby's. Atwood turned him down flat. Following the unfruitful exchange, Foulkes called his client somewhat shamefacedly to tell him that he had made history. Atwood calmly reassured his British proxy not to worry. "I already had the opportunity to double my money," he told him.

Uncommonly kind, Atwood possessed a quiet ruthlessness when it came to collecting. After identifying his desires, he patiently waited them out. "I saw things I wanted in private collections that people had no intention of selling," he once told the *New York Times*. "I do not bargain, and I am not a pressure buyer or seller. I simply asked them what they thought was a fair price. And when they told me, I agreed. I think bargaining is for the birds."

Atwood always had the last word. He relied on people and trusted them until he didn't, as his curator Andrewes explained. "If they crossed him he simply never did business with them again."

Uninterested in amassing the largest collection of timepieces, Atwood pursued the most pivotal instruments. He acquired one of the earliest existing pendulum clocks, made by the Dutch clockmaker Salomon Coster circa 1657, and one of the oldest timekeepers, a Greek-Byzantine sundial from AD 450 with mathematical markings used for telling time in different latitudes. In an act of astonishing fortitude, he obtained the Thomas Mudge "Green" chronometer, one of only three of Mudge's marine timepieces known to exist, produced during the competition among clock and instrument makers during the eighteenth century to determine longitude at sea. The earliest, known as Mudge No. 1, sits in the British Museum.

For years it was thought that the Mudge Green had been lost at sea. Inexplicably it surfaced at a London auction and was sold to a Swiss dentist named Eugen Gschwind. An avid collector of watches and enamels from the sixteenth and seventeenth centuries, Gschwind enjoyed both a personal and a professional relationship with the Stern family and often commissioned Patek Philippe to make special watches for him. Gschwind also consulted with several important Swiss collectors.

Atwood approached Gschwind, told him to name his price, and gained possession of the famous timekeeper. The transaction not only secured for Atwood the celebrated chronometer but launched an important relationship between Atwood and the dentist. With his many contacts, Gschwind became Atwood's eyes and ears, ensuring that in the future he would be offered the choicest timepieces.

When a piece was found to be unobtainable, Atwood engaged experts to make costly replicas. An example of John Harrison's marine timekeeper stood high on his list. Harrison had won the Longitude Prize, and Atwood was insistent that Harrison's triumph of time measurement be represented in his collection. Unfortunately for Atwood, all four surviving examples of Harrison's history-making device were housed at the National Maritime Museum in Greenwich, England. Considered national treasures, they were as likely to be sold as an original copy of the Declaration of Independence.

However, Atwood knew that Harrison and his brother, James, had made three high-precision tall-case regulators in which he experimented and tested his sea clocks. Atwood discovered that a British man named Colonel Quill owned one of the three. The regulator case surfaced after a strange turn of events that ran the course of nearly two hundred years. The long case had remained in the Harrison family for decades. At some point it is believed that they sold it to pay off a family debt. Then, during the 1860s, it mysteriously turned up in the cellar of the Old Malt Shovel, an inn in the village of Hull, in terrible shape. Quill bought it in 1954, by which time it was in even worse shape. The case was painted an awful pink-brown color and the dial redecorated, while a number of its parts had been replaced, although James Harrison's name still appeared in gold on the arch above the dial. Quill restored the case and the movement to what he believed was its original condition and sold it to Atwood in 1980. Atwood then commissioned a skilled London clock maker to produce an exact replica of Harrison's first marine timekeeper, complete with the movement's large oak wheels seen through the glass side-panels, and Harrison's invention, the grasshopper escapement made in brass, with an ebonized case and gilt-capped hood pilasters. It stood six feet, eleven and a half inches tall.

Over the years, Atwood became acquainted with other great collectors. Later he would say that, as a result, his life had been "tremendously enriched with new and lasting friendships." He came to enjoy strong ties with Patek Philippe, which, on occasion, produced special orders (after first being approved by the Sterns) for an elite circle of patrons.

In 1972 Atwood decided he wanted to own a secular perpetual calendar.

Only an extremely limited number of these instruments were known to exist. Unlike the perpetual calendar that accounts for leap years, the secular version also automatically adjusts for non-leap years that, according to the Gregorian calendar, stop during three consecutive centuries (2100, 2200, 2300) before resuming once again at the start of the year 2400. Atwood commissioned Patek Philippe to craft one especially for him. It was the first time that the house incorporated a secular perpetual calendar into one of their watches. The eighteen-carat gold double-dial keyless lever pocket watch was signed *Made for Seth G. Atwood*. Thirty-five years later, this very piece went under the hammer at Christie's in Geneva, fetching $1,042,338.

Although largely forgotten by the broader public, among serious connoisseurs the great watches commissioned by Henry Graves, Jr., and James Ward Packard continued to hold sway. Aurel Bacs, who became the co-head of watches for Christie's in Geneva after working at Sotheby's, grew up on fine mechanical watches. As a teenager in Zurich he had accompanied his father, an architect and watch devotee, to auction houses, antique stores, flea markets, and private collectors in search of them. Bacs had also come under their spell. "As soon as anyone enters the watch market, you come across Graves and Packard as much as one comes across Napoleon and Churchill when you are studying history and politics," he said. "They are the two most relevant, stellar names, the most influential in the world of collecting."

But the Graves and Packard watches posed a thorny problem for all interested parties. Although they were two of the most famous collections, they were also among the least well documented. Outside of the thirty-one Packard watches owned by the American Watchmakers Institute, the successor organization to the Horological Institute of America, nobody, not even Patek Philippe, had a clear idea of just how many watches were out there.

Following the end of World War II, the Packard collection was briefly split up. Twenty-six pieces left the Chicago Museum of Science and Industry and returned once again to the Smithsonian Institution, where they were housed at the National Museum of History and Technology (later the National Museum of American History). Five timepieces went directly to the U.S. Naval Observatory, also in Washington.

After Elizabeth Packard died in 1960, the bulk of her estate went to her family, the Gillmers. What was left of Ward's watches had been casually dispersed. In a fifty-page catalogue of the public auction of her Lakewood estate, among the more than eight hundred items listed, not one was a watch or a clock.

Scattered among family and others, some who viewed them as treasured heirlooms and others whose enthusiasm in many cases never matched the original owners', the Graves and Packard watches simply became an after-thought. They remained in private hands, were placed in safety deposit boxes, or were passed on and in some cases misplaced or lost. Some who owned these little mechanical marvels had no clue or understanding of their importance, let alone their monetary value.

In short, most of both men's watch collections seemed to vanish, except for a group of choice timepieces. Chief among the Graves watches was the Supercomplication. Just as she vowed that day on the lake, Gwendolen ensured that the piece never left the family. This Patek Philippe Depression-era masterpiece had catapulted to the top of Seth Atwood's wish list. Of course, he was not alone in his desire. And he was not the only interested party who knew just where to find it.

Since acquiring Patek Philippe in 1932, the Sterns had continued to procure uncommon timepieces for the family's private collection. Among these, 120 antique pieces, including enamel portraits from the seventeenth-century French master Jean Petitot, obtained by Henri Stern, sat in the firm's conference room on the Rue du Rhône. Despite the industry's vicissitudes, the family remained exceptionally dedicated to the craft, viewing themselves as defenders of the watchmaking tradition, a vocation that spanned all three generations. "My grandfather and then my father infected me with it," Philippe Stern once mused, "and inspired the idea of safekeeping some of the most interesting pieces crafted by Patek Philippe since the company was founded in 1839."

Once Alan Banbery joined the *maison* in 1965, the Sterns began scaling up their acquisitions. On occasion Banbery was asked to appraise important private collections, where he usually came across significant Patek Philippe pieces, which, he believed, should be back with the company. More often than not, these collections would be broken up and sold at auction. After cataloguing the four hundred–piece collection of a Swiss actor that contained five beautiful Patek Philippe pocket watches, Banbery attended the auction and bought all five. "I went to Henri Stern with the pieces," he later recalled. "Mr. Stern was a man of few words, and those he did speak were full of importance. He asked me how much I had spent and then went off in a cloud of his pipe smoke. He returned saying, 'That was a bit expensive.'" Banbery took this as a sign that he had carte blanche.

As Banbery spun around the globe on sales trips, he scheduled his calendar to overlap with watch auctions. "I took the latest catalogues on flights," he said, passing them back and forth with Philippe Stern, the company's heir apparent, who often traveled with him on business. Groomed from childhood to take on the Patek Philippe mantel, the younger Stern began working at the Henri Stern Watch Agency in New York in 1963, learning the watchmaking métier as he studied business, before returning to Geneva in 1966. While in New York, he impressed his colleagues with his work ethic and democratic attitude. As one of them recalled, he had once sat for hours, tying suede cords on ladies' wristwatches in such a way that they wouldn't scratch the wrist.

Back on the Rue du Rhône, Stern and Banbery discreetly probed retailers about their unsold stock. When rare pieces came back for servicing they courted the owners, gently inquiring as to whether they might sell their pieces or trade them for others.

Once while on a business trip to the Henri Stern Watch Agency in New York, Banbery recalled, he was asked to authenticate a watch for a woman sitting impatiently in the reception area. "I opened it up and assessed it quite rapidly," he remembered, somewhat amused. It was a lady's pendant watch completely set in diamonds and rubies with a beautiful red enamel dial. "On top of that it was a minute repeater that struck on the hour, quarter, and minute on two gongs. I thought it was quite something, like seeing a lady in a very beautiful dress." Engaging the woman in small talk, Banbery quickly realized that she didn't seem particularly attached to the piece. She was in a hurry, she explained, on her way to the car dealership to pick up her new Cadillac. "I joked that that was quite an expensive car and we might be interested in acquiring the watch, which would help pay for the Cadillac. We agreed on a price. That was that, and we both went off very happy." As it turned out, Princess Henriette of France, the twin sister of Louise Elisabeth and the eldest child of King Louis XV, had once owned the lovely pendant watch.

Together, Banbery and Philippe Stern quietly put together a gargantuan collection. Within five years Banbery was named its curator. The idea of erecting a museum had yet to take root, and the collection remained well out of public view. "I squirreled them away in the depths of the Patek building," joked Banbery.

In 1973 Stern began adding significant wristwatches to the collection. Because of the popularity of cheap quartz watches, he practically had the field to himself. Mechanical wristwatches were considered largely undesirable relics, and most collectors wanted antique pocket watches and clocks. Stern quietly acquired pieces directly from private clients. "It was rather easy to make a nice collection," he noted, "because you didn't have that many collectors."

In putting together his collection, Stern came up with three criteria: "The first was the case and the overall quality, the second was something interesting from a technical standpoint, like a new movement, and the third was artistic interest, such as engraving or enameling," To aid him, Stern had exclusive access to Patek Philippe's comprehensive archives, which documented the production of each piece: its movement and case numbers, any maintenance performed, and the name of the purchaser, stretching back to the company's founding. As a result, Stern's pursuit became more treasure than hunt.

In some circles, people whispered that the Sterns maintained a "hot list" of significant pieces, culled from the ranks of its exhaustive leather-bound records. Along with pieces of spectacular technical caliber, Banbery and Stern also prized timekeepers with exceptional provenance. Through the years Patek Philippe had attracted a diverse set of celebrated admirers. Eventually the pair would obtain the gold split-second chronograph wristwatch that Duke Ellington bought during his European tour in 1948 at the Geneva salon and Joe DiMaggio's chronograph wristwatch. This was the piece that "Joltin' Joe" acquired around the time he signed a record-breaking $100,000 contract with the Yankees in 1949.

When it came to prominent patrons of the modern era, perhaps none loomed larger than James Ward Packard and Henry Graves, Jr., And when it came to timepieces, none was more prominent than the Graves Supercomplication, which Patek Philippe itself had crowned as "representing the pinnacle of watchmaking achievement."

In this rarefied clique of connoisseurs, many had fantasies of owning this bluest of ticking bluebloods, now known simply as the Graves watch. But nearly all of them had become disheartened about the prospects of its ever coming on the market. In 1960 Gwendolen Graves Fullerton had given the lavishly constructed watch to her son, and he had never given any indication that he could be enticed to sell.

However, few pieces that had come into Seth Atwood's sights had escaped joining one of his two collections: the museum collection, funded through his companies, and his private collection that he personally financed. In the years before he opened his Time Museum, he had become acquainted with Reginald Fullerton, Jr. The retired industrialist met with the descendant of one of America's greatest watch patrons on a semiregular basis, usually in Manhattan, and often over a martini lunch.

Over time they drifted into an amiable friendship. Erudite and person-
able, Atwood held forth on a great number of subjects, but with a subtle
tenacity remained locked on the Supercomplication. "Name your price,"
Atwood generously invited his companion. Fullerton never did.

Fullerton had been born six months after his grandfather took receipt
of the Supercomplication. As a boy, he often watched his grandfather wind
and care for his collection of timepieces, and he developed a deep appreciation for the older man's horological passion. Fullerton grew up in the
family apartment at 1030 Fifth Avenue and on his parents' gray-shingled
estate in the wealthy Old Money enclave of Watch Hill, Rhode Island. He
had gone to the right schools—the Westminster School in Simsbury, Connecticut, and Phillips Academy in Andover, New Hampshire—and had
memberships in the right gentlemen's clubs: the exclusive Links, the Racquet and Tennis Club, and the University Club. After graduating from
Yale in 1957, he joined the family business, becoming a vice president
at Bankers Trust. In 1968, the same year that Atwood began his feverish
watch collecting, Fullerton married a pretty girl with the right pedigree:
Kathleen Knudson of Coral Gables, Florida, and Litchfield, Connecticut.
Her father, Shirley D. Knudson, was the president of the investment firm

Henry's grandson
Reginald "Pete" Fullerton, Jr.,
inherited the Supercomplication.
Photograph courtesy of Sotheby's,
Inc. © 2012.

Eskay Corporation and had also founded the greeting card manufacturer Treasure Masters.

In August 1953, five months after Henry died, Fullerton had purchased his first Patek Philippe, an eighteen-carat gold wristwatch with a subsidiary seconds dial. His acquisition happened to coincide with the twentieth anniversary of the Supercomplication, and following in his grandfather's footsteps, Fullerton began assembling his own superb collection of timepieces. Through the years, Gwendolen had given her son several of her father's watches that she had inherited. In the fall of 1960, the pair traveled to Switzerland, bringing with them a number of Henry's watches for servicing at Patek Philippe's Geneva salon. Looking to his grandfather for inspiration and guidance, Fullerton later wrote about conducting séances in which he convened with "the ghost of Henry Graves, Jr.," about his acquisitions.

The Supercomplication drew any number of suitors, but Fullerton showed no interest in parting with his grandfather's most important pocket watch until March 26, 1969. Six days earlier his mother had died. (Reginald Sr. had passed away the previous year.) It was a Wednesday, and once again Atwood and Fullerton had gotten together. In their previous casual summits Atwood had demonstrated his unusual depth of knowledge of the watch, the magnitude of craftsmanship and the enormity of dedication it took for such an undertaking. "He was obviously in love with it" was how one of his intimates described Atwood's intense interest.

On this occasion, Fullerton and Atwood stayed up until two in the morning drinking, discussing everything except the watch. Then, without prompting, Fullerton turned to Atwood and blurted out that he would sell this most coveted treasure. Fullerton's terms: $200,000 in cash. Atwood agreed.

Soon after the deal was sealed, Atwood walked into a branch of the United Bank of Rockford and told the teller that he wanted to withdraw $200,000 in cash (the equivalent of more than $1.2 million today). Alarmed, the bank teller informed him that he needed to see the manager. With such an unusually large request, the manager invited Atwood to come to his office, where Atwood amiably explained that he was buying something with a large price tag and needed the cash. Not entirely convinced, the manager agreed. But first, he said, he needed to confirm that Atwood had the funds.

Atwood not only consented, but helpfully offered to speed the process along, giving him the telephone number of one of the bank's executives. "I have a Mr. Atwood here who wants to take out $200,000," the manager cautiously spoke into the phone. "Is he good for the money?" Asked to describe Atwood, the bank executive then informed the manager, "Oh, yes, he owns this bank and seven others." With that, Atwood stuffed the cash into a suitcase and walked out into the sunlight.

The news of the sale shocked the insular world of collectors. Enraged, one of Fullerton's friends confronted him directly. "I was very angry," he recalled. "I told him, What are you doing that for? You are going to regret it."

The Stern family was caught off guard. "They never assumed that he would sell it to Seth," someone close to all of the players noted. "Patek Philippe was sad. They didn't understand what had happened. There was a tacit understanding in their mind, if Fullerton was going to sell it, it would be to them." Alan Banbery too was surprised. "I must confess, that on finally learning that the Graves watch had been sold—without Patek Philippe having the possibility of first refusal—I was somewhat disappointed, to say the least." Yet Hank Edelman, president of Patek Philippe USA, who began working for the watchmaker in New York in 1961, knew, as did many, that "if Atwood was after something, there was no point in bidding. He wanted originals and the first of everything. For the most part, if it was out there he got them."

CHAPTER TWENTY

❦

Back in Time

In the face of the quartz onslaught, among those who still believed in a future that included mechanical watches was Henry B. Fried. "People want to see the art of the mechanics," the seventy-eight-year-old dean of American watchmakers mused brightly. "They want to see that the watch has a heart. They want to hear it tick." But Fried was also a pragmatist. The son and grandson of watchmakers and the author of fourteen books and hundreds of articles on watchmaking, he had taught watch and clock making in New York City high schools for thirty-five years. Appointed technical director of the American Watchmakers Institute, he had numerous titles and certificates, including president of the New York City Horological Society, the New York State Watchmakers Association, and the Horological Institute of America. Fried was the first American to receive the Silver Medal of the British Horological Institute.

Having spent much of his life loupe to eye, examining and writing about the Packard collection, it was conceivable that outside of Packard himself or Patek Philippe, no one was more intimately familiar with the automobile pioneer's watches. In his eighth decade, Fried still traveled the country giving lectures on science and time. When discussing the intricacies of a watch's movement, he was known to break into an explanation of how a sailor could calculate his position at sea using a decent watch and dividing up the 24,000 miles around the Equator. He was punctual to a fault but spoke of watches like a romantic. Once, when describing his preferences for pocket watches over wristwatches, he offered, "They are watches with a soul." Fried believed that the Packard watches supplied a great public service. In a world of automated conveniences, such magnificent mechanical engineering, he wrote, proved "that the skilled horologist is a technician, a scientist, a mathematician, an artist and even a musician."

But the decline of the mechanical watch had thinned out the ranks of men like Fried, and watchmaking schools had difficulty attracting students. By the mid-1980s highly skilled watchmakers appeared to be going the way of blacksmiths. "In ten to twenty years," Samuel Beizer, chairman of the jewelry department at the Fashion Institute of Technology, told the *New York Times*, "there will be no one around to make or repair these watches."

In the spring of 1986 a wild rumor hit the watch world like a thunder-clap: Patek Philippe had decided to abandon the mechanical watch. Over-night, perceptions changed. All of a sudden the mechanical wristwatch, considered both obsolete and too contemporary to rate as a collectible, had transformed into an *objet d'art*. Although the rumor turned out to be false, the gears had irretrievably shifted. Mechanical watches were now a market.

Just a few auction seasons earlier, Sotheby's watch expert, Daryn Schnipper, had wondered if she should tell those hopefuls turning up with vintage wristwatches that they weren't salable. Now she found herself in the middle of a frenzy. "The rumor alone incentivized people," she said, still stunned by the quick turn. "The Swiss were leaders in luxury watches, and Patek Philippe led the Swiss in mechanical timepieces. If they ditched making mechanical watches, it signaled the end of an era."

This was a time when Americans deep-dived into one of the most con-spicuous periods of wealth culture since the 1920s. The demand for luxury goods soared, everything from a $245 nine-ounce tin of Beluga caviar to a $32,000 DeLorean sports car. *U.S. News & World Report* declared, "Wealth is back in style." As the soap-operatic celebration of greed and consump-tion played out weekly by the Ewings on *Dallas* and the Carringtons on *Dynasty* made for must-see television, *Forbes* magazine launched its inau-gural list of the wealthy, "The *Forbes* Four Hundred." Notably, in the maga-zine's paean to the affluent, Old Money had dwindled to roughly a quarter of the scorecard, while the rest was composed largely of the self-made. Flush with money, the newly minted moguls flaunted their toys with abandon, the shinier and bigger, the better.

Luxury was fast becoming associated with products and with mass affluence. Almost anybody could purchase the once elite status symbols; Gucci purses, Mercedes-Benz cars, and cell phones were all becoming commodities available to the masses. The commodification of status drove the wealthy elite in search of even rarer treasures. They collected art. Cor-porate raiders and private equity princes, such as Henry Kravis, the British advertising magnate Charles Saatchi, and Japanese CEOs, battled it out for Van Goghs and Rothkos. The desire for expensive toys and the dwindling number of hand-finished watches sent prices for mechanical watches sky-rocketing. At auction houses across the globe they became the subject of intense attention. In May 1986 a 1955 Patek Philippe split-second wrist chronograph with perpetual calendar, one of only three such wristwatches produced, made history when it realized $158,889 at Christie's in Geneva. During the same auction several other wristwatches also raised the mercury. A Patek Philippe chronograph with perpetual calendar wristwatch once owned by the late King Talal of Jordan fetched $48,889. The very pieces that had received only glancing curiosity now broke records. In 1980 Sotheby's held an auction at which it sold six wristwatches, which together realized $16,000; six years later, during the 1986 season, an auction of mechanical timepieces brought in $1 million.

As the market exploded, members of the board of the American Watch-makers Institute proposed that they sell the Packard watches to save the Institute and put the money toward its Perpetuation Fund, 40 percent of which covered the organization's operating expenses. Although in 1971 the AWI had established a trust with the purpose of funding a museum, by 1986 its realization remained as elusive as ever.

Among the strong proponents of the sell-off was Henry Fried, now an honorary member of the board, but not everybody agreed. "It was a shame that they wanted to sell them," recalled Philip Poniz, a watch expert with an encyclopedic knowledge and a specialty in sniffing out fakes. "They were magnificent watches, and nobody wanted to see them leave the Institute, to say nothing of the country. But I do understand why the Institute decided to part with them." Poniz, who viewed the watches as "a part of the glory days when America was gaining strength," said they were "part of America's heritage."

Facing down the opposition within the ranks of the AWI, those in support of a sale remained insistent that the sell-off was the best way to ensure that the Institute had a future. The Packard watches remained the AWI's crown jewels, but over the years the collection had also become less than a sterling asset. The organization couldn't even afford to insure the pieces in order to keep them on public display and had exhibited them only once in thirty years, in 1985, to celebrate the twenty-fifth anniversary of its merger with the Horological Institute of America.

The most valuable of the collection was Packard's astronomical watch. Discreetly, the AWI board put out feelers. One member with deep ties to the Swiss returned with word that a European auction house was interested in the consignment and had offered a $2.2 million minimum bid for the sale. Although an auction would provide the highest price for the watch, the idea that the Institute would have absolutely no control over who would own such a prestigious timepiece left some on the board skittish. Chief among their concerns, the Packard might get caught up in the South American drug wars. At the time the Medellín and Cali drug cartels had perfected an illicit money-laundering chain through the sale of art, jewelry, real estate, and cars. As Jim Lubic, the AWI's executive director, later recalled, "They feared the collection would end up in some drug lord's sock drawer." Spooked, the AWI declined the offer and tabled the idea for the time being.

In the early fall of 1988 Patek Philippe was in the midst of preparations for the firm's 150th anniversary the following spring. Nearly five hundred watches, about half of the total that Alan Banbery and Philippe Stern had skillfully acquired over the past twenty years, would be exhibited at the Museum of Watchmaking and Enamels in Geneva. As part of the jubilee celebrations, a special auction, "The Art of Patek Philippe," the first thematic sale devoted to a single subject, was also being readied. Spearheaded by Osvaldo Patrizzi of Habsburg Feldman (which later became Antiquorum), three hundred Patek Philippe watches were going under the hammer at the Hotel des Bergues.

Sitting in his office on the Rue du Rhône surrounded by photographs of his prizewinning Labrador retrievers, a white marble bust of Count Antoine Norbert de Patek, and some of the many antiques he had collected over the years, including the rare opium pipe made from the branch of a Chinese lemon tree, Alan Banbery received an unexpected inquiry from Marvin Whitney at the American Watchmakers Institute in Ohio, who had called with a delicate matter to discuss. The AWI was considering selling the Packard, and Whitney wanted to gauge Patek Philippe's interest in acquiring the last stunning *grande complication* that the automaker had received during his final days at the Cleveland Clinic. With its beautiful celestial chart that navigated the night sky over Packard's beloved hometown of Warren, this was the piece said to have helped pave the way for the Supercomplication, in more ways than one.

Banbery had crossed paths with both Marvin Whitney and the Packard before. Roughly six years earlier he had photographed the watch while working on the first of his two seminal volumes on the history of Patek Philippe watches. At the time, he had requested taking the watches back to Geneva to better facilitate the various stages of dismantling necessary to photograph the movements in detail. Reluctant to allow such rare and valuable watches out of the country, Whitney instead organized the shoot from the security of the Smithsonian Institution, where several of the pieces remained.

At the Smithsonian, Banbery was appalled at the state of the collection. "They were badly displayed," he sniffed. "Unless you knew who Packard was or what they were, you wouldn't have taken a second look." Furthermore the watches themselves were in bad shape. "The oil was dry and the gold was dull," he said. "The entire display was dusty."

Rolling up his cuff-linked sleeves, Patek Philippe's refined curator went to work in a makeshift photo studio in the basement of the former National Museum of History and Technology building. Over nine days in cramped and dark quarters Banbery's small team shot each piece with surgical precision, showing the watches' movements to their anatomical advantage. Despite conditions that he later described as "pretty horrendous," Banbery not only obtained "fantastic photos," but he came to a mutual understanding with Whitney.

After the Supercomplication had fallen to Seth Atwood, the chance to acquire the Packard was not one to let pass. Banbery recalled his phone conversation with Whitney: "I asked if they set a price, and he had. And more or less we said yes."

As the two men were about to end their conversation, Banbery said, "I don't mind saying if you ever consider selling the others . . ." He was, he recalled, given every assurance that should the AWI go that route again, he would be at the top of their call list.

Soon after their telephone conversation, Larry Pettinelli, an executive at the Henri Stern Watch Agency, was sent to meet representatives of the American Watchmakers Institute in a private airport lounge in Cincinnati. "I really had no idea what I was picking up," he recalled. "I had a check in one hand. I was supposed to make the exchange, look at the watch, and make sure it was this watch." In that moment, Pettinelli, who would be named president of Patek Philippe USA eight years later, said that he "had no idea of its historical significance."

That changed quickly. On September 12, 1988, Patek Philippe held a press conference announcing that it had paid a staggering $1.3 million for the James Ward Packard astronomical pocket watch, displaying the watch for photographers. The price was more than eighty-one times the $16,000 that Packard had originally paid for the piece. "It created quite a flurry around the world," Banbery recalled. "We got over a million dollars in free publicity."

So did James Ward Packard. Knowledgeable collectors began sniffing around for other pieces from the celebrated collections that had been largely forgotten.

In something of a coup, five of Henry Graves, Jr.'s prizewinning Patek Philippe pocket watches went under the hammer as part of "The Art of Patek Philippe" auction, organized by Osvaldo Patrizzi. Circling the globe, Patrizzi had developed close ties with private collectors and all of the major watch manufacturers. In a business built on relationships, Patrizzi's served him exceedingly well, offering a steady stream of prime customers and magnificent pieces that his auction house regularly put on the block. Philippe Stern for one was said to have purchased hundreds of Patek Philippe watches for his museum from Patrizzi's sales, often at record prices.* Patrizzi's connections would eventually deliver the *maison* more Graves watches perhaps than any other auction house.

At "The Art of Patek Philippe" sale on April 9, 1989, Graves's rare platinum tourbillon repeater, no. 198311, realized $260,000. It went to a private collector. Three others went under the hammer that evening as well: the platinum chronometer with up-and-down indicator, no. 198050-1913; the gold chronometer, no. 191000; and no. 191016, a second platinum Geneva Observatory prizewinning chronometer. The Patek Philippe Museum acquired all three. An unknown buyer took the fifth, a gold chronometer, no. 190977, that received second prize at the 1924–25 Geneva Observatory trials.

Coincidentally, just months later, Sotheby's unearthed the open-face, dual-time-zone pocket watch in silver with gold details and a hinged engine-turned case that Patek Philippe made especially for James Ward Packard. In January 1990 the piece, no. 190757, realized $19,800. Packard had originally paid $200 for it in 1920.†

While the sales of these watches were as exciting as they were notable, as it turned out, they were also exceedingly rare.

In anticipation of Patek Philippe's sesquicentennial, Philippe Stern made a number of decisions aimed at safeguarding the *maison* into the next cen-

---

*In 2007 Antiquorum's board ousted Patrizzi, accusing him of misappropriation of funds, among other things, allegations he denied. The same year the *Wall Street Journal* aired allegations that Patrizzi intervened with watchmakers to pump up prices at auction, with the manufacturers themselves feverishly bidding on pieces. Again Patrizzi denied the charges.

†In 1998 Christie's sold a trio of James Ward Packard's early pocket watch acquisitions, two open-face key-wound verge watches and a pivoted detent chronometer (Chevalier, Piguet, and French), for $1,495.

tury. Following his father, in 1977, Stern became managing director and looked at his son Thierry to one day sit at the same table. "It's my obsession to keep Patek Philippe in the family," he was known to proclaim. Even as longtime competitors were swallowed up, he remained dedicated to the company's traditions. A few offers to buy the firm had all been politely refused.

As the only independent family watchmaker in Geneva, Patek Philippe also continued to manufacture nearly all of its own components (except for hands, dials, and sapphire crystals) and decorate and hand-assemble each watch. The meticulous process of producing a complicated wristwatch required at least 1,200 steps and six hundred hours of quality-control testing and adjustments. In an interview with *Cigar Aficionado* Stern explained, "Today, exceptional watchmaking is about creating innovative, useful complications that may be based on old or historic technology but still fit in with our present-day lives."

Stern himself listened to each minute repeater for tonality and purity of sound before it left the manufactory. The process came down to "know-how and not mathematical rule." As he explained, "The gong should not be too close to the case to allow some space and air. The beauty of it is that no two minute repeaters have the same sound even of the same model reference number."

Stern realized that in order to remain independent, Patek Philippe also needed to sustain its profitability. The firm had always produced its timepieces in limited quantities. By 1980 the house manufactured fewer than ten thousand watches annually. Stern decided that, having survived by hewing close to its founding principles, the firm needed to do something to further ensure its survival.

In June 1979 Stern met with several of his executives to discuss how best to mark the firm's anniversary, still a decade away. One of the ideas was to reproduce the Graves Supercomplication. Max Studer, Patek Philippe's technical director, suggested a better idea: If the Graves symbolized the final frontier beyond which further complications were not technically possible in 1933, why not design a pocket watch now that had even more complications? Both sweeping and shrewd, the project would commemorate Patek Philippe's 150 years by showcasing its unrivalled command of watchmaking.

Stern assembled a team led by Paul Buclin, Patek Philippe's "master of complications," and engineer Jean-Pierre Mussy, for what they dubbed the Calibre 89. The watchmakers faced many of the same issues that their predecessors had in building the Graves. For a watch of such layered intricacy, each complication had both to work independently and to operate in concert with the others without disrupting the whole. An error of hair-thin fractions could be multiplied throughout the mechanism. Despite the accumulated knowledge amassed over decades, the watchmaking trade still relied on the individual skill and memory of the men of the bench. There was no repository holding manuals, drawings, or descriptions, so Buclin's team took apart complicated timepieces as a compass. They drafted components, coming up with 1,660 detailed parts drawings.

However, the team behind the Calibre 89 had advantages over their predecessors, for when Seth Atwood acquired the Supercomplication, he sent it to Patek Philippe for a complete overhaul and cleaning. Alan Banbery asked if he might photograph the masterpiece, and Atwood had agreed, under the condition that Banbery personally hand-carry the watch on a flight from Chicago to Geneva with an armed police escort. Along with photographs of the intricate Graves movement, by the time Stern approved the project, his watchmakers had the benefit of computers. The entire set of component drawings were scanned and uploaded, then automatically calculated and recalculated according to different positions in the movement. The computer also calculated the geometrics of the parts and indicated how to cut and assemble them. In Graves's time every step, every process was done through trial and error—by hand.

After nine years, the Calibre 89 was complete. It weighed two pounds, nine ounces and measured five inches high, four inches wide and one and a half inches deep. Encased in eighteen-carat gold, all 1,728 parts were hand-finished, -polished, and -washed. The watchmakers assembled the 332 screws, 415 pins, 429 mechanical components, 184 wheels, 129 rubies, 61 bridges, and 68 springs by hand. After exhaustive tests, the team gathered on July 1, 1988, for its unveiling with bottles of Champagne. Philippe Stern arrived to witness the final wind-up. Its creators wound the watch and put its astounding thirty-three functions through their paces. Aside from timekeeping, the Calibre 89 indicated the time of sunrise and sunset and the time in a second time zone. It featured a perpetual and secular calendar; a lunar calendar showing phases and age of the moon; an astronomical calendar, including houses of the zodiac; a chronograph; and various chimes. It was described as the world's "most complicated portable timepiece." With the aid of computer technology, Patek Philippe had broken its own record.

On February 8, 1989, Patek Philippe displayed the Calibre 89 for the press. For the photographs, the celestial chart was presented, showing the constellation of the eagle, the ascendancy of Altair—Philippe Stern's lucky star.

A month later, with a full slate of anniversary celebrations on tap, the watchmaker sent the Calibre 89 on a world tour, with stops in Tokyo and Los Angeles, in anticipation of a special auction for the monumental watch to take place back in Geneva on April 9 (part of "The Art of Patek Philippe"). In New York the *maison* organized an invitation-only dinner at Tiffany & Co. Over dessert, the select group viewed a film placing the Calibre 89's importance alongside Stonehenge. Patek Philippe planned to craft three additional Calibre 89's: one in rose gold, a second in white gold, and the third in platinum. "Occasionally an object or work of art of such importance comes up for auction," Patek Philippe announced, "that the attention of the whole civilized world is focused on the drama of its disposal."

At the Hotel des Bergues, an anonymous syndicate of buyers won the bid for the Calibre 89 with $2.97 million, a record for any timepiece sold at auction. The sale of three hundred pieces from "The Art of Patek Philippe" broke records from start to finish, fetching $15.1 million, smashing all previous watch auctions. Patek Philippe had solidified its standing in history and positioned itself for success in the twenty-first century.

The anniversary exhibition of nearly five hundred Patek Philippes from the *maison*'s private collection at the Museum of Watchmaking and Enamels in Geneva was also a tremendous success. Visitors from all over the world flocked to the exhibit, causing the museum to begin limiting the number of entrants at a viewing. In the Villa Bryn Bella, built during the first half of the nineteenth century, the ancient creaking floors could not carry the weight of all the people.

Flush with victory, the Sterns began erecting a formal museum to house the family's private collection and to present the history of watchmaking. "Watches are part of human cultural development," Philippe Stern explained. "They are very much influenced by the sciences—physics, mathematics, astronomy—as well as art—painting, sculpture, design. You can see the evolution of those disciplines in watches, because they were always adapted to their time."

On June 17, 1989, Sotheby's shattered records for a wristwatch sold in the United States with a 1947 Patek Philippe gold and enamel world watch that showed the time in forty-one cities; the cost was $275,000. Two months earlier, Patek Philippe had established the world record for a wristwatch when a tourbillon made in 1960 went under the hammer in Geneva for $433,290.

The vintage market for mechanical wristwatches and rare complicated pieces, especially those made by Patek Philippe, expanded. In the space of five years a Patek Philippe wristwatch, reference no. 1518, with a perpetual calendar, chronograph, and moon phase display, jumped in price at auction from $14,000 in 1984 to $100,000 in 1989.

Buoyed by the leap in disposable income and the growth in connoisseurship, Swiss watch exports reached a record high of $5.4 billion in 1990, up 12 percent over 1989, itself a record year. While Switzerland made up just 6 percent of the watches manufactured, it also accounted for half of their value. The pendulum had swung back.

Enabled by the massive wealth that accompanied the global economic boom, a new breed of collector emerged. Although perhaps short on historical perspective and technical understanding, most thought nothing of purchasing a watch the price of which was, in some cases, comparable to the price of a house.

One of New York's well-known princes of finance came to own fifteen complicated Patek Philippe wristwatches that he kept in a safe in his penthouse apartment. When deciding on which one to wear he said, "I am an investor. I wear a watch until I lose money, and then I switch." Although his father owned a Patek Philippe Calatrava wristwatch, this collector had not become serious about timepieces until he had a lunch meeting with a Mexican investor who showed up wearing a minute repeater. "I didn't know what it was. He opened my eyes to *grandes complications*." The financier began traveling regularly to Geneva, always arriving at the salon on the Rue du Rhône with the same question: "What do you have that I can't live without?" Each piece he ordered required approval from Philippe Stern himself, and all took a year or two to complete. He sometimes slipped away from his office to an auction at Sotheby's or Christie's, not to bid but rather "just to see the value of my watches."

Once, while traveling to Europe, this collector had forgotten that his latest wristwatch, purchased more than a year before, was on its way to New York, until his secretary called. "[She] told me there was an armored truck at the office with a delivery from Patek Philippe." The watch, his most complicated piece to date, contained the most coveted complications: a tourbillon, a minute repeater, a perpetual calendar with a retrograde date hand, and a moon phase display. Encased in platinum, it had a price tag of $650,000.

Around the time that the *New York Times* predicted in a 1982 article that within eight years, "modern quartz will have nearly silenced the tick," a group of Swiss banks approached a Lebanese-born business consultant named Nicolas Hayek, who had launched Hayek Engineering in Zurich in the early 1960s. His consultancy firm had been involved in a number of major corporate restructurings, including those of AEG-Telefunken and Swiss Railways. It had also advised a broad sweep of global outfits, among them Nestlé, Siemens, and U.S. Steel, raising Hayek's profile while making him a very wealthy man. The banks asked Hayek to draw up a plan to liquidate its two main manufacturers, ASUAG and SSIH, makers of such venerable brands as Omega and Longines.

Instead of liquidating the hidebound giant Swiss flagships that produced movements and watches, he merged them, forming the Société Suisse de Microélectronique et d'Horlogerie (SMH) and purchased a majority stake. In 1983 he introduced a plastic quartz watch with novelty dials and flamboyant colors. Hayek called his timepiece Swatch, a contraction of *Swiss* and *watch,* and revived the industry.

Significantly, Hayek redirected the intense interest in Swatch toward high-end mechanical Swiss watches and SMH-controlled brands, such as Omega, Longines, and Rado, doubling prices to make them more exclusive. He snapped up dying brands with golden names such as Blancpain and Breguet, reinvesting in innovation and producing component parts, becoming the industry's sole powerhouse. Hayek became known as the man who saved the Swiss industry from death's door.

In contrast, in the spring of 1993, the American Watchmakers Institute once again decided to sell off more pieces from the Packard collection, among them, the beautiful Patek Philippe pocket alarm clock that played the *Jocelyn* lullaby and the skeletonized Vacheron Constantin framed in rock crystal and ringed in sapphires. Again they chose to forgo a major auction consignment and instead notified a select group of collectors and dealers around the world that the Institute (later renamed the American Watchmakers-Clockmakers Institute) in Cheviot, Ohio, would be entertaining silent, sealed bids for twenty-one watches from the Packard collection. The request for bids moved swiftly and informally. Oddly, neither Alan Banbery nor anyone else at Patek Philippe was given a heads-up.

Andrew Crisford, a partner in the London firm Bobinet Limited, a dealer of antique clocks, watches, and scientific instruments, had gotten wind of the impending sale. Established in 1973, Bobinet maintained a select list of clients, including Seth Atwood, whose passion for timepieces aligned with their enormous spending power. A Breguet specialist, Crisford focused mainly on clocks and pocket watches made between the 1800s and the 1930s—the golden age of mechanical watchmaking.

Crisford called Milton Stevens, the AWI's executive secretary, to confirm that the rumor of the silent auction was true. Not long after Patek Philippe had acquired the Packard, Crisford and his partners had flown to Ohio to purchase three of the *grandes complications* from the Packard collection on behalf of one of their well-heeled clients. Apparently, in 1990, two years after Banbery and the Sterns made headlines paying $1.3 million for Packard's astronomical watch, they were, as were many, unaware that Bobinet had scooped up three pieces from the automaker's collection (reportedly for $2.25 million).

This time the AWI decided to sell the remaining twenty-one watches, keeping five French and American pieces that the original assessor in 1928 deemed not of museum quality. Henry Fried himself had written about this particular group of timepieces, saying they were "not unlike the first gatherings of the average inexperienced watch collector. Mr. Packard probably yearned for these, and like every other beginner, acquired gifts and probably bought a few." The lot included all of the precious watches produced from the English workshops of Smith, Benson, and Dent and the outstanding one-of-a-kind Patek Philippes crafted to Packard's discerning specifications.

Crisford jumped on the first plane from London. "All of the top collectors knew about this collection," he recalled later. "Some of the most complicated Patek Philippe watches were included. One or two of the watches would've been enough to cause excitement, but the idea of acquiring them all . . ." Borrowing a hotel typewriter, Crisford punched out his bid and put it in an envelope. The AWI accepted only sealed written tenders. Crisford submitted his in person but found the episode puzzling. "It was an extremely unusual method of disposal for a public body to choose this over a public auction." Still, he won the day (reportedly with an offer of $3.55 million). The twenty-one pieces joined the first three from the collection that Crisford bought on behalf of someone he called "an established collector."

Crisford wasn't the only person baffled by the transaction. "I was not informed of the second 'sale,'" noted Alan Banbery, despite the assurances he had received earlier from Marvin Whitney, regarding the future sale of additional pieces from the Packard collection.

After the considerable shock surrounding the sale began to quiet came the rumors of who bought the pieces. The decorous silence that ruled the gentlemen's club of collectors and dealers, not to mention the amount of potential future commissions at stake, all but guaranteed anonymity for the new owner of James Ward Packard's pocket watches. The majority of speculation was that a Middle Eastern prince had snapped up the watches.

Not surprisingly, Crisford refused to divulge the name of his client, except to say, "The collection is with a royal head of state. It's not British, it's in the Middle East, but it's not Brunei, and not Saudi Arabia. It is a royal collection housed in a special room in the palace."

As was true of pieces owned by Henry Graves, Jr., the appearance of a James Ward Packard watch on the market had become almost as rare as a total lunar eclipse. The AWI's bizarrely orchestrated sale seemed to shut the door on the possibility that Packard's magnificent watches—which also happened to include the most complicated and rare cache—might surface. As Crisford said, "It is extremely unlikely they will ever go on the market."

But there were other doors leading to other timepieces.

~

CHAPTER TWENTY-ONE

The Comeback

The rarer the object, the bigger the desire to possess it. Immediately after the American Watchmakers Institute's private sell-off of the Packard watches, both Sotheby's and Christie's, along with the smaller Antiquorum, led a streak of blockbuster mechanical watch sales of Patek Philippes, Vacheron Constantins, Rolexes, Audemars Piguets, Cartiers, and Jaeger LeCoultres.

As interest in vintage wristwatches mounted, the Italian collector and expert Jader Barracca cowrote the seminal *Le Temps de Cartier*. Martin Huber and Alan Banbery's books on Patek Philippe pocket and wristwatches became bibles for collectors. For the first time, enthusiasts had precise information on models, including the exact quantities manufactured. These books became an index with which to navigate the inscrutable labyrinth of more than a century's worth of exclusive and rare timepieces and a concrete way to determine their value.

Men dusted off their grandfathers' old watches and widows scavenged through their garages, reaching behind old bureaus and into safety deposit boxes for long-forgotten timepieces. Thanks to the popularity of mass-market vibrating quartz timepieces, the 1970s and 1980s had severely depleted mechanical watch inventories. By the 1990s the appearance at auction of a Patek Philippe chronometer or a Rolex Daytona not only sent prices soaring in the rarities market; it catapulted the contemporary market. In 1994 Blancpain, founded in the eighteenth century—a name barely on life support toward the end of the twentieth—announced that it had sold nine of its "1735" models, including an $800,000 *grande complication* equipped with a perpetual calendar, split-second chronograph, minute repeater, and tourbillon. Such staggering numbers enticed a great deal of press coverage, drawing new collectors into the heat.

Perhaps the only thing that attracted more interest than price records was a particularly rare object with spectacular provenance. In this narrowest of universes, the watches once owned by Henry Graves, Jr., and James Ward Packard became some of the most coveted. Over the years, scarcely more than a handful—not including the AWI sales—had turned up in privately brokered deals. During the sixty-year sweep since Packard's death, fewer than ten of the gentlemen's watches had come to light at auction. After five of Graves's pocket watches sold in 1989 as part of "The Art of Patek Philippe" jubilee sale, and the Packard in 1990 at Sotheby's, a sixth watch turned up eight years later, on October 31, 1998, during the first day of Antiquorum's two-day sale, "Important Watches, Wristwatches and Clocks in Geneva." It was Graves's prizewinning, eighteen-carat gold pocket lever chronometer with thirty-six hours power reserve indication. The Patek Philippe (no. 178448), completed in 1926 for the financier, fetched 46,200 Swiss francs ($34,709). This piece is now on exhibit at the Patek Philippe Museum in Geneva.

Watch collectors were a competitive lot, and the contest between the top-tier players soon became heated. Among them, Patek Philippe was the most formidable, with Alan Banbery shattering a number of records both at auction and privately. The astonishing $1,715,000 that Patek Philippe paid in 1996 for the one-of-a-kind Calatrava with an astronomic minute repeater, perpetual calendar, and moon-phase indication, the classic round wristwatch named for the watchmaker's symbol, made in 1939, not only raised the bar and made headlines, it earned an entry in the *Guinness Book of World Records* for the highest price ever paid at auction for a wristwatch. Two years later, Banbery paid $950,000 for yet another record breaker, a 1953 eighteen-carat pink-gold World Time wristwatch with double crowns and a cloisonné enamel dial representing a map of North America, its outer dial ringed with forty-one major cities.

As the market took off and the stampede broadened, those watches commissioned for Graves and Packard became something of a Holy Grail. With so few of their timepieces in circulation and so much damage done by the quartz crisis, outside of a small group of collectors, their appearance quickly became a race against time. As Aurel Bacs of Christie's put it, "These watches rarely come on the market. One day they will no longer be available. They will be put away in bulletproof cases. And then it's *finito*. Even if you have all the money in the world, it is the highest level of agony and pain for a collector who has the money and it is not for sale. It is a knife in a collector's heart."

At the start of 1999, Sotheby's promoted its two-day February sale of "Important Watches, Wristwatches and Clocks," bringing together more than five hundred timepieces, including two of Graves's Patek Philippe wristwatches that had been consigned to the house. The first of the two was the watchmaker's first perpetual calendar chronograph, reference no. 1518, crafted in 1947. The watch, including the buyer's premium, fetched $96,000, comfortably above its high-end estimate of $80,000. The second was his pink gold rectangular wristwatch, reference no. 2425, also produced in 1947, with a subsidiary seconds dial. It brought in only $7,475, just below the low-end estimate of $8,000.

The entire sale realized $5,082,386, the highest total in a decade for wristwatches at Sotheby's. At the time, Graves's collection had yet to become known outside an elite circle of the most knowledgeable bidders. Yet a rival watch expert snickered, thinking the auction total had fallen short. "Sotheby's missed the train," he said.

Within weeks, that changed as another opportunity presented itself. And Sotheby's did not blink.

Approaching his ninth decade, Seth Atwood surprised the watch world just as he had when he began collecting, and with little discussion, put the gears in motion to shut down his Time Museum and with it, his longtime horological passion. On March 12, 1999, when the Museum in Rockford, Illinois, closed its doors for good, Sotheby's watch department was already negotiating a deal with Seth Atwood to auction off its best pieces. Although he kept the reasons for this turn of events to himself (many believed he did not want to burden his family with maintaining the museum), it was Atwood's great desire to keep the more than three thousand instruments together as a collection. Daryn Schnipper pushed hard to secure the commission, though she had her eyes on the biggest prize of all: the Graves Supercomplication.

When Atwood struck upon the shocking decision to sell his collection, assessed at $33 million, the city of Rockford stepped up to buy it. Michael Lash, the director of Chicago's public art program who had grown up in Rockford, wanted Atwood's timepieces to remain in Illinois, but the nostalgia bid never got off the ground, as his hometown lacked the necessary money. Mayor Richard Daley of Chicago and Governor George Ryan of Illinois also made something of a crusade of retaining Atwood's collection in a national time museum.

The great Chicago collector Justice Warren Shepro had also lobbied hard to keep Atwood's legacy in Chicago. "There has never been another collection as comprehensive as his," he exclaimed, "nor one comprising individual pieces of such caliber and importance." Atwood and Shepro had become great friends, and in fact Shepro moved from Chicago to Roscoe, Illinois, simply to be closer to Atwood's Time Museum.

In May Atwood held a silent auction. The city of Chicago came up with only $25 million, well below several other bids that exceeded $40 million. Yet in the end, Atwood awarded his watches to that city because it was the only bidder to agree to keep the majority intact. Several successive deals were struck to shore up the considerable difference. Chicago's Adler Planetarium and Astronomy Museum privately raised funds to buy nine of the museum's most important astronomical instruments. Atwood agreed to display 1,550 of his clocks and watches at the private Museum of Science and Industry while the city raised the funds to pay for them and to cull eighty-one pieces and sell those at auction to make up the difference between the city's low bid and the higher ones that he had turned down. In the end, however, Chicago's deal fell apart. The entire Time Museum would eventually be split up into four sales and sold at Sotheby's in New York. The last of the sales, held over three days in October 2004, would bring $18,210,690.

Of all of the magnificent timepieces that had come into Atwood's possession over the years, he insisted upon retaining just one: a gold, one-minute tourbillon pocket watch with a coaxial escapement that also incorporated a perpetual calendar, equation of time, phases of the moon, thermometer, and a power reserve indicator. Atwood had commissioned the piece from George Daniels.

Six years later, after the final sale on February 21, 2010, Atwood died. He was ninety-two.

With the Graves Supercomplication, which Daryn Schnipper had taken to calling "the Big Kahuna," within its grasp, Sotheby's received an astonishing phone call from a woman claiming to be a close friend of Marilyn Preston Graves. The daughter of Duncan Graves, Marilyn had died on July 3, 1998, and the caller, representing her estate, explained that she had stumbled upon a cache of remarkable pocket watches and sought an appraisal. While going through Marilyn's Fairfield, Connecticut, home, she had discovered a shoebox stuffed under a bed with four pocket watches. Perhaps they were valuable?

There were three Patek Philippe tourbillon minute repeaters, two encased in gold and the third in platinum. The fourth, also platinum, was by the watchmaker Jules Jürgensen. All four were engraved with a coat of arms, an eagle issuing out of a coronet, and a banner with the Latin motto *Esse Quam Videri*. During the past forty years, little more than a handful of Graves watches had surfaced. Now not one but four, it seemed, had suddenly reappeared.

Distant relations and interested parties often streamed through Sotheby's glass doors seeking confirmation for objects after happening upon long-lost gems. It was the desire of every auction house to come across some hidden treasure that would break records, splashing across international headlines, raising the bar on prices for years to come. The odds, however, rarely tended to favor the scavengers. Stunned, Schnipper turned over in her mind the likelihood of stumbling upon unknown Graves timepieces while Sotheby's was in the midst of commissioning his masterpiece.

She dropped everything and traveled to Connecticut.

Marilyn was the last of Duncan Graves's surviving children. He had died in 1977 and his wife, Helen, in 1962. Marilyn's twin sister, Helen Mitchell Graves, had died in 1968 at thirty-six, and her disinherited brother, Duncan, Jr., in 1970 at forty-three. Duncan and Helen's children were said to have been stricken with a congenital disease, and as a result none had married or had children of their own. At sixty-six, Marilyn had lived the longest of her siblings. At the deaths of Harry Graves 3rd and George Coe Graves II, Henry's family tree had been sheared.

A society beauty with a taste for nightlife, Marilyn had devoted much of her time to volunteer work at the Memorial Sloan-Kettering Cancer Center and the Junior League of New York and remained active in the Colonial Dames in the State of New York. After Duncan took up full-time residency at his two hundred–acre Connecticut estate, Breeze Hollow Farm, she divided her time between Fairfield and her father's Park Avenue duplex, bought in 1953, the year Henry Graves, Jr., died. As her father's sole heir, Marilyn Preston Graves inherited her father's entire estate, along with several of her grandfather's watches.

While the important players in the watch world had come to know the name Reginald H. "Pete" Fullerton, Jr., and his association with his grandfather, the name Marilyn Preston Graves scarcely registered. But as Schnipper discovered, the watches were unmistakably Henry Graves, Jr.'s: complicated, interesting, and beautiful. They also happened to be some of the most important pieces that the American financier had commissioned.*

All of the timepieces were made especially for Henry. At least two had

---

*In 1998 Marilyn Graves donated an oak long-case clock made by Thomas Tompion circa 1675–78 to the Metropolitan Museum of Art. The clock, veneered with oystershell-cut olivewood and marquetry panels of stained bone, ivory, and gilt brass mounts, was very likely one of her grandfather's handed down to her after her father died.

arrived during the 1920s, at the height of Henry's and Ward's pursuit of complicated timepieces. One gold watch, no. 198052, was the piece equipped with eleven complications; another, no. 174961, had twelve complications. The latter was the glorious pocket watch that Henry had commissioned in 1919 and received just days before his daughter's wedding in 1926; it was also one of the pieces that signaled his return to the game. The third, received in 1948, was a rare platinum, Observatory prizewinning, open-face one-minute tourbillon, no. 198427. It was one of three platinum Patek Philippe tourbillons known to exist, all of which were made exclusively for Graves. The fourth watch, the Jules Jürgensen pocket chronometer, no. 15954, acquired in 1940 after his four-year collecting break, did not match the Patek Philippes in caliber, but it had earned the first-class certificate at the Neuchâtel Observatory.

The watches went on sale on June 15, 1999, at Sotheby's in New York. Although it had been just four months since the house sold the pair of Graves wristwatches, word moved through the fickle watch world that previously unknown *grandes complications* from the important twentieth-century American Patek Philippe patron had surfaced with impeccable provenance, technical daring, and overall rarity.

At the auction, Patek Philippe made quick work of the first of the two gold pocket watches, acquiring the piece with eleven complications for the museum for $794,500. But their elation was brief; the second gold watch, featuring twelve complications, went to an anonymous bidder for $640,500. The rare platinum tourbillon realized $453,500; it too slipped the watchmaker's grasp, going to an anonymous collector. The Jules Jürgensen fetched only $24,150. Together, the three Graves Patek Philippes tallied up nearly $2 million, breaking the American auction record for mechanical watches. Six months later, Sotheby's broke all world records, selling the Supercomplication for more than $11 million. By the time the hammer dropped, the Graves name was electric.

Once the auction room applause died down, interest in acquiring a watch commissioned by Henry Graves, Jr., exploded, and with it renewed attention was given to the collection of James Ward Packard, touching off a new chase.

Although the appearance on the market of pieces owned by the two great horological patrons of the twentieth century was now widely anticipated, they remained extremely rare. And it was largely those previously sold watches that recirculated on the market. When, on May 18, 2004, the Graves platinum tourbillon (no. 198427), first sold in 1999 at Sotheby's as part of Marilyn Preston Graves's consignment to a private collector, went under the hammer once again, this time through Christie's in a Geneva sale, it fetched $1,763,780, more than double the pre-sale estimates and nearly a 300 percent gain over its sale five years earlier. This time the Patek Philippe Museum held the winning bid. A month later, when the Packard dual-time pocket watch (no. 190757) appeared at Christie's, fourteen years after it first sold at Sotheby's, Patek Philippe swooped down and bid $59,750, winning the gold and silver timepiece, paying more than three times what it had first earned under the hammer. Two years earlier, the watchmaker scored another victory when it toppled the bid for Graves's exceptional silver desk clock at an Antiquorum auction in Geneva for 1,103,500 Swiss francs ($729,297). Alan Banbery would deliver yet another coup de grâce, obtaining the clock's near identical twin, first made for James Ward Packard in 1923, in a private transaction.

The journey of Packard's Paperweight to the Patek Philippe Museum had a bit of kismet. Before she died in 1960, Elizabeth Packard had asked her niece Katherine to donate the desk clock to a museum. But her nephew Dick Gillmer had long had his eye on the spectacular silver and gold embellished clock and told the estate's executor that he wanted it. Outside of her diary, where Elizabeth expressed her abject disappointment with Gillmer, she made it clear that she intended to cut her nephew out of her will. Unfortunately, there were no legal instructions, and a small tiff broke out, with Katherine refusing to relinquish the clock. One of Mrs. Packard's executors stepped in with a solution. He suggested the two flip a coin for it. When Katherine refused, the executor sold it to Dick for $250, perhaps a tenth of what Ward had paid for it in 1923.

Both the Graves and Packard Paperweights now sit in the Patek Philippe Museum in Geneva.

On November 14, 2005, buoyed by the manic prices going for Graves pieces, the private collector who had first acquired the gold Graves pocket watch with twelve complications (no. 174961) consigned Christie's to sell it. Again the Patek Philippe Museum won the sale, paying $1,980,200, nearly triple what it fetched six years earlier. "An absolutely new chapter in the market price for watches and the Graves name had opened," declared Aurel Bacs.

A year earlier, during Christie's spring watch sale on May 18, the Jules Jürgensen sold in 1999 had also resurfaced. A private European collector snapped it up for 167,300 Swiss francs ($130,724). Perhaps the least technically important of the cache found hiding under Marilyn Preston Graves's bed had jumped 481 percent from its prior sale.

Others were stirred to test the market. After sitting undisturbed in a bank vault for twenty years, Graves's platinum tourbillon (no. 198311), which was sold to a private collector at "The Art of Patek Philippe" sale in 1989, was back on the market at Christie's fall watch sale held on November 17, 2008, in Geneva. It hit the market just as the global economy began its descent into the slough, yet it still fetched $821,910, a more than 200 percent increase over its first sale.

Patek Philippe's aggressive pursuit soon delivered the greatest number of Graves watches and, outside of the anonymous collector who snapped up the majority of the American Watch Institute's Packard pieces, perhaps the largest number of the automaker's watches as well.

Spread over four floors and spanning five centuries, the Patek Philippe Museum featured the Graves and Packard pieces on the first floor, surrounded by more than one thousand watches. The Sterns' collection of antique pieces dating back to the sixteenth century occupied the second floor.

Each of the Graves and Packard collections sat inside of its own custom-made eucalyptus and glass case, presented to its best and most glittering advantage, just steps apart from the other. The showstopper, the Packard, shimmered under the soft lights with its five hundred stars rotating in the celestial chart directly across from the Graves pieces. Underscoring the heft of these two collectors in historical context, Philippe Stern's concession from the anonymous bidder, the mysterious figure who changed the market with the single bob of a paddle, was that arrangements were made to temporarily loan the Supercomplication to the museum. There collectors, tourists, and hobbyists could see for themselves the spoils of a most uncommon victory under bulletproof glass, just a few feet away from the watches of the man who started it all with the first tick. "Hence at the beginning of the 20th century, for Patek Philippe the supremely complicated watch was put at stake in a fascinating duel between two gentlemen," reads the accompanying plaque. "These two masterpieces are even today among the most complicated watches ever to have been made."

Yet the trail of the Packard pieces that had elevated the Graves collection had grown cold through the years. Every so often one would pop up, usually in serendipitous fashion. In 2003 a gold Wittnauer pocket watch owned by Packard went up on eBay. The seller, a septuagenarian dairy farmer living in a trailer in Chautauqua, New York, where Ward had maintained his summer home, claimed that he scoured local flea markets for antiques and collectibles and then sold them on the online auction in order to help pay for his health insurance. Under a seller's handle he advertised a "Repeater Gold Watch John [sic] Packard Auto Giant." The cuvette of the eighteen-carat gold Wittnauer pocket watch was engraved *James Ward Packard, November 5, 1916, Warren Ohio*. The back case featured the automaker's monogram in stylized blue enamel, with the *P* chipped. The repeater setting was broken and it needed a cleaning. It sold for $1,576.51.

Four years later the retired dairy farmer turned up Packard's eighteen-carat gold Patek Philippe wristwatch with iridium hands and engine-turned cushion case. The owner described the watch as "one of a kind." Again he put it up for sale on eBay, where he incorrectly stated that most of the auto tycoon's collection remained on exhibit in the Smithsonian. Among the six bidders, the winner came in at $47,100; it was the Patek Philippe Museum. After the sale, the Museum dispatched a representative to this small, westernmost corner of New York to pick up the watch, examine it for authenticity, and hand over a check.

A few Graves pieces also trickled in from other hidden corners. In early 2009 a Connecticut couple contacted the New York office of Antiquorum. They had come into possession of a cushion-shaped platinum Patek Philippe minute-repeater wristwatch crafted in 1927, with the movement no. 198095. The watchmaker had made only a very limited number of such wristwatches in this period, perhaps half a dozen, and Henry Graves, Jr., owned at least two of them. After a dealer they found on the Internet offered them $30,000, the couple decided to try their luck with an auction house. Julien Schaerer, the watch director at Antiquorum USA, inspected the wristwatch and found that it had a broken crystal and missing minute hand. When Schaerer turned it over, to his shock he saw the unmistakable Graves emblem with the motto *Esse Quam Videri*. Antiquorum checked with the Patek Philippe archives and learned that Graves had acquired the wristwatch on August 16, 1928, during his extended European summer. It went under the hammer in March 2009 and sold for $630,000.

A year later an old Main Line WASP who described himself by saying, "I used to be on top of Wall Street until I found myself at the bottom of a bottle of gin," decided to see how much he might get for a gold Patek Philippe pocket watch manufactured in the 1920s. Discovered among the discarded contents left in a newly purchased Manhattan apartment, the watch was given to him as a gift, but the gentleman had tucked it away and forgotten about it for years. Several known watch dealers in New York City offered him an appraisal. One, he said, gave it a value of $8,000, another between $10,000 and $12,000. None of them made any mention of the significant crest emblazoned across the back, and he decided to do some investigating on his own. He typed the words *Esse Quam Videri* and *Patek Philippe* on his computer and the Internet search spit out links to the sale of the Graves Supercomplication. Floored after reading that the stunner had fetched over $11 million, he finally took the piece to Christie's. First manufactured in 1913, the chronometer with a thirty-six-hour up-and-down reserve indicator had the movement no. 177483 and had earned a Geneva Observatory prize in 1920. The watch was purchased on October 14, 1921, and afterward the family crest was engraved on the case. It appeared that another of Graves's watches had turned up quite unexpectedly.

Although not a highly complicated piece, the period of the watch certainly encompassed Graves's collecting years, and it possessed a first prize from the Geneva Observatory timing competition that he had ardently pursued. The luxury apartment in which the pocket watch turned up had once been owned by Duncan Graves. Christie's put the watch on the cover of its catalogue for the New York "Important Watches" sale on December 14, 2010, where it sold for $242,500.

Many Graves and Packard timepieces were known, but considerably more remained mysteries. With a reference and case number, one gained admittance to that history, but the converse was not true. The name of an owner did not necessarily unlock the secrets of a watch's past. Without a watch's specific numbers, one ended up searching for treasure without a map.

While the Graves and Packard watches denoted a high level of prestige to those collectors privileged enough to obtain one, for a watchmaker to have manufactured timepieces for this storied pair conferred a validation of a different sort. Throughout history, a watchmaker's name was made on the timepieces it produced, but its reputation was established by the patrons who favored it.

Since the majority of those Graves and Packard watches that had surfaced over the years were manufactured by Patek Philippe, most had come to believe that the two men had battled it out over their *grandes complications* at the seat of that firm. It was largely unknown to the contemporary horology market that both men, though faithful clients of Patek, had in fact acquired timepieces from other watchmakers. And the watchmakers of Switzerland had long wished to discover that one or more of these long-forgotten timepieces might come to light at market, offering the kind of invaluable endorsement, like that of a royal warrant, that a Graves and Packard patronage implied.

In 2006 a sheaf of yellowing letters was discovered deep in the archives of Vacheron Constantin on the Quai d'Ile in Geneva. Lying undisturbed for some eighty years, the correspondence had been conducted between Henry Graves, Jr., and the watchmaker. Written during Henry and Florence's European idyll, the letters spanned the summer and fall of 1928, during which time Henry communicated his desires for a particular Vacheron Constantin pocket watch, culminating in a meeting at the Hôtel de Crillon in Paris with Charles Constantin. It was there that he received his tourbillon chronometer, awarded the First First Prize at the Geneva Observatory.

The letters unearthed tangible confirmation that Graves had strayed into the orbit of another Swiss watchmaker, even as Patek Philippe was in the midst of producing his horological tour de force, the Supercomplication. Short of the actual timepiece, it was just the kind of evidence that Swiss watchmakers had hoped for, and it was just the kind of news known to ignite the market. For Vacheron Constantin, it was incontestable proof that Henry Graves, Jr., had patronized its *maison* on the Quai d'Ile.

As it so happened, a curious story regarding a Vacheron Constantin timepiece had emerged four years earlier. A spectacular eighteen-carat gold pocket watch, movement no. 415946, equipped with ten complications, including a tourbillon split-second perpetual calendar, with an enamel face and manufactured in 1932, was sold at Sotheby's in New York on October 22, 2002. The *grande complication* had earned first prize at the Geneva Observatory in 1934 and five years later was presented at the Swiss National Exhibition in Zurich. It was part of the Esmond Bradley Martin estate. A grandson of the Pittsburgh steel magnate Henry Phipps, Martin was a brilliant chess player, wildlife conservationist, orchid breeder, and one of the most important and prodigious watch collectors of modern times. At Sotheby's the piece realized $295,000.

Following the auction, however, the trail of the watch was said to have taken a murky and somewhat controversial turn. Within weeks the timepiece was flipped among dealers, according to one watch expert, with the price doubling at each sale, before it was finally sold for a reported $2 million to Vacheron Constantin. Almost immediately there were questions about this odd turn of events; chief among them was why the price for this watch had jumped so remarkably high in such a short time. According to this watch expert, a story soon spread that the pocket watch had been part of Henry Graves, Jr.'s collection, which only led to more questions. Why didn't Vacheron purchase the piece at auction? Did someone in the very small world of horology have information about the watch's origins that hadn't been shared at the time of sale? Though the timepiece was indeed superlative, the Sotheby's auction did not describe its provenance as being part of the Graves collection, maintaining that it was never given such information. Furthermore, the expert who said that he examined the watch before the auction did not recall seeing the Graves coat of arms. A stunner in its own right, had the watch been pronounced as one of Graves's, the bidding at auction would have pushed the price well beyond the $295,000 it fetched.

Certainly the news that Vacheron Constantin had discovered a previously unknown Graves watch elicited a great deal of excitement. It was the Graves provenance, according to one of the biggest watch dealers in the world, that brought about the astonishing price tag just weeks later. For Vacheron it was a horological triumph; one of the most important watch collectors of the twentieth century had patronized this august watchmaker. It was said that the *maison* planned to display the piece in the small museum maintained at their flagship salon in Geneva.

The news also prompted a measure of scrutiny. The Graves crest became a point of contention and dispute. Within weeks after it had exchanged several pairs of hands the famous eagle and *Esse Quam Videri* were rumored to have appeared on the case and then disappeared. Another major expert who attended the sale did not recall seeing the coat of arms, nor did his own extensive records note the connection between Graves and this particular watch. As speculation mounted, there were rumblings that specialists were called in to authenticate the watch, only to have the piece quietly withdrawn and the matter dropped. One of the watch experts mulling over the matter later posited that the process had been manipulated in order to justify its $2 million price tag.

When asked if the watch had indeed been one of Henry Graves, Jr.'s, a representative for the *maison* offered the following reply: "This piece was destined for Henry Graves, Jr., but Vacheron Constantin finally kept it for communication/promotion purposes. The watch was sold to a private collector in the [19]50s." The *maison* also noted that this particular pocket watch, now part of a private collection, did not contain Graves's famous coat of arms.

With the matter seemingly unresolved, wrapped inside the puzzle surrounding this particular watch was yet another mystery. Henry Graves, Jr., did indeed purchase at least one remarkable pocket watch from the house of Vacheron Constantin, the gold Observatory winner (movement no. 401562), in 1928. However, this pocket watch remains at large, its whereabouts unknown. Should it ever emerge, if this episode is any indication, it is sure to attract an enormous purse.

CHAPTER TWENTY-TWO

≈

The Final Gavel

On October 25, 2010, the local Tennessee newspaper the *Chattanoogan* ran the obituary of an eighty-four-year-old man who had died the day before. A decorated veteran, he had served in both the Korean War and World War II, earning the rank of first lieutenant. Raised in Detroit, he graduated from Michigan State University and went to work as an engineer with the Spun Steele Division in Birmingham, Michigan. A father of three, he enjoyed traveling the world, sailboat racing, alpine skiing, and fox hunting. It was not his many laudable accomplishments that rated the column space, however; it was his name: Warren Packard III. He was the grandson of William Doud Packard and the great-nephew of James Ward Packard.

Fragile and in ill health for quite some time, he had delighted Packard motorcar enthusiasts just a month earlier when he traveled to Warren, Ohio, accompanied by his nurse, to attend the reopening of the National Packard Museum. While saddened by the news of his passing, few outside a fringe of interested parties made much of his death. This soon changed.

In the course of going through Warren Packard III's estate, his family uncovered four pocket watches. The first pair, one in silver and the other gold, were simple pieces made by the American Watch Company during the late 1800s, with personalized engravings on the cuvettes, lovely family trinkets but not valuable heirlooms.

The second pair, however, were spectacular: a Patek Philippe and a Vacheron Constantin, the watches Elizabeth Packard had given her nephew Warren Packard on July 26, 1928, after her husband's death. They were the very watches that, after Warren Packard died in a plane crash the following summer, had passed to his son Warren Packard II, and finally to his grandson. In mint condition, they had remained in their fitted presentation boxes, the Patek with a spare mainspring and crystal and the Vacheron accompanied by handwritten instructions. Ninety-two years after their manufacture, the pieces had hardly been touched.

• • •

As was becoming clear with each new find, the Packard watches originally bequeathed to the Horological Institute of America represented only a portion of the auto magnate's collection. Previously unknown and unrecorded in the literature, these watches had never changed hands outside of the family, nor had they come up for sale. Packard acquired the two, which were completed in 1918 and 1919, during his most prolific collecting period. The Patek Philippe (movement no. 174907) was one of the only known minute repeaters made by the firm to feature a power reserve and an uncommon eighteen-carat gold Murat-style case. Of the two, the Vacheron Constantin (movement no. 375551) was even rarer. Encased in twenty-carat gold and featuring a gold dial, it was the only timepiece with a trip repeater, *grande* and *petite sonnerie,* chronograph, and half-quarter repeater that the watchmaker is known to have crafted in its history.

Shortly after the New Year, a representative for the family inquired at the major auction houses about selling the watches. Ultimately Christie's landed the consignment and decided to include the pieces in its "Important Watches" sale set for June 15, 2011, in New York. The watch specialists came up with estimates for the pocket watches. Christie's placed the two American Watch Company pieces between $2,000 and $3,000. The Patek Philippe was given an estimate between $200,000 and $400,000, while the Vacheron Constantin was bracketed between $250,000 and $500,000. Insiders believed the pieces would immediately blow through these numbers. After all, six months earlier, when the previously unknown Graves pocket watch unexpectedly emerged from one of the former apartments of his son, it had quadrupled its median estimate to more than $200,000. While lovely, that piece had more provenance than rarity. The Packard timepieces, on the other hand, were extraordinary, and over the previous twenty years fewer than a handful of the automaker's watches had come on the market. Moreover the superlative Vacheron Constantin was the only example of the watchmaker's from Packard's considerable collection ever to go under the hammer.

The final decade of the twentieth century had ended with the spectacular sale of the Graves Supercomplication. The first decade of the new millennium would close with the spectacular discovery of two previously unknown James Ward Packard watches.

The appearance of the Packard watches happened to coincide with the global economy's worst performance since the Great Depression. In 2008, when American credit card debt hit $972.7 billion, the stock market went into free fall, dropping 778 points on September 28—the largest single point loss in history—knocking out $1.2 trillion in market value. As the venerable financial institutions Lehman Brothers and Bear Sterns imploded and the auto industry went to Congress hat in hand, asking for a bailout, vintage mechanical watch sales began another remarkable trajectory. As the year 2008 rolled into a new decade, the world looked disturbingly similar to that of 1929.

Along with those of paintings, jewels, and wine, the price of a mechanical watch seemed boundless. An Antiquorum auction of five hundred watches in New York brought in $7.2 million, a 105 percent jump over the house's initial estimates. During a Sotheby's sale of pocket watches and wristwatches held in the fall of 2010, nearly 81 percent of the lots sold, bringing in more than $5 million. In November 2009, Christie's in Geneva put up a collection of ten Patek Philippe wristwatches dating from 1938 to 1982 that realized $5.7 million, twice the amount of their presale estimate. At the end of 2010, Christie's International announced that it had ended the year with its highest annual total ever achieved for watches, realizing $91.2 million in sales. By the time 2011 drew down, Christie's would report breaking its own record, with its year-end global tally for clocks and watches reaching $115.6 million.

The mechanical watch had emerged once again as the coveted status symbol of kings.

At the June sale, it was assumed that both Patek Philippe and Vacheron Constantin would make aggressive plays for the Packard watches.

A sense of anticipation began to permeate Christie's mezzanine-floor auction room in Rockefeller Center. The sale of 416 watches was divided into two sessions, with the Packards coming up toward the middle of the first set. By now, most buyers guarded their privacy and increasingly bid via the Internet and telephone, although, as with the sale of the Supercomplication a decade earlier, the room filled up quickly, bringing in both heavy hitters and spectators lured by the "fairy-tale," as the sale's catalogue described the discovery of the two Packard watches. A hive of dealers from the Fuller Building on Madison Avenue emerged along the back wall like a firing squad. A well-known Italian dealer dressed in scruffy beard and loafers without socks took a seat close to the door. Watch specialists from rival houses showed up to view the proceedings as well.

Arriving from Switzerland, Dominique Bernaz, a striking figure in a gray pinstripe suit, chunky black eyeglasses, an Hermès belt, and his unmistakable buzz of close-cropped silver hair, folded his lanky frame into one of the middle-row seats just next to the phone bank on the left side of the room. Bernaz had run Patek Philippe's Geneva salon for seven years before taking his position as head of retail at Vacheron Constantin. His presence attested to the powerful draw of the Packard pieces. He spent the auction with his mobile phone pressed to his ear, undoubtedly connected to the *maison* on the Quai de l'Ile in Geneva.

Aurel Bacs, conducting the auction, swept his gaze over the room. Those already sitting shifted in their seats, tapping on their BlackBerrys and iPads. Several significant pieces came up early. A great many broke through their estimates, offering a prelude to the obvious showstoppers. Lot 62 set the pace. A 1965 Patek Philippe pocket watch with a cloisonné enamel dial depicting a boat on Lake Geneva opened at $42,000 and jumped to $80,000. After volleying rapidly between the room and telephone banks, it jumped to $200,000, $250,000, and north, with several pauses that broke the momentum. When Bacs called "Fair warning," the watch had reached $800,000; with the buyer's premium it came to $962,500. The room broke into long and loud applause. The bidder requested anonymity.

As the lots edged closer to the Packard pieces, fresh waves of spectators crowded into the room. Bacs made quick work of the two "nostalgia" Packards made by the American Watch Company, which each realized less than $3,000. By the time he announced Lot 99, the Patek Philippe, the overflow crowd had spilled out through the door onto the mezzanine.

Bacs began, "Ladies and gentlemen, this is the historically important Patek Philippe made for James Ward Packard in 1918." The room fell silent. Bidding opened at $100,000 and within seconds jumped to $160,000. From there the bids shot up from $200,000 to $600,000, ricocheting back and forth between three of the phones, when a new phone bidder entered and pushed the velocity forward. At $720,000, Bacs broke the tension, quipping, "I'll take it with pleasure." Another phone bidder joined, helping to ramp up the price to $820,000. With the buyer's premium, the winning paddle held by the Patek Philippe Museum, took the watch that hammered under for $986,500.

Introducing Lot 100, the highly anticipated Vacheron Constantin, Bacs told the crowd, "This is the momentous eighteen-carat gold James Ward Packard. I assure you it is the most unusual, most important Vacheron Constantin I have ever handled." The bidding opened fierce at $250,000, racing past its low-end estimate within seconds. Phone bids came in fast and furious, six coming in from China and across Europe. They carried the watch to $380,000, then $500,000, and then $550,000. When the price hit $880,000, Dominique Bernaz, his phone pressed against his ear, raised his paddle and called out "$900,000." All eyes turned to the silver-haired gentleman from Vacheron Constantin. Without breaking his tempo, Bacs chirped, "Brand new in the room!"

Then once more a phone bidder called it: $920,000. Bernaz pierced the air with his finger and countered: "$1 million." The room let out a collective sigh. One of the dealers perched along the back wall, an Orthodox Jew wearing a black skullcap, let out a small prayer, "Baruch ha-Shem." Wringing every bit of theater available to him, Bacs announced, "This is now already the most expensive Vacheron Constantin ever sold on the U.S. continent!"

Five phone bidders dropped out, leaving just Bernaz and one buyer on the phone, who brought the room back to the contest by offering $1.15 million. With a glance, Bacs acknowledged that it was Bernaz's move next. Playing back the race to Geneva over his phone, Bernaz broke the silence, "Wait, I'm on the line."

"Just don't lose the line," retorted Bacs. "You will regret it very dearly." The room roared. Bernaz came back with a bid of $1.3 million. The phone buyer returned fire with $1.35 million, and Bernaz responded with $1.4 million. "He's not giving up yet," shot Bacs as he surveyed the crowd. Neither was the collector on the phone, who threw out a $1.45 million bid. Bernaz appeared to recede into his cell phone. A hush took over, and the room hung suspended in time for several long seconds. Bernaz emerged and shouted out, "$1.5 million!" The entire back row broke into cheers. With that the phone bidder withdrew, Bacs dropped the hammer, and the room erupted in ballpark-style applause. With the buyer's premium the gold pocket watch came to $1.8 million against a $500,000 top estimate.

In September Vacheron Constantin had opened its first stateside boutique on Madison Avenue in New York, and, to celebrate the acquisition, the watchmaker launched a month-long "American Heritage of Vacheron Constantin" exhibit there. It displayed the Packard along with several pieces from its Geneva museum and those never before seen by the public, including the *grande complication* created in 1935 for King Farouk of Egypt.

Vacheron Constantin had their Packard. A Graves remained out there somewhere.

Then the nearly unthinkable happened. Henry Graves, Jr., came roaring back once again to best his rival. It appeared that, at auction, this gentleman's duel had an afterlife.

On April 17, 2012, ten months after the remarkable pair of Packard watches had been discovered, Sotheby's announced it would auction off thirteen previously unknown and unpublished watches owned by Henry Graves, Jr. The largest assembly of Graves pieces ever to go on the block in history, all Patek Philippe's, they included his gold coin-form watch, an eighteen-carat gold Tonneau-shaped minute-repeater wristwatch, an eighteen-carat gold Observatory prizewinning pocket watch, and his massive rectangular platinum wristwatch crafted during World War II, when platinum was declared a strategic metal. For decades the watches had remained in the bedroom of his grandson Reginald Fullerton, Jr., who had wound each piece daily until the day he died, on March 15, 2012.

The discovery came nearly thirteen years after the Graves Supercomplication went under the hammer and sent a stir through the watch world. John Reardon, Sotheby's international senior specialist, had come full circle. In 1999 he had traveled the country with the Graves to show collectors; now he likened this latest episode to "finding a dozen Picassos from the finest Picasso collector sitting in his closet."

Since the sale of the Supercomplication, however, much had changed. Demand and prices for watches had exploded. On May 10, 2010, Christie's in Geneva broke the record for the sale of a wristwatch, bringing the hammer down on an eighteen-carat gold Patek Philippe perpetual calendar chronograph made in 1943 for $5.7 million. Watch auctions across the globe had soared. In 2011 the four major auction houses—Antiquorum, Bonham's, Christie's, and Sotheby's—reported $257.2 million at sales in Geneva, Hong Kong, New York, and London. Led by new Chinese collectors, the buying fervor was driven in part by the shift in global wealth to the East. In 2011 luxury watch imports jumped 49 percent in China. Collectors from Asia and Russia had become the dominant players, supplanting the Americans, the Europeans, and the Middle Eastern petro-sheiks. Characterized by their voracious acquisitiveness, the new players shared a zeal to brandish their newfound wealth with their timekeeping trophies.

Just four months earlier, on November 14, one of Graves's extremely rare platinum Patek Philippe minute-repeater tourbillon pocket watches went back on the block at Christie's in Geneva. It was the third time that this particular piece, movement no. 198311, had surfaced in twenty-two years. The Observatory prizewinning pocket watch, first manufactured in 1932, had fetched a stunning $1,273,265.

Yet the news of the unknown Graves pieces, along with forty belonging to his grandson, hitting the market sent the watch world, now accustomed to epic sales, into a lather. Called a "landmark" and "historic," the sale became one of the increasingly rare occasions that lured collectors from all over the globe to New York to make their presence known.

On the morning of June 14, Sotheby's auction room filled up quickly, leaving a number of individuals to stand at the back. Propped up by the sizzling watch market and the major timepieces on offer, the crowd seemed anything but subdued.

Sotheby's auctioneer Tim Bourne launched the sale with the first time-piece of the collection, the eighteen-carat gold Patek Philippe minute repeater that Henry had given to his son-in-law, Reginald Fullerton, on his wedding day in 1926. It opened at $28,000 and sold for $134,500. A Graves silver Patek pocket watch fetched $56,250,* nearly double its high-end estimate. The room watched with piqued interest as the telephone banks to the left of the auctioneer fought it out among their bidders and the momentum swelled with each succeeding lot. The financier's eighteen-carat gold Observatory prizewinning pocket watch, also made in 1926 and engraved with the Graves coat of arms, estimated to realize between $40,000 and $70,000, went under the hammer for $302,500.

None of the lots failed to find a buyer. With money and emotions in the mix, buyers jumped at what many viewed as perhaps the last chance to own a part of history. All of Graves's timepieces vaulted in multiples well above their estimates. The gold coin-form watch made for him brought $92,500, more than six times its $15,000 high-end estimate. The winning bid for the immense Patek platinum rectangular wristwatch, expected to go for anywhere from $30,000 to $50,000, hit $362,500.

The undisputed star of the sale was Graves's eighteen-carat gold Ton-neau-shaped minute repeater, movement no. 97589, one of three made for the financier by Patek Philippe and engraved with the Graves crest. Its mirror image in platinum, movement no. 198212, sits in the Patek Philippe Museum. One of the watchmaker's earliest minute repeaters, Patek had yet to designate a reference number for the model. Chatter among horophiles was that this was the piece to watch. The wristwatch's high-end estimate stood at $600,000, and while the auctioneer opened the sale at $450,000, the bidding gingerly climbed to $1 million. With the price rising, the bidders expanded from the telephones to the room. At $2.2 million a new phone bidder entered the contest. At $2.3 million, the room began to waver, and the bids slowed down even as the price tag soared. There was clamoring at the phone banks, hesitations, and a few false stops before the piece moved up to $2.4 million. A smiling woman positioned at the back of the room shot her arm straight ahead and called

---

*All of the prices quoted from the June 14, 2012, auction include the hammer price and the buyer's premium.

the last bid, $2.6 million. Applause broke out. When the hammer fell, the watch, including the buyer's premium, went for $2,994,500, making it one of the most expensive wristwatches sold at auction and a record for Sotheby's. In an odd twist of auction-house theater, the woman, Sabine Kegel, a watch specialist at Sotheby's rival, Christie's Geneva, had bid on the wristwatch on behalf of an anonymous client.

The entire sale brought $8,339,813, well above the $4.5 million high-end estimate, with Graves's pieces realizing $5,003,125. His grandson's collection brought in an additional $3,336,688. In 1999 Graves's exceptional pocket watch had captivated the market; more than a dozen years later it was one of his wristwatches that stirred collectors. The spectacle, over in the blink of an eye, would likely not be repeated anytime soon.

Sixty years after his death, Henry Graves, Jr., had once more bested James Ward Packard, this time at auction.

The Graves-Packard duel left a legacy of some of the most coveted objects ever crafted. The men began their contest during gilded times. When it was over the American and global economy had crashed. There would never again be patrons like Henry Graves, Jr., and James Ward Packard, two men who pushed each other to the limit while challenging the best watchmakers to go beyond the edge of what was technically possible.

But both men left behind a number of mysteries, horological buried treasures. The total number of watches made for Graves and Packard is unknown. Some remain in private hands, safeguarded as precious possessions. Others are unidentified even to their owners; they may be sitting in the drawer of some unwitting person, just waiting for the right moment to surface. If the past is anything to go by, it is highly probable that several more may show up sometime in the future. It's just a matter of time.

In 1928, just before the economy crashed and burned, Henry Graves, Jr., traveled to Geneva as Patek Philippe's watchmakers crafted his Supercomplication. A story later circulated that the financier had commissioned a variation on his magnum opus, a second Supercomplication. In illustrations that were drawn up, it too contained a celestial chart; even a movement number had been designated, 198025. Few, however, knew about the watch, and most of the particulars had been lost over the years. To date, this spectacular watch, which also contained a perpetual calendar, phases of the moon, sunrise, sunset, true and mean time, and winding indicator, has yet to surface. Until Henry's grandson died, it was thought that perhaps this piece may not have gone beyond the initial discussion phase.

A drawing of Henry Graves, Jr.'s mysterious second Patek Philippe sky chart pocket watch. Its whereabouts remain unknown. Photograph courtesy of Sotheby's, Inc. © 2012.

But when Reginald Fullerton, Jr., died, he left behind not only his grandfather's watches, but a pair of fitted Patek Philippe presentation boxes, their interior lids each concealing a certificate of origin. One was for the ship's striking pocket watch that he had received some five years after James Ward Packard took possession of his. The other held Henry's second potential Supercomplication, another sky chart pocket watch. Both boxes opened the June 14, 2012, auction, selling for a total of $56,845; however, each remained empty, offering no hint of the watches' whereabouts, and yielded nothing of Henry's secrets. Efforts to discover their location persist, but so far have led nowhere.

This may be Henry Graves, Jr.'s most enduring mystery. While a photograph of his ship's bell exists, the mystery watch remains just that, a mystery. All of those with primary knowledge have long passed. Were this second Supercomplication to be found it would be like finding a previously unknown Michelangelo. It would upend the market and introduce a new flurry of speculators and questions. Primary, of course, is why, after achieving horological immortality, Graves would commission another Supercomplication.

One interesting theory being floated offers yet another twist on both the Supercomplication and the Graves-Packard collecting joust. It begins with Graves receiving this sky chart, movement no. 198025, only to discover that Packard had acquired his luminous astronomical watch, just before the automaker died. Competitive to a fault, Graves resolved to return his timekeeper to Patek, dispatched with the instructions to craft him a watch that was "certainly more complicated than that of Mr. Packard!" The watchmakers, working with Graves, transformed the existing ébauche mechanism, evolving over several years into the piece known as the Graves Supercomplication. For now, this premise and all others dwell in the realm of speculation.

Collectors will have to wait until time decides to reveal her secrets. Most believe the pieces from the AWI's Packard collection, the Supercomplication, and those watches now sitting in the Geneva museums of Patek Philippe and Vacheron Constantin will never return to the market. And perhaps they are correct. These watches, made for the archetypes of American success, the self-made entrepreneur and the heir of a Wall Street fortune, no longer reside in the United States.

Much of the Packard collection now sits in a private room in one of the palaces of Sultan Qaboos bin Said of the oil-rich nation of Oman that borders the United Arab Emirates, Saudi Arabia, Yemen, and the Arabian Sea. The sultan, who came to power in 1970 after deposing his father in a bloodless coup, has been called "Oman's Renaissance man." As for the Graves Supercomplication, the magnificent watch came to settle in another Middle Eastern kingdom flying under a different flag, its owner, Sheikh Saud bin Mohammed bin Ali Al-Thani, a second cousin of the ruling Emir of Qatar Sheikh Hamad bin Khalifa Al-Thani, had become something of an open-secret in the upper echelons of cloistered horological circles. Named as head of Qatar's National Council for Culture, Arts and Heritage in 1995—the same year his cousin toppled his own father from the Qatari throne—Sheik Saud Al-Thani had over a decade disgorged as much as $2 billion from the royal coffers filling up five national museums in an effort to transform this desert kingdom into a preeminent cultural titan. This endeavor had in fact turned the sheikh, who also maintained a private collection, into the world's greatest collector and the emirate, the world's largest buyer of art. The sheikh's horological desires had apparently begun with the Graves Supercomplication and ended with select choice acquisitions later.

Few thought it rather unlikely that these instruments, the pinnacle of an era and the embodiments of the passions of two men, would ever leave their royal fortresses, just as they were not likely ever to return to American shores.

But if the odyssey of the two men's collections has proven anything, it is that time is a constant, and possession is not. Fortunes are won and lost, nations change, and people die. When asked whether these extraordinary timepieces might once again surface, one longtime horological veteran pursed his lips, bringing the tips of his fingers together, considered the question, and replied, "Never say never."

EPILOGUE

Never indeed. Those fateful words hit like a gong of truth the first week of November 2012, just as the watch world converged upon Geneva for the year's important fall sales. Spiking an already febrile climate, a startling piece of news made international headlines: Sheikh Saud bin Moham-med bin Ali Al-Thani of Qatar had pledged the Graves Supercomplica-tion along with an additional 240 *objets d'art* to Sotheby's to cover debts owed to the auction house. According to those suing him, the sheikh, now publicly unveiled as the anonymous bidder of the legendary watch, had during the course of the previous eighteen months left eleven auction houses and art dealers in the lurch. At Sotheby's alone he was reported to have accumulated a staggering debt of some $42 million.

The sheikh, a man that *The Art Newspaper* once described as some-one who "moved through markets like a whirlwind," was now accused of leaving a scorched trail among its glittering circles. In September, a group of three coin dealers in New York, London, and Washington, D.C., filed suit against the sheikh, accusing him of defaulting on the $19.8 million he owed (a sum that would balloon to $25 million with interest) for the prized Prospero collection of ancient Greek coins sold on January 4, at a New York auction. On November 9, London High Court judge Justice Haddon-Cave, ruled to freeze $15 million worth of Al-Thani's assets over the dispute. Putting a fine point on what some had believed of the sheikh's longtime perfidy of excess, the judge labeled his behavior, "discreditable, dishonorable and disturbing." The sheikh's lawyer Stephen Rubin denied the accusations, telling the *Daily Mail* newspaper in England, "There are no doubt timing issues which go to why he cannot pay at the moment, but that's not a reason to make a freezing order against someone."

It was stunning turn of events for the Qatari royal who had only recently emerged in spectacular fashion from the ignominy of his house arrest in 2005, following accusations that he had misled the Qatari government about the value of his decade-long spending spree, using a London dealer to create false invoices. Although stripped of his title as head of the nation's culture council, Al-Thani had quickly re-established his position as the art world's King Midas, whose bids turned everything he touched into auction house gold.

Aside from the frisson of excitement that the sheikh's most recent troubles stirred, they offered the sensational specter that horology's most prized timepiece would soon be back in play.

Al-Thani had long cut an intoxicatingly if not bewildering figure across auction houses, museums, and among dealers around the world. His acquisitions on behalf of the Qatari museums and his own personal collections encompassed an astonishing range of interests: Islamic art, eighteenth-century French furniture, photography, jewels, ancient artifacts, antique libraries, vintage bicycles, and even entire fossilized dinosaurs. His dominance brought him to best both London's Victoria and Albert Museum *and* the French government at auction. Al-Thani frequently astonished the market not simply for the volume of his purchases but for the vast sums he spent to acquire them. In 1999, the feverish collector set a world record when he purchased Gustave Le Gray's mid-nineteenth-century seascape *Grande Vague: Sète* for £507,500 ($840,370). A year later he parted with $15 million to buy 136 vintage photographs, including works by Man Ray and Alfred Stieglitz, emptying his purse of another $8.8 million to possess a complete set of Audubon's *Birds of America*. During an Islamic art sale held at Christie's in London in 2004, he dropped £901,250 ($1,615,400) on a Mughal fly whisk inlaid with rubies, 113 times above its high-end estimate of £8,000 ($14,339). On a visit to Qatar in 2004 to interview Al-Thani, a reporter for *The Art Newspaper* described warehouses packed floor to ceiling and auction tags and dealer labels still hanging from the countless number of pieces.

It was precisely Al-Thani's profligate and unpredictable spending habits that had lifted the sale of the Supercomplication to stratospheric heights in 1999. The art world's royal celebrity had simply changed the game of horology, sideswiping all known quantities of serious watch collectors in an unprecedented sale that kicked up incomparable levels of uncertainty.

Al-Thani's latest fall from grace and his mounting legal woes set the stage for the very distinct possibility that Sotheby's would once again offer the Supercomplication at auction. While the auction house remained extraordinarily tight-lipped on the situation, one prominent dealer noted that representatives had quietly asked him to offer his take on what price the treasured pocket watch might realize. After calling the sheikh an "embarrassment to Qatar," and "absolutely incorrigible," he dismissed the inquiry out of hand as "completely uncharted territory and simply impossible to speculate over."

Indeed, this bizarre turn of events created an interesting dilemma that paralleled some of the same issues that attended the Supercomplication's sale in 1999. Primarily, who might step up to buy it and where might the price land. Outside of the Patek Philippe Museum and the Stern family, with Thierry Stern, who succeeded his father as president of the watchmaker in 2009, anyone coming up with a list of known suitors with the kind of money needed to win the day would find it visibly thin. To date, no other watch had come close to the record breaking $11 million bestowed on Graves's masterpiece. Just as the auction drew a wild card in the form of Sheikh Saud bin Mohammed bin Ali Al-Thani, another turn under the hammer would likely attract someone outside of the normal solar system of watch collectors. With the market now flush with new billionaire connoisseurs from China and Russia in the game, there is also the question of whether the watch would once again command such a towering prize.

The story of the Graves Supercomplication is far from over.

ACKNOWLEDGMENTS

I began writing this book three years ago, intrigued by this horological arms race that ran the course of nearly thirty years, culminating with the Graves Supercomplication. The little-known story behind the lives and watches of Henry Graves, Jr., and James Ward Packard, shaped by obsession and fueled by passions, took hold of me completely. Soon I found myself following in the footsteps of these men across America and Europe, digging through archives, reading yellowed letters and diaries, sorting through cracked, leather-bound photo albums, uncovering hidden treasures, and peering into the microscopic world of complicated watch movements.

Part treasure hunt, part voyage of discovery, this was a journey made all the more interesting by the number of extraordinary people I had the great good fortune to meet along the way. Each, in one way or another, brought me closer to unraveling this tale. First, William Andrewes, a man for whom to call horology a passion is inadequate; it is a subject on which he holds forth as if reciting poetry. At an early and difficult stage, Will took the time to ring me with the kindest, most encouraging words. Next, John Reardon, a talented horological historian and the author of *Patek Philippe in America*; his fascination for Henry Graves, Jr., the collector was indescribably infectious. Not only did John open numerous horological doors, guiding me through the pivots and escapements of mechanical watches with patience and enthusiasm, but he also took great delight in my sleuthing.

During the course of writing this book I prevailed upon the collected wisdom housed in a number of libraries and institutions. They were an unending source of historical gems, and as a result, I was able to pore over letters and old newspaper clippings, examine everything from art catalogues to hotel menus from the 1880s, view historical photograph collections, learn about the beginnings of both electricity and automobiles, and become familiar with the manners and rituals of New York society at the turn of the century. I am eternally grateful to the dedicated librarians and curators who allowed me access to their archives, including the Brooklyn Historical Society, the New-York Historical Society, the Thomas Watson Library at the Metropolitan Museum of Art, the Brooklyn Museum of Art Library, the New York Stock Exchange Archives, the Arthur W. Diamond Law Library at Columbia University, the New York Public Library, the Brooklyn Public Library, and the Archives of the National Packard Museum in Warren, Ohio. In particular I would like to thank Sheldon Steele, executive director of the Larz Anderson Auto Museum, who showed me the Roderick Blood Collection Papers in Brookline, Massachusetts; Ilhan Citak at the Special Collections, Lehigh University Libraries, Bethlehem, Pennsylvania; Flavia Ramelli at the Patek Philippe Archives in Geneva, Switzerland; Laura Ronner and Anna Marie Sandecki at the Tiffany & Co. Archives Collection in Parsippany, New Jersey; Carlene Stephens at the Smithsonian Institution in Washington, D.C.; Jim Lubic at the American Watch and Clockmakers Institute in Harrison, Ohio; Jim Moske at the Office of the Secretary Records, The Metropolitan Museum of Art Archives in New York; Gregory August Rami at the American Natural History Museum in New York; Dorthea Sartain, curator of archives at the New York Explorers Club; and Michele Tucker, curator at the Saranac Lake Free Library in Saranac, New York.

While the primary players in this gentleman's duel have long since passed, they did leave a remarkable collection of mechanical watches in their wake. Where the formal record of their lives was thin, a number of scholars, family members, collectors, and experts generously offered their own anecdotes, thoughts, and time, and in some cases shared correspondence, diaries, photographs, and artifacts that went a long way in illuminating the men behind the names Henry Graves, Jr., and James Ward Packard. I am indebted to Terry Martin, a true Packard aficionado, who graciously allowed me to sit in his barn as he patiently restored an old Packard motorcar while I sifted through thousands of Packard Motor Car Company documents. Among the many generous souls who shared their time and knowledge, I must offer tremendous thanks to Alan Banbery, Sharon Collinson, Andrew Crisford, Helen Ebersole, Cheryl Graves, Buz Graves, Dorcas Hardy, Chris Hildebrand, Mary Hotaling, Ian Irving, Airell Jenks, Paul Johnson and the dedicated men at the Packard Proving Grounds, Wendell Lauth, Judy Meagher, Elizabeth Pyott, Betsy Solis, Myron Summers, and Wendy Shupe.

My efforts to understand the history and technology of mechanical watches was greatly aided by Aurel Bacs and Daryn Schnipper, who shared their deep understanding of horology and the auction world. I would also like to particularly thank Isabelle Ah-Tec, Blair Hance, Maria Kelly, Yianbo Liu, Julien Marchenoir, Erin McAndrew, Aude Pittard-Campanelli, Nicolas Poupon, and Darrell Rocha.

During my travels while researching this book, I always found refuge with good friends at my various stops along the way: in London, the Vassallos, Mary, Peter, Alex, Ellie, and Mira; Gedi and Offri Hampe and Chana and Naftali Arnon in Jerusalem; and in San Francisco, Kathy and David Rosenberg-Wohl, who also helped guide me through the issues of family trust law and the mazelike Graves family trusts.

The MacDowell Colony gave me the chance to immerse myself in the early writing of this book with an incomparable view of the New Hampshire woods, as well as the great luck to meet the wonderful Alvin Singleton and Mamiko Otsubo, both of whom I now call friend.

I offer thanks to Bill Saporito, who graciously sent me on completely unrelated reporting assignments for *Time* that kept me in the world of twenty-first-century journalism, even as my mind was occupied in turn-of-the-century America and eighteenth-century Europe.

For the production of this book, I would like to thank my editors: Leslie Meredith, Dominick Anfuso, and Donna Loffredo; also Jennifer Weidman, Elisa Rivlin, Kathryn Higuchi, and Leah Johanson. I am especially grateful to Marty Beiser, who first saw its potential, and my agent, Michelle Tessler, for her unflagging support.

Finally, thank you to my many lovely friends and family who provided support in a variety of instances and who listened as I excitedly described my "Nancy Drew moments" (or at least pretended to). Bettina Chavanne also translated French texts, Michelle Freedman's historical knowledge helped me with period fashion, and Duke Sherman provided digital mastery. Leslie Frishberg, Jeremy Quittner, Karyn McCormack, Esther Perman, Sue Ruopp, and Susan Scandrett kept me in laughs and encouragement. Thank you, thank you all!

NOTES

All quotations and descriptions of events not cited here or in the text are drawn from interviews that I conducted or come from personal observations and other sources. Owing to several factors, research on the watches owned and commissioned by Henry Graves, Jr., and James Ward Packard revealed discrepancies regarding dates of manufacture and/or purchase; in these cases, where possible, I used the date from the earliest archival source.

CHAPTER ONE: LOT 7

1 *Along with his wife Gerdi*: Nancy Wolfson, "Time after Time," *Cigar Aficionado*, December 1, 1999.
1 *"to go fast and take risks sometimes."*: Ibid.
4 *the brilliant seventeenth-century British clockmaker*: David S. Landes, *Revolution in Time: Clocks and the Making of the Modern World* (Harvard University Press, 2000), 219–21.
4 *First commissioned in 1925*: "Masterpieces of the Time Museum," Sotheby's, December 2, 1999; Patek Philippe, *Star Calibre 2000* (Editions Scriptar, 2000), 21.
4 *Elaborately constructed*: "Masterpieces of the Time Museum," Sotheby's, December 2, 1999.
4 *Graves paid 60,000 Swiss francs*: Ibid.
5 *King Farouk I of Egypt*: For further information on the life of King Farouk I, see Hugh McLeave, *The Last Pharaoh: Farouk of Egypt* (McCall Publishing, 1970).
6 *the cigar-chomping John Pierpont Morgan spent*: "Catalogue of the Collection of Watches, the Property of J. Pierpont Morgan" (Chiswick Press, 1912).
6 *Morgan eventually bequeathed*: "J.P. Gives $7,500,00 in Art," *New York Times*, December 18, 1917.
6 *acquired the Supercomplication on March 26, 1969*: "Setting the Standard," *Patek Philippe*, no. 8 (Autumn/Winter 2000): 73.
8 *Alan Bond shattered*: "Who Buys a Painting for $104m?," *BBC*, May 6, 2004.
9 *Bond was unable to come up with the funds*: Michael Kimmelman, "Getty Buys Van Gogh 'Irises' but Won't Tell Price," *New York Times*, March 22, 1990.
9 *the Japanese multimillionaire Ryoei Saito*: This account is drawn from Terry McCarthy, "The Last of the Big Spender," *Independent*, November 16, 1993; Carolyn Kleiner, "Van Gogh's Vanishing Act," *U.S. News and World Report*, July 24, 2000.
9 *In 1997, an anonymous bidder*: "The 30 Most Expensive Paintings of All Time," *tech-*

nology.am, November 9, 2009; Holland Cotter, "Picasso in Lust and Ambition," *New York Times,* October 23, 2008.

9 *Bill Gates purchased:* "Gates Pays More Than $30 Million for Painting by Winslow Homer—Price for Seascape Reportedly Is Record for American Art," Associated Press, May 5, 1998.

9 *In 1985 the Forbes:* Howard G. Goldberg and Adam Lechmere, "World's Most Expensive Bottle Claimed Fake as Renowned Collector Sued," *Decanter.com,* September 6, 2006.

9 *Star of the Season:* Scott Reyburn, "Diamonds Attract Funds as Largest Gem Prices Surge 76% in Year," *Bloomberg News,* July 14, 2008.

9 *the watch fetched $1,715,000:* Nancy Wolfson, "Time after Time: Patek Philippe Watches," *Cigar Aficionado,* November/December 1999.

10 *"The watch, for me":* Matthew Malone, "Time Bomb," *Portfolio,* November 18, 2007.

11 *Banbery first joined Patek Philippe:* "A Shared Passion for Exquisite Timepieces," *Patek Philippe Museum,* Tribune Des Arts, Le Tribune de Genève, undated, 8–9.

11 *Banbery purchased the one-of-a-kind Calatrava:* Nancy Wolfson, "Timekeeper," *Cigar Aficionado,* November/December 1999.

11 *In 1989, as the firm celebrated:* Philippe Stern, "The World's Most Beautiful Museum of Horology," *Patek Philippe Museum,* 4.

12 *"an epic poem to time":* "A Voyage through Watchmaking History," *Patek Philippe Museum,* 7.

13 *An Egyptian limestone:* Sotheby's sale results, "Masterpieces of the Time Museum," December 2, 1999.

13 *English brass-and-silvered mechanical clock:* Ibid.

13 *grossing $25.3 million:* "Sotheby's Sale: Property from the Andy Warhol Collection," April 23 to May 3, 1988, Sotheby's Historical Timeline.

16 *"We're out":* Laurie Kahle, "Keeping Time," *Robb Report,* February 1, 2002.

17 *the entire sale realized $28,285,050:* Sotheby's sale results, "Masterpieces of the Time Museum," December 2, 1999.

18 *"adversaries in pursuit":* Lucien F. Trueb, "Battle of the Super Collectors," Special Patek Philippe, *Watch Time,* undated.

18 *"driving force behind the complications":* Arthur Lubow, "Complicated Collectors," *Patek Philippe,* no. 12 (Autumn/Winter 2002): 36.

CHAPTER TWO: THE FIRST TICK

19 *On the evening of March 30, 1922:* Diary of Elizabeth Gillmer Packard, March 30, 1922.

19 *Measuring about one inch long:* Private collection, Patek Philippe Archives.

20 *"Ward & I were very prosaic":* Diary of Elizabeth Gillmer Packard, January 4, 1901.

21 *"If I ever marry":* Ibid.

21 *"It was a complete surprise":* "Early Morning Nuptials, J. W. Packard and Miss Elizabeth Gillmer Quietly Married Today," Warren wedding announcement, August 31, 1904.

21 *"Wedding":* Diary of James Ward Packard, August 31, 1904.

21 *a special walking stick:* Diary of James Ward Packard, June 1, 1917; Patek Philippe Archives.

22 *beautiful polished ebony cane:* Private collection, James Ward Packard diaries 1917–18, Patek Philippe Archives.

22 *the first U.S. president*: "Packard Twin Six (President Warren Harding)," *CNET*, February 18, 2011.

22 *record sales of $38 million*: "Packard Motor Car Profits," *New York Times*, November 10, 1922.

22 *a 100 percent dividend*: "Packard Doubles Stock," *New York Times*, November 21, 1922.

22 *In the Allies' air corps*: Kalamazoo (Michigan) Aviation History Museum.

25 *The DNA of several of these designs*: "Packard's First 50 Years: A Salute to a Pioneer Enterprise," *Kiplinger Magazine*, November 1949, 17.

25 *A visitor to Lakewood*: Alvan Macauley, address at the Packard Memorial Banquet, Lehigh University, June 9, 1927, published in *Inner Circle*, July 2, 1927, Special Collections, Lehigh University.

26 *"Not a dish was broken"*: Helen G. Ebersole, "The Packards in Lakewood," Village of Lakewood Archives.

26 *bought up twenty-five lakeside acres*: Joseph Green Butler, *History of Youngstown and the Mahoning Valley, Ohio* (American Historical Society, 1921), 2:233; Ebersole, "The Packards in Lakewood."

26 *The first Packard in America*: For the early history of the Packard family, I drew upon the accounts in Terry Martin, *Packard: The Warren Years* (Historical Publishing Network and the National Packard Museum, 2009), 13; Butler, *History of Youngstown and the Mahoning Valley, Ohio*, 232–34.

27 *a population of sixteen hundred people*: "Ohio History Central," Warren, Ohio, *Encyclopedia of Ohio*, online.

27 *Warren's blue skies*: For the early history of Warren, Ohio, I relied primarily on *History of Trumbull and Mahoning Counties* (H. Z. William & Bros., 1882), 1:258.

27 *"broad vision and optimistic views"*: Butler, *History of Youngstown and the Mahoning Valley, Ohio*, 233.

28 *"jolly"*: Cited in remarks made by August Parker-Smith, circa 1928, Special Collections, Lehigh University Libraries, Bethlehem, Pa.

28 *a regular churchgoer*: For Mary Doud Packard's biographical information, I drew upon material from the Archives of the National Packard Museum, Warren, Ohio.

28 *when Ward was eleven"*: James Ward Packard diaries, November 24, 1874–May 22, 1875.

29 *"Weather was very hopeless"*: Ibid.

29 *"who insisted on doing things"*: H. F. Olmstead, "Down through the Years with Packard," *Packard Magazine*, Autumn 1927, 4.

30 *Sir Humphry Davy first demonstrated*: "Humphry Davy Demonstrating the Arc Light, 1809," Collection of Dr. Bayla Singer, Rutgers University.

30 *One evening Ward persuaded his brother*: From remarks made by Alvan Macauley, Packard Memorial Banquet, Lehigh University, June 9, 1927, published in *Inner Circle*, Packard newsletter, July 2, 1927.

30 *All of Warren emptied*: Olmstead, "Down through the Years with Packard," 4.

30 *the Lee collection*: Annual Report of the Essex Institute, Salem, Massachusetts, 1913, 9.

32 *Over a twenty-four-hour period*: "Watches: These Are the Best Built in the World," *Life*, December, 23, 1940.

32 *Every part of the mechanism*: Paul Moore, "Mysteries of Time," circa 1928, Roderick Blood Collection, Larz Anderson Auto Museum, Brookline, Massachusetts.

32 *"a pencil mark on a scrap of paper"*: Ibid.

32 *"Crudeness and imperfections"*: From remarks made by Alvan Macauley at the Packard Memorial Banquet, published in *Inner Circle*, July 2, 1927.

33 *a rare eighteen-carat gold Victor Kullberg chronometer*: Henry B. Fried, *The James*

W. Packard Collection of Unusual and Complicated Watches (Horological Institute of America, 1959), 43.

34 *his first watch from Patek Philippe*: Henry B. Fried, *The Museum of the American Watchmakers Institute* (AWI Press, 1993), 58; Patek Philippe Archives.

34 *his two greatest* grandes complications: Fried, *The Museum of the American Watchmakers Institute*, 79–80; Patek Philippe Archives.

35 *the Leroy No. 1*: Patek Philippe, *Star Caliber 2000* (Editions Scriptar, 2000), 29.

36 *designated with the movement no. 198023*: Fried, *The Museum of the American Watchmakers Institute*, 76–77; Patek Philippe Archives; Patek Philippe Museum, Geneva.

CHAPTER THREE: A SHINING LIGHT

37 *he had traveled to the Philadelphia Centennial Exposition*: Joseph Green Butler, *History of Youngstown and the Mahoning Valley, Ohio* (American Historical Society, 1921), 2:233. For background on the Philadelphia Expo, I drew from reports found in the Centennial Exposition Digital Collection, http://libwww.library.phila.gov/CenCol/exhibitionfax.htm.

37 *In his room at Saucon Hall*: Special Collections, Lehigh University Libraries.

38 *"A handsome little fellow"*: Remarks made by August Parker-Smith, circa 1928, Special Collections, Lehigh University Libraries.

38 *"Design of a Dynamo Electric Machine"*: Special Collections, Lehigh University Libraries.

40 *Brooklyn in 1884 offered Ward*: There were a number of sources that were helpful in researching Brooklyn in the 1880s, including the Brooklyn Historical Society Library; Stephen M. Ostrander, *A History of the City of Brooklyn and Kings County* (published by subscription, Brooklyn, 1894); dispatches from the *Brooklyn Daily Eagle*.

40 *Joseph Pulitzer*: For an account of the role of Pulitzer and the Statue of Liberty, the National Park Service provides a summary in its digital history.

40 *Ward had traveled from Bethlehem*: James Ward Packard diaries, August 1884, Archives of the National Packard Museum, Warren, Ohio; Terry Martin, *Packard: The Warren Years* (Historical Publishing Network and the National Packard Museum, 2009), 19.

41 *nearly fifteen hundred arc lamps*: Richard Dennis, *Cities in Modernity: Representations and Productions of Metropolitan Space* (Cambridge University Press, 2008), 132.

41 *Thomas Edison installed six twenty-seven-ton dynamos*: Sean Dennis Cashman, *America in the Gilded Age* (New York University Press, 1993), 16–17; see also Theresa Mary Collins, Lisa Gitelman, and Gregory Jankunis, *Thomas Edison and Modern America: A Brief History with Documents* (Palgrave Macmillan, 2002), 17.

41 *In the first camp*: For the war on currents and the early days of commercial electricity, I drew upon the accounts found in Jill Jonnes, *Empires of Light: Edison, Tesla, Westinghouse, and the Race to Electrify the World* (Random House, 2003), and the PBS documentary *Tesla: Master of Lighting*, originally broadcast in April 2004.

41 *"Westinghoused"*: Craig Brandon, *The Electric Chair: An Unnatural American History* (McFarland, 1999), 9.

41 *A former Boston newspaperman*: For the history of Sawyer-Man, I primarily drew upon Charles D. Wrege and Ronald G. Greenwood, "William E. Sawyer and the Rise and Fall of America's First Incandescent Electric Light Company, 1878–1881," Rutgers University, GMI Engineering Institute, undated; "Mr. Edison Challenged by Mr. Sawyer," *New York Herald*, August 13, 1880; "Mr. Sawyer on the Warpath," *New York*

Times, March 26, 1882; "Sawyers Eastern Electric Manufacturing Company," *New York Times*, March 30, 1882.

42 *While Sawyer-Man proposed a plan*: "The Triumph of Science," *New York Times*, June 29, 1884.

42 *conducted their feud publicly*: "A New Electric Light," *New York Times*, October 30, 1878; "Carbon in Electric Lamps," *New York Times*, July 29, 1883.

43 *"the only institution for the insane"*: "The Sawyer-Man Electric Light," *Brooklyn Eagle*, October 27, 1884.

43 *"I took charge"*: James Ward Packard diaries, November 25, 1884, Archives of the National Packard Museum, Warren, Ohio.

43 *"I took charge of the whole factory"*: Ibid.

43 *Ward paid $4.50*: Ibid., October 13, 1884.

43 *"ideal home for gentlemen and families"*: Berkshire advert in the *Brooklyn Daily Eagle*, circa 1884.

43 *Ward traveled to Philadelphia*: James Ward Packard diaries, September 16, 1884.

43 *"scene of bustle and brilliancy"*: "A Dazzling Exhibition," *New York Times*, September 2, 1884.

43 *an instrument that measured time*: "The Electrical Exhibition," *New York Times*, September 6, 1884.

43 *"Started lamps at the mill"*: James Ward Packard diaries, August 7, 1884.

44 *Jay Gould had once summoned*: Butler, *History of Youngstown and the Mahoning Valley, Ohio*, 232. There are many accounts of Jay Gould's life, notably Edward J. Renehan Jr., *Dark Genius of Wall Street: The Misunderstood Life of Jay Gould, King of the Robber Barons* (Basic Books, 2005).

44 *"You can weigh the scrap yourself"*: Butler, *History of Youngstown and the Mahoning Valley, Ohio*, 232.

44 *the two Packard men arranged to meet*: James Ward Packard diaries, November 23, 1884.

44 *America's wealthiest man*: Among the many sources recounting the life of John Jacob Astor, see Axel Madsen, *John Jacob Astor: America's First Multimillionaire* (Wiley, 2001).

44 *epic table d'hôte*: Among the treasure trove of information contained at the New-York Historical Society is a series of period menus, including those from the Astor House Hotel.

45 *Ward accompanied his father*: James Ward Packard diaries, November 23, 1884.

45 *an enormous granite and limestone edifice*: *The Electrician and Electrical Engineer* 2 (1884–87): 272.

45 *"A noble outgrowth"*: "The New Building of the Mutual Life Insurance Company," *New York Times*, June 26, 1884.

46 *"The bride-to-be"*: James C. Young, "Maiden Lane of the Jewelers," *New York Times*, February 17, 1924.

46 *fashionably adorned with enamel portraits*: Nick Foulkes, *High Society History of America's Upper Class* (Assouline, 2008), vi.

46 *Founded by a Genevese cabinotier*: Company history of Vacheron Constantin.

47 *"The American pieces you referred to"*: Letter from François Constantin, September 24, cited in Foulkes, *High Society*, v.

47 *As early as 1847*: John Reardon, *Patek Philippe in America: Marketing the World's Foremost Watch* (Cefari, 2008), 10.

47 *Over three months Patek journeyed*: Ibid.

48 *he purchased a pair of overalls*: James Ward Packard diaries, February 17, 1885.

48 *a handful of Swiss and French watches*: Fried, *The James W. Packard Collection of Unusual and Complicated Watches*; inventory of Packard watches taken in 1928.

48 *"No establishment"*: *Illustrated New York: The Metropolis of Today,* merchant catalogue, 1888, 204.

49 *In 1888 alone*: "Our Illustrated Record of Electrical Patents," *Electrical World,* January
7, 1888, 260; Archives of the National Packard Museum, Warren, Ohio.

49 *an improved incandescent electric lamp*: Ibid.; U.S. Patent Office, No. 426.055, serial
no. 332,884, filed December 7, 1889, approved April 22, 1890.

50 *Ward was now the second superintendent*: "New York Notes," *Electrical World,* January 7, 1888, 9.

50 *"for the care and judgment"*: Ibid., 27; "The Electric Light," *Electrical World,* June 2,
1888.

50 *all but one American light manufacturer*: Norman R. Ball and John N. Vardalas,
Ferranti-Packard: Pioneers in Canadian Electrical Manufacturing (McGill-Queens,
1994), 40.

50 *The Park Hotel*: *Electrical World,* January 7, 1888, 69.

50 *County Jail in Brooklyn*: "County Money," *Brooklyn Daily Eagle,* September 21,
1888.

50 *"bring out the beauties"*: *Electrical World,* January 7, 1888, 180.

50 *429,813 lamps*: "The Electric Light," *Electrical World,* June 2, 1888, 281.

50 *Ward attended the tenth annual clam bake*: *Electrical Review,* September 1, 1888, 5.

50 *Thomson-Houston Electric Company purchased*: Wrege and Greenwood, "William E.
Sawyer and the Rise and Fall of America's First Incandescent Electric Light Company, 1878–1881," 46.

51 *$16 a day*: James Ward Packard diaries, 1890.

51 *Warren boasted linseed oil*: "Ohio History Central," Warren, Ohio, *Encyclopedia of
Ohio,* online.

51 *Ward received the patent*: U.S. Registered Patent No. 426,055, "The Packard Brothers
and the Packard Lamps," http://home.frognet.net/~ejcov/packard.html; "Our Illustrated Record of Electrical Patents," *Electrical World,* May 8, 1890, 809.

53 *Following his graduation in 1882*: Butler, *History of Youngstown and the Mahoning
Valley, Ohio,* 233.

53 *half a dozen local investors*: Ball and Vardalas, *Ferranti-Packard,* 41.

53 *"lamps of all candle power"*: Packard Electric Company advertisement, circa 1891,
Archives of the National Packard Museum, Warren Ohio.

53 *On October 6, 1890, Packard Electric held a board meeting*: Ibid.

54 *"an expert in lamp manufacture"*: *Electrical Review,* May 8, 1891 cited in ibid., 40.

55 *"as money, long experience"*: "A New Transformer," *Electrical World,* no. 23 (1894): 97.

55 *the company offered repairs*: Ball and Vardalas, *Ferranti-Packard,* 64.

55 *it would soon manufacture more lamps*: The Federal Writers' Project, *Ohio Guide*
(North American Book Distributor, 1940), 403.

55 *the first city in the country to install electric incandescent lamps*: Ibid.

55 *manufactured by Jules Jürgensen*: Invoice, Cross & Beguelin, for James Ward Packard,
November 1891.

CHAPTER FOUR: *ESSE QUAM VIDERI*

57 *On the evening of April 8, 1892*: This account is drawn from several sources including
"A Great Sale of Pictures," *New York Times,* April 9, 1892; the archives of the Thomas
Watson Library at the Museum Metropolitan Museum of Art; Annotated Catalog:
American Art Association Paintings, April 7, 8, Chickering Hall, New York, Sothe-
by's, et al.

57 *Alexander Graham Bell*: Patrick Bunyon, *All Around the Town: Amazing Manhattan Facts and Curiosities*, 2nd ed. (Fordham University Press, 2010), 234.

57 *Oscar Wilde had stood*: *The Complete Works of Oscar Wilde* (Oxford University Press, 2005), 389.

57 *a well-regarded collection of paintings*: For the art collection of Henry Graves, Sr., I drew from a number of sources, notably the Thomas Watson Library at the Metropolitan Museum of Art; the Library and Archives at the Brooklyn Museum of Art; Henry Graves, *Deluxe Illustrated Catalogue of the Notable Art Treasures Collected by the Late Henry Graves, Orange, New Jersey* (J. J. Little & Ives, 1909).

58 *Earlier, at an auction*: "The Detmold Treasures," *New York Times*, February 28, 1890.

58 Cypripedium Henry Graves Jr.: *Journal of Horticulture, Cottage Gardener and Home Farmer Journal of Horticulture Office* (Journal of Horticulture Office,1895), 80.

58 *one of the leading collectors*: George B. Warren and Thomas Benedict Clarke, *Catalogue of Antique Chinese Porcelains* (Merrymount Press, 1902), 13.

58 *"foolish extravagances"*: "In the Grand Chinese Porcelain Collection Tradition: Kangxi Porcelain," *Antiques & Fine Art*, Autumn, 2001; see also Raymond B. Fosdick, *John D. Rockefeller, Jr.: A Portrait* (Harper & Brothers, 1956).

58 *countless priceless pieces*: For the Chinese porcelains collected by Henry Graves, Jr., I relied on a number of sources, notably the Thomas Watson Library at the Metropolitan Museum of Art; the Library and Archives at the Brooklyn Museum of Art; "Property of the Estate of the Late Henry Graves, Jr., Chinese Art Jades Porcelains, Sculpture," Parke-Bernet Galleries, October 14 and 15, 1959.

58 *the sale of seventy-four canvases*: "A Great Sale of Pictures," *New York Times*, April 9, 1892.

59 *Palmer had single-handedly reshaped*: A lively description of Potter Palmer's influence on Chicago and retail can be found in the PBS broadcast "Chicago: City of the Century," *American Experience*.

59 *Also on the hunt was Stanford White*: Paul R. Baker, *Stanny: The Gilded Life of Stanford White* (Free Press, 1989); Department of Drawings and Archives of Avery Architectural and Fine Arts Library, Columbia University, New York.

59 *"rich men who did not want their names revealed"*: "A Great Sale of Pictures," *New York Times*, April 9, 1892.

59 *"two most conspicuous"*: Ibid.

59 *"The prices realized"*: Ibid.

60 *"Europe had plenty of art"*: Meryle Secrest, *Duveen: A Life in Art* (Knopf, 2004), 43.

60 *A cartoon*: Ibid., 95.

60 *The wealthy Boston philanthropist*: Ibid.

60 *As a boy of ten*: Kenneth Whyte, *The Uncrowned King: The Sensational Rise of William Randolph Hearst* (Counterpoint, 2009), 18.

60 *"In the past, dominant nations"*: Sean Dennis Cashman, *America in the Gilded Age: From the Death of Lincoln to the Rise of Theodore Roosevelt*, 3rd ed. (New York University Press, 1993), 175.

61 *"altogether extraordinary unity"*: "Graves Art Collection on Exhibition," *New York Times*, February 22, 1909.

61 *"of distinguished Colonial ancestry"*: "Henry Graves Dead," *New York Times*, September 1, 1906.

61 *In 1635 his forebears arrived in America*: Kenneth Vance Graves, *John Graves 1635 Settler of Concord, MA and His Descendants*, (K.V. Graves, 2002), part 1 of 2.

62 *In 1859 Henry Sr.*: *Proceedings of the New Jersey Historical Society* (New Jersey Historical Society, 1909), 163–64.

62 *Maxwell & Graves*: For the history of Maxwell & Graves, I chiefly used sources from the New York Stock Exchange Archives; *Finance and Industry: The New York Stock Exchange Banks, Bankers, Business Houses, and Moneyed Institutions: The Great Metropolis of the United States* (Historical Publishing, 1886), 79.

62 *"every convenience"*: Finance and Industry, 79.

62 *The railways joined*: Among the many good histories of the railroads in America is Richard White, *Railroaded: The Transcontinentals and the Making of Modern America* (Norton, 2011).

63 *With starting capital of $500,000*: "Liberty National Bank of New York," *Bankers Magazine* 91 (1915): 689.

63 *Liberty National tallied up*: Banker's Magazine 63 (1901): 435.

63 *Henry Sr. purchased capital stock*: "Manhattan Beach Hotel Company," *New York Times*, November 4, 1890.

63 *the Manhattan Beach Marine Railway*: A Complete History of the New York and Brooklyn Bridge from Its Conception in 1866 to Its Completion in 1883 (S. W. Gree's Son, 1883), 71.

63 *a Spanish émigré*: For the background and history of Atlas Portland Cement and the association between de Navarro and Maxwell & Graves, I relied primarily on the account found in Earl J. Hadley, *The Magic Powder* (G. P. Putnam & Sons, 1945).

63 *he organized the first line of steamers*: "José F. de Navarro Dead at Eighty-Six," *New York Times*, February 4, 1909.

64 *Atlas had a market capitalization of $10 million*: José F. de Navarro, *Sixty Six Years Business Record* (L. H. Bigelow, 1904), 71.

64 *"I don't believe the Central Railroad"*: Cited in a company history written for the de Navarro family.

65 *"Probably no one"*: "The Oranges of New Jersey," *New York Times*, July 1, 1894.

65 *"not a back-slapper"*: "Setting the Standard," *Patek Philippe*, no. 8 (Autumn/Winter 2000): 73.

65 *"brilliant reception"*: "In the Oranges," *New York Times*, January 4, 1892.

66 *the "dollar princesses"*: An account of this matrimonial phenomenon can be found in Cassandra Jardine, "The Dollar Princess Who Changed History," *London Telegraph*, April 3, 2007.

66 *The passel of brides*: "Duchesse Tallyrand Is Dead, Youngest Daughter of Jay Gould," *New York Times*, November 30, 1961.

66 *Gould and Boniface divorced in 1906*: Ibid.; "Marriage," *Time*, July 21, 1924.

66 *Mary Leiter married Lord George Curzon*: F. C. Kiessling, "The Heiress and the Viceroy," *Milwaukee Journal*, March 8, 1978.

66 *Florence Isabelle Preston*: For Preston's family history, I drew upon *The New York Genealogical and Biographical Record*, vol. 53 (New York Genealogical and Biographical Society, 1922); Daughters of the American Revolution; John S. Wurts, *Magna Charta*, part 5, *A Collection of Baronial Pedigrees* (Brookfield, 1946), 1233.

68 *"At least 1,500 cards"*: "The Social World," *New York Times*, January 21, 1896.

68 *New York had become*: "The New York Produce Exchange," *New York Times*, September 22, 1901.

68 *had speculated heavily*: "A Heavy Failure," *New York Times*, February 10, 1882.

68 *"unable to meet our engagements"*: Ibid.

68 *"Few subjects have recently arisen"*: "The Margins to Be Paid Back," *New York Times*, February 18, 1882.

69 *The family descended*: Wurts, *Magna Charta*, 1223.

69 *a $2.5 million dowry*: Arthur T. Vanderbilt, *Fortune's Children* (HarperCollins, 1991), 173.

69 *Reverend Dr. John Wesley Brown*: "A Day's Weddings, Graves-Preston," *New York Times*, January 22, 1896.

70 *"costly and numerous"*: "Weddings in Society," *New York Times*, October 5, 1882.

71 *"palace of jewels"*: Christopher Gray, "Before Tiffany & Co. Moved Uptown," *New York Times*, July 6, 2006.

71 *In addition to receiving royal appointments*: George Frederic Heydt, *Charles L. Tiffany and the House of Tiffany & Company* (Tiffany & Co., 1893), 39–40.

72 *Henry purchased a magnificent*: Deed of Shadowbrook between Charlotte C. Graves and Henry Graves, Jr., July 25, 1900, State of New York, County of Westchester.

72 *a wealthy Irishman of unknown origins*: Lisa Kreisberg, "A Sumptuous Relic, Castle for a Musical King," *New York Times*, February 13, 1977; deeds, State of New York, County of Westchester.

73 *"Well, I must arrange my pillows"*: Randy Nelson, *The Almanac of American Letters*, (William Kaufman Inc., 1981) 179.

73 *In 1926 the composer*: Kreisberg, *New York Times*, February 13, 1977.

73 *Stan Getz and his wife*: Ibid.

74 *the scandalous divorce*: "Divorce for Robert Graves," *New York Times*, December 29, 1900.

CHAPTER FIVE: MR. PACKARD'S HORSELESS CARRIAGE

77 *On Christmas morning in 1899*: Mary Doud Packard diaries, 1899, Archives of the National Packard Museum, Warren Ohio.

77 *Warren amused his grandmother*: Ibid.

78 *he purchased the boat*: Beverly Kimes, *Packard: A History of the Motorcar and the Company* (Automobile Quarterly, 2005), 23.

78 *"Took launch engine apart"*: James Ward Packard diaries, 1892, Archives of the National Packard Museum, Warren, Ohio, cited in ibid., 23.

79 *drew up blueprints for a carriage*: H. L. Olmstead, "Down through the Years with Packard," *Packard Magazine*, Autumn 1927, 4.

79 *President Grover Cleveland turned to J. P. Morgan*: Daniel Alef, *America's Greatest Banker* (Titans of Fortune Publishing, 2009), 6.

79 *a Canadian manufacturing subsidiary*: Norman R. Ball and John N. Vardalas, *Ferranti-Packard: Pioneers in Canadian Electrical Manufacturing* (McGill-Queens, 1994), 44.

79 *The legal fight dragged on*: Kimes, *Packard*, 23.

80 *Westinghouse undercut its rival's million-dollar bid*: The story of Westinghouse and the Chicago Fair can be found in Hubert Howe Bancroft, *The Book of the Fair; an Historical and Descriptive Presentation of the World's Science, Art, and Industry, As Viewed through the Columbian Exposition at Chicago in 1893*, vol. 1 (Bancroft, 1895); see also Henry G. Prout, *A Life of George Westinghouse* (Cosimo Classics, 1921), 134–41.

80 *Inside the Great Hall of Electricity*: There are a number of accounts of the Chicago Expo. I drew upon several, notably *World's Columbian Exposition of 1893*, Paul V. Galvan Library Digital History Collection, Illinois Institute of Technology.

80 *"There will be no occasion for lighting a match"*: Julian Ralph, *Harper's Chicago and the World's Fair* (Harper & Brothers, 1893), 195.

81 *The striking building designed by Louis Sullivan*: *World's Columbian Exposition Official Catalogue*, vol. 7–12, Transportation Exhibits Building Annex, Special Building and the Lagoon, Dept G. Transportation Exhibit Railways, Vessels, Vehicles (W. B. Conkey Publishers, 1893), 782–84.

82 *Henry Ford*: Clay McShane, *Down the Asphalt Path: The Automobile and the American City* (Columbia University Press, 1995), 107.

82 *So did Charles Duryea*: Steven A. Riess and Gerald R. Gems, *The Chicago Sports Reader: 100 Years of Sports in the Windy City* (University of Illinois Press, 2009), 91.

82 *"ran no faster"*: Chaim M. Rosenberg, *America at the Fair: Chicago's 1893 World's Columbian Exposition* (Arcadia, 2008), 22.

82 *By then the country*: "Motorists on the Road," *New York Times*, April 8, 1928.

82 *The* Chicago Times-Herald *provided the catalyst*: Alfred Heitmann, *The Automobile and American Life* (McFarland, 2009), 30.

83 *Chugging through the wet*: Jeffrey Steele, "On the Past Track of Auto Racing," *Chicago Tribune*, September 15, 1996; "J. Frank Duryea, Auto Maker Dies," *New York Times*, February 16, 1967.

83 *more than five hundred*: Steven Watts, *The People's Tycoon: Henry Ford and the American Century* (Random House, 2006), 46.

83 *the* Horseless Age: Terry Martin, *Packard: The Warren Years* (Historical Publishing Network and the National Packard Museum, 2009), 23.

83 *Booth had hoped to enter*: Kimes, *Packard*, 24.

83 *"made a commotion among the horses"*: Albert Rufus Baker and Samuel Walter Kelley, "Horseless Carriages For Doctors," *Cleveland Medical Gazette*, May 1897, 444; "Dr. Booth's Cab in Completed Form," *Horseless Age*, July 1896, 21–22.

83 *"motoring with driving"*: "Dr. Booth's Cab in Completed Form," *Horseless Age*, July 1896, 21–22.

83 *Cowles was given a salary*: James Ward Packard diaries, May 16, 1896, cited in Kimes, *Packard*, 23.

84 *"The 'Great River' Thames"*: William Doud Packard letter to Mary Packard, October 10, 1889, Archives of the National Packard Museum, Warren, Ohio.

84 *"rather crude cars"*: Warren Packard essay, unpublished circa 1924, Roderick Blood Collection, Larz Anderson Auto Museum, Brookline, Massachusetts.

84 *the quadricycle became a familiar sight*: H. F. Olmsted, "Down through the Years with Packard," *Packard Magazine*, Autumn 1927, 5.

84 *giving it to his nephew Warren*: Warren Packard essay, unpublished, circa 1924, Roderick Blood Collection.

84 *Ford took the machine out*: Watts, *The People's Tycoon*, 40–41.

84 *the first Oldsmobile*: "Motorists on the Road," *New York Times*, April 8, 1928.

85 *by 1904 Durant was manager*: Ibid.

85 *Ward traveled to Cleveland*: James Ward Packard diaries, 1898, cited in Kimes, *Packard*, 25.

85 *Winton set the world record*: Watts, *The People's Tycoon*, 66.

85 *an 800-mile endurance run*: "Winton Motor Carriage Co.," Cleveland State University Department of History, Center for Public History + Digital Humanities.

85 *"To Cleveland tried Winton Motor carriage"*: James Ward Packard diary, 1898, cited in Kimes, *Packard*, 25.

86 *"Dispense with a horse"*: Winton advertisement, circa 1898, *Scientific American*.

86 *nearly eleven grueling hours*: Martin, *Packard*, 33.

86 *he dispatched his own foreman*: Ibid.

86 *Whenever the "motor"*: Ibid.

86 *"I must go up sometime this week"*: Letter from James Ward Packard to George Weiss, April 11, 1899, Archives of the National Packard Museum, Warren, Ohio; also cited in Martin, *Packard*, 38.

87 *"If you are so smart"*: Arthur W. Einstein and Steven Rossi, *Ask the Man Who Owns One: An Illustrated History of Packard Advertising* (McFarland, 2010), 11.

CHAPTER SIX: ASK THE MAN WHO OWNS ONE

88 *Inside the factory*: "The Name We Conjure With," *The Packard,* April 1, 1911, 2.

88 *Mary Packard invested*: Mary Packard diaries, 1899, Archives of the National Packard Museum, Warren, Ohio.

88 *"I believe that this report"*: Letter of James Ward Packard to George Weiss, April 11, 1899, cited in Terry Martin, *Packard: The Warren Years* (Historical Publishing Network and the National Packard Museum, 2009), 38–39.

88 *On June 17 Weiss sold*: Ibid., 39.

89 *a salary of $100 a month*: Ibid.

89 *"a carriage in want of a horse"*: Michael G. H. Scott, *Packard: The Complete Story* (Tab Books, 1985), 8.

89 *Ward's own experience*: Warren Packard essay, unpublished, circa 1924, Roderick Blood Collection, Larz Anderson Auto Museum, Brookline, Massachusetts.

89 *Even petrol*: "The Packard in Packard Today," *Packard Inner Circle,* July 2, 1927, Special Collections, Lehigh University Libraries.

89 *A one-cylinder, nine-horsepower engine*: Beverly Kimes, *Packard: A History of the Motor Car and the Company* (Automobile Quarterly, 2005), 34.

89 *"Early manufacturing"*: Packard Motor Car Sales Educational Course, circa 1925, Roderick Blood Collection.

90 *gray iron cylinder castings*: Arthur Einstein and Steven Rossi, *Ask the Man Who Owns One: An Illustrated History of Packard Advertising* (McFarland, 2010), 13.

90 *"greatest contributing influences"*: "Packard Sales Educational Course," February 15, 1926, vol. 3, no. 20; see also F. F. Beall, "Requirements in the Treating of Materials," *Steel Processing and Conversion,* vol. 3, no. 5 (1917): 159–62.

90 *an underwhelming 7.1 horsepower*: Martin, *Packard,* 43.

91 *until the automatic transmission was introduced*: Scott, *Packard: The Complete Story,* 2; Einstein and Rossi, *Ask the Man Who Owns One,* 13.

91 *various models would earn some 454 patents*: Einstein and Rossi, *Ask the Man Who Owns One,* 73.

91 *at least a dozen runaway horses*: Warren Packard essay, unpublished. Roderick Blood Collection, Larz Anderson Auto Museum, Brookline, Massachusetts.

92 *"First carriage got running"*: James Ward Packard diary, 1899, cited in Dates in History of Packard from Notes of J. W. Packard, Warren, Ohio, February 21, 1921, Roderick Blood Collection.

92 *"proved satisfactory in every particular"*: *Warren Tribune,* cited in Einstein and Rossi, *Ask the Man Who Owns One,* 12.

92 *the direct costs to build the first Packard*: Ibid.

92 *nearly double the annual income*: *Engineering and Contracting* (Gillette Publishing, 1922), 91.

92 *paid $1,250*: Martin, *Packard,* 51.

92 *America had more blacksmiths than doctors*: For an account of these changes at the turn of the century, see "America 1900," David Grubin Productions, *American Experience,* PBS, which contains some of the turn of the century stats.

92 *128 pioneering auto firms*: Scott, *Packard: The Complete Story,* 10.

93 *"We could allow"*: Warren Packard essay, unpublished. Roderick Blood Collection, Larz Anderson Auto Museum, Brookline, Massachusetts.

93 *"it will never"*: *Literary Digest,* October 14, 1899, cited in *Packard: A History of the Motor Car and the Company,* 92.

93 *"If God had designed"*: Packard Sales Education Course, early Packard history, circa 1925, Roderick Blood Collection.

93 *The town council passed a law*: Packard: A History of the Motor Car and the Company, 52.

94 *When the dogs scrambled*: Warren Packard essay, "Back in 1902," unpublished, circa 1924, Roderick Blood Collection.

94 *the boy had a variety of functions*: Ibid.

94 *rough-and-tumble test track*: Evan P. Ide, *Packard Motor Car Company* (Arcadia, 2003), 48–49.

94 *The invention of the steering wheel*: Warren Packard essay, unpublished, circa 1924, Roderick Blood Collection.

95 *"Solidly built"*: *Horseless Age*, May 16, 1900, cited in James A. Ward, *The Fall of the Packard Motor Car Company* (Stanford University Press, 1995), 11.

95 *"stood the test in fine shape"*: "Good Record," *Warren Daily Tribune*, May 18, 1900.

96 *no "suitable rooms to let"*: Untitled essay, unpublished, dated May 26, 1900, Roderick Blood Collection.

96 *The automobile came home in a boxcar*: Warren Packard's childhood memories recounted in the *Packard Inner Circle*, vol. 12, no. 13 (circa 1927).

96 *"new-fangled foreign contraption"*: Martin, *Packard*, 69–71.

96 *"uninterrupted whirr"*: "Society Sees Motor Show," *New York Times*, November 6, 1900.

96 *"curious contrivances"*: Ibid.

96 *if only to help alleviate*: "Cantankerous Combustion," *Petroleum Age*, March 2008.

97 *"We have no million dollar factory"*: Martin, *Packard*, 69.

97 *William Rockefeller*: Ibid.

97 *thirty-three cars worth $40,000*: Ibid., 71.

97 *A total of five Packards*: There are several accounts of the New York–to-Buffalo run, notably ibid., 77–83; "On the Report of Endurance Contest," *Automobile Topics Illustrated*, October 19, 1901.

98 *"Expositions are the timekeepers"*: From the speech by President William McKinley on September 6, 1901, at the Temple of Music, Pan Am Expo, Buffalo, New York.

98 *the organizers chose to cut the race short*: "All Pan-American Automobile Events Called Off," *Cycle and Automobile Trade Journal*, October 1, 1901.

98 *the first transcontinental race*: There are several accounts of this endurance run. I primarily drew from "Old Pacific," *Chicago Tribune*, January 29, 1922; "Fetch Recollection," June 7, 1943, Roderick Blood Collection; "Pacific to Atlantic by Automobile," *Automobile*, June 27, 1903; Marius C. Krarup, "Across America in an American Automobile," *Automobile*, July 25, 1903.

99 *"Thank God, it's over"*: "Fetch Recollection," June 7, 1943, Roderick Blood Collection.

99 *the London watchmaker E. Dent & Company*: The information regarding Packard's early commissions comes primarily from Henry B. Fried, "James Ward Packard Collection of Watches," in *The Museum of the American Watchmakers Institute* (AWI Press, 1993), 82–85, 88–91.

101 *"The Packard is a car for a patrician"*: Cited in Peter Davies, *American Road: The Story of an Epic Transcontinental Journey at the Dawn of the Motor Age* (Macmillan, 2003), 24.

CHAPTER SEVEN: EAGLE ISLAND

105 *In 1903 Levi P. Morton*: Among the biographical information available on Morton, I primarily drew upon Mark O. Hatfield, with the Senate Historical Office, *Vice Presidents of the United States, Levi P. Morton (1889–1893), Vice Presidents of the United*

States, 1789–1993 (Washington: U.S. Government Printing Office, 1997), 269–74; Thomas Lannon, *Levi P. Morton Papers, 1818–1920,* New York Public Library, Humanities and Social Sciences Library, Manuscripts and Archives Division, June 2008.

105 *purchasing it in the spring*: Eagle Island lease, 1910.

105 *"Of the many camps"*: "Wawbeek," *New York Times,* June 12, 1910.

105 *Alfred Gwynne Vanderbilt's colossal Camp Sagamore*: Howard Kirschenbaum, *The Story of Sagamore* (Sagamore First Raquette Lake, 1996).

106 *"The great forest"*: "Fine Homes Displacing Real Camps in the Adirondacks," *New York Times,* May 31, 1903.

106 *society gilded their camps*: Among the histories of the Great Camps, I drew primarily from Craig Gilborn, *Adirondack Camps: Homes Away from home, 1850–1950* (Adirondack Museum/Syracuse University Press, 2000).

106 *Maitland De Sormo wrote*: Maitland C. De Sormo, *Summers on the Saranacs* (Adirondack Yesteryears, 1980), 174.

107 *The Graveses rubbed elbows*: For background on the Great Camps on Upper Saranac Lake, I chiefly drew from De Sormo, *Summers on the Saranacs*; Gilborn, *Adirondack Camps*.

107 *He had motored up to the Saranac Inn*: "Saranac Inn, Edwin Gould Guest of Mr. and Mrs. Henry Graves, Jr., at Their Camp," *New York Times,* July 17, 1910.

107 *His main rival*: "Wawbeek; Two School Boys Upset Old River Drivers in Boat-Tilting Match," *New York Times,* September 3, 1910; "Saranac Water Carnival," *Tupper Lake Herald,* September 1, 1911.

108 *it became widely known*: "Upper Saranac Speed Boats Put in Order for the Coming Races," *New York Times,* July 9, 1911.

108 *At the pop of the starting pistol*: De Sormo, "The 'Eagle's' Great Leap," in *Summer on the Saranacs,* 267.

108 *"of which only $345,622"*: "Henry Graves's Estate," *New York Times,* March 3, 1907.

109 *Henry Graves, Sr., drew up his will*: The Last Will and Testament of Henry Graves, Sr. December 20, 1904, in *Atlantic Reporter,* "Fidelity Union trust Co. v. Graves et al. 147/282, Court of Chancery of New Jersey," 1947, 750–58.

109 *"Inasmuch as my sons"*: Ibid., 751.

109 *he was entrusted to manage*: "Lackawanna Railroad Stock Control Held by a Few Men," *New York Times,* March 4, 1914.

110 *the country's wealth quadrupled*: Nick Foulkes, *High Society: History of America's Upper Class* (Assouline, 2008), 32.

110 *the canine arrived*: Jean Zimmerman, *Love, Fiercely: A Gilded Age Romance* (Houghton Mifflin, 2012), 82.

110 *Cornelius Kingsley Garrison Billings*: "High Society's High Jinks," *Life,* January 2, 1950.

111 *an "intimate" St. Valentine's Day bal poudre*: "A St. Valentine's Dance," *Town & Country,* February 24, 1906.

111 *"They control the people"*: Louis D. Brandeis, *Other People's Money and How the Bankers Use it* (Frederick A. Stokes Company, 1913).

111 *"The man of great wealth"*: President Theodore Roosevelt, State of the Union Address, December 3, 1906.

111 *nine thousand workers went on strike*: "Ludlow Massacre," Digital Records, Colorado State Archives, http://www.colorado.gov/dpa/doit/archives/digital/Ludlow.htm.

112 *an "unreasonable monopoly"*: Standard Oil Co. of New Jersey v. United States, 221 U.S. 1 (1911).

112 *thirty rare Old Master prints*: File 2, June 15, 1920, George Coe Graves Correspondence, Office of the Secretary Records, The Metropolitan Museum of Art Archives.

112 *the entire contents of his Osterville estate*: Correspondence of Gifts, October 28, 1930, George Coe Graves Correspondence, Office of the Secretary Records, The Metropolitan Museum of Art Archives.

112 *His collection of furniture*: Art News Annual, vol. 31, item notes v1–1878, Art Foundation, New York.

112 *gave unremarkable donations*: "Red Cross Fund to Be $110,000,000," *New York Times*, June 24, 1917.

113 *donating a triangular wedge of land*: Deed of land, June 19, 1925, filed Westchester County, New York; "Gets Irving Memorial," *New York Times*, May 31, 1934.

113 *Rather than take up his father's seat*: New York Stock Exchange Archives.

CHAPTER EIGHT: THE COLLECTOR

114 *"the highest average"*: "High Prices Bid at Graves Sale," *New York Times*, February 27, 1909.

114 Sheep Shearing *topping the lot*: Ibid.

114 *Over the previous two days*: Notable Art Treasures Collected by the Late Henry Graves, February 25, 26, 27, 1909, Mendelssohn Hall American Art Galleries catalogue.

114 *"The number of prominent persons"*: "High Prices Bid at Graves Sale," *New York Times*, February 27, 1909; "Gallery Notes, Graves Art Collection on Exhibition," *New York Times*, February 22, 1909.

114 *realizing $281,452*: "High Prices Bid at Graves Sale," *New York Times*, February 27, 1909.

115 *only four were originals*: Michael Kimmelman, "Havemeyer Collection," *New York Times*, March 26, 1993.

115 *Isabella Stewart Gardner bought*: Meryle Seacrest, *Duveen* (Knopf, 2004), 213.

115 *bronze doors*: Ibid., 88.

116 *Under rather opaque circumstances*: Among the descriptions of Henry Graves, Jr.'s collecting habits, see "Henry Graves Jr. 1868–1953," in *Notable American Prints: The Collection of Henry Graves, Jr. May 7–31 1959*, Kennedy Art Galleries Catalogue, American and Naval Prints, 1959.

116 *From the Friederich von Nagler and Berlin Museum collection*: Masterpieces of Engraving and Etching Collection of Henry Graves, Jr., American Art Association, Anderson Galleries, catalogue, 1936, 38.

116 *he scooped up a brilliant selection*: "Henry Graves Jr. 1868–1953," in *Notable American Prints*.

116 *"Mr. Graves aimed for the best"*: Ibid.

116 *"That his acumen was seldom at fault"*: Ibid.

116 *Henry owned nearly two hundred paperweights*:P.J. Curry Co. Appraisers Report of the Estate of Henry Graves Jr., August 19, 1954.

116 *he owned Colonials*: Ibid.

117 *"People were largely unaware of these collections"*: "Setting the Standard," *Patek Philippe*, no. 8 (Autumn/Winter 2000): 73.

117 *"And if you're not going to buy the best"*: Ibid.

117 *Alfred G. Stein, a watchmaker*: "Alfred G. Stein Dies, Noted Watchmaker," *New York Times*, February 21, 1934.

117 *the only American juror*: John Reardon, *Patek Philippe in America: Marketing the World's Foremost Watch* (Cefari, 2008), 43.

118 *"The World's Best Watch"*: Advertisement, cited in ibid., 43.

118 *"There is a watch which"*: Advertisement, cited in ibid, 44.

118 *The Geneva Observatory contests*: William J. H. Andrewes, "Pursuing Precision," *Patek Philippe,* no. 12 (Autumn/Winter 2002): 14.

119 *"taste for purity of design"*: "Important Watches, Wristwatches & Clocks, Including Property from the Collection of Henry Graves, Jr.," Sotheby's, New York, catalogue, June 15, 1999, 86.

119 *when the average American's annual income*: Scott Hollenbeck and Maureen Keenan Kahr, "Ninety Years of Individual Income and Tax Statistics 1916–2005," *Statistics of Income Bulletin,* Winter 2008, table 1.

119 *"Croesus's" . . . four times as many millionaires*: *North American Review,* January 1929, cited in Larry Samuel, *Rich* (Amacom, 2009), 45.

120 *"plumed, pearled"*: Larry Samuel, Rich, 46.

120 *"Millionaire's Row was doomed"*: R. L. Duffus, "Upper Fifth Avenue to House the 4,000," *New York Times,* May 24, 1925.

120 *The Great War continued to rage*: Among the many significant accounts of World War I, see J. M. Winter, *The Great War and the British People* (Macmillan, 1986); Paul Fussell, *The Great War and Modern Memory* (Oxford University Press, 1975).

120 *a German U-boat*: Diana Preston, *Lusitania: An Epic Tragedy* (Bloomsbury, 2002).

120 *Equipped with* foudroyante: Henry B. Fried, *The Museum of the American Watchmakers Institute* (AWI Press 1993), 79–80.

121 *Carnegie's imposing three-story*: Christopher Gray, "A Mansion for Me, Another for My Cars," *New York Times,* June 14, 2009.

122 *"one of the most remarkable"*: "Packard Challenges Graves, through Patek Philippe," *Patek Philippe Museum* Tribune Des Arts, Le Tribune de Genève, undated, 63

CHAPTER NINE: ACCELERATION

Unless noted, details and descriptions of James Ward Packard's watches in this chapter come from a variety of sources, including Henry B. Fried, *The Museum of the American Watchmakers Institute* (AWI Press 1993); Patek Philippe Archives; Sotheby's; Christie's; private collections. As well, unless otherwise indicated, details and descriptions of Henry Graves, Jr.'s timepieces come from Patek Philippe Archives; Patek Philippe Museum, Geneva; Tiffany & Co. Archives Collection; Sotheby's; Christie's.

124 *The son of James Frederick Joy*: Pete Davies, *American Road: The Story of an Epic Transcontinental Journey at the Dawn of the Motor Age* (Macmillan, 2003), 18.

124 *"Princes of Griswold Street"*: James Arthur Ward, *The Fall of Packard* (Stanford University Press, 1995), 11.

124 *"jump in"*: Beverly Kimes, *Packard: A History of the Motorcar and the Company* (Automobile Quarterly, 2005), 52.

124 *"New Automobile Which Harry Joy Has Ordered"*: *Detroit Journal,* November 7, 1901, cited in Davies, *American Road,* 20.

124 *he sank $25,000*: Davies, *American Road,* 22.

124 *a new 32,000-square-foot factory*: Terry Martin, *Packard: The Warren Years* (Historical Publishing Network and The National Packard Museum, 2009), 109.

125 *"Joy here"*: James Ward Packard diary, July 1, 1902, Archives of the National Packard Museum, Warren, Ohio.

125 *ratcheted up his investment*: Davies, *American Road,* 22.

125 *Total capitalization*: Arthur Einstein and Steven Rossi, *Ask the Man Who Owns One: An Illustrated History of Packard Advertising* (McFarland, 2010), 19.

125 *"I want you and we all want you to be"*: as cited in Ibid., 20.

125 *Initially Ward gave his assurances*: Letter from Henry Bourne Joy to Russel A. Alger, December 20, 1928, Roderick Blood Collection, Larz Anderson Auto Museum, Brookline, Massachusetts.

125 *The pioneering industrial architect Albert Kahn*: Among the many sources on the life and work of Albert Kahn, I primarily drew from David L. Lewis, *Ford and Kahn* (Michigan History, 1980); Roger Matuz, *Albert Kahn: Builder of Detroit* (Wayne State University Press, 2002); Louis Bergeron and Maria Teresa Maiullari-Pontois, "The Factory Architecture of Albert Kahn," *Architecture Week,* November 1, 2000.

126 *Initially he told Joy*: Letter from Henry Bourne Joy to Russell A. Alger, December 20, 1928, Roderick Blood Collection.

126 *"More than one cylinder"*: Martin, *Packard,* 103.

126 *By July 1918*: Ibid., 137.

126 *"He worried himself sick"*: Elizabeth Gillmer Packard letter to Mr. and Mrs. Joy, April, 21, 1928, Roderick Blood Collection.

127 *"one of the best in the country"*: Martin, *Packard,* 97.

127 *George Weiss broke all ties to the company*: Martin, *Packard,* 119.

128 *Within ten years of moving to Detroit*: Davies, *American Road,* 23; see also Walter Boynton, "Packard Grows from 2-Acre to 75-acre Plant," *Automotive Industries,* December 8, 1920.

128 *In 1902, the year before the company moved*: Ibid., 101.

128 *By 1912 the company had orders worth $14 million*: Davies, *American Road,* 23.

128 *would not experience financial woes*: Martin, *Packard,* 149.

128 *"the connecting link"*: Davies, *American Road,* 21.

128 *At the twelfth annual Automobile Club of America banquet*: Ibid., 24.

128 *raising the minimum wage*: Steven Watts, *The People's Tycoon: Henry Ford and the American Century* (Vintage Books, 2006), 178.

128 *William C. Durant pronounced*: David Lewis, "The Geniuses Who Shaped an Industry," *New York Times,* January 25, 1976.

129 *Packard "was the most valuable name"*: "Packard," *Fortune,* January 1937.

130 *William K. Vanderbilt bought his first Packard in 1911*: "Prominent Packard Owners of Long Standing," Packard Motorcar Company, Roderick Blood Collection.

131 *At the push of a slide button*: For a description of how a minute repeater functions, see Jack Forster, "Big Time," *Forbes Life Magazine,* December 14, 2009.

131 *"matters of selection"*: Paul Moore, "Mysteries of Time," circa 1928. Roderick Blood Collection, Larz Anderson Auto Museum, Brookline, Massachusetts.

132 *"mark of sissies"*: "Watches: These Are the Best Built in the World," *Life,* December 23, 1940.

132 *Alberto Santos-Dumont*: Cartier company history.

135 *Henry purchased at least five pocket watches*: Tiffany & Co. Archives Collection.

135 *Charles Tiffany himself*: For an early history of Tiffany, I drew upon a number of sources, notably George Frederic Heydt, *Charles L. Tiffany and the House of Tiffany & Company* (Tiffany & Co., 1893).

135 *The Graves family file*: Tiffany & Co. Archives Collection.

136 *Henry's son Duncan had purchased*: Ibid.

136 *The pair wed in a lavish ceremony*: "Graves-Dickson," *New York Times,* July 7, 1918; "Miss M. A. Dickson Becomes Bride of Henry Graves, 3d," *New York Tribune,* July 7, 1918.

136 *Henry Jr. laid out $80,000*: "Westchester Buyers," *New York Times,* June 2, 1920; deed of the house, filed Westchester County, New York.

CHAPTER TEN: VALLÉE DE JOUX

For the early history of clocks and time, I drew upon several sources, notably David S. Landes, *Revolution in Time: Clocks and the Making of the Modern World* (Belknap Press of Harvard University Press, 1983); Seth Atwood and William Andrewes, *The Time Museum,* monograph (Johnson Press, 1983); Jo Ellen Barnett, *Time's Pendulum: From Sundials to Atomic Clocks, the Fascinating History of Timekeeping and How Our Discoveries Changed the World* (Houghton Mifflin Harcourt, 1999).

140   *"The clock did not create"*: Landes, *Revolution in Time,* 58.
140   *in the year AD 1086 the Chinese emperor*: There are several accounts of Su Sun and his astronomical clock, in particular, ibid., 17–19; Hong-Sen Yan and Marco Ceccarelli, paper presented at the International Symposium on History of Machines and Mechanisms, Spring 2009, 73.
140   *The caliph of Baghdad*: Thomas F. Glick, Steven John Livesey, and Faith Wallis, *Medieval Science, Technology, and Medicine: An Encyclopedia* (Psychology Press, 2009), 129.
141   *the clock tower at Norwich Cathedral*: Michael Naumann, "Rhythm of the Drum," *Patek Philippe,* no. 12 (Autumn/Winter 2002): 4.
141   *A power source such as weights*: Atwood and Andrewes, *The Time Museum,* 4.
143   *In 1370 King Charles V of France*: Patek Philippe, *Voyage to the End of Time* (Patek Philippe SA, Geneva, 1989), 17.
143   *In 1518 François I of France*: Landes, *Revolution in Time,* 87.
143   *In England craftsmen hoping to impress*: "Early English Watch Making," *New York Times,* October 3, 1897.
143   *a German locksmith named Peter Henlein*: *Britten's Old Clocks and Watches and Their Makers: Being an Historical and Descriptive Account of the Different Styles of Clocks and Watches of the Past in England and Abroad to Which Is Added a List of Eight Thousand Makers* (B. T. Batsford, 1899), 57.
144   *Mary, Queen of Scots commissioned*: Ibid., 94–95.
144   *Hans Holbein's famous portrait of Henry VIII*: Catherine Cardinal, "Costume Drama," *Patek Philippe,* no. 12 (Autumn/Winter 2002): 69.
144   *"in the likeness of an apple"*: Ibid., 72.
144   *During Queen Elizabeth's forty-five-year reign*: *Britten's Old Clocks and Watches and Their Makers,* 90–93.
146   *Bürgi built a clock*: Landes, *Revolution in Time,* 104–5.
146   *Rudolph raised court patronage to heights*: "Prague During the Rule of Rudolph II (1583–1612)," in *Heilbrunn Timeline of Art History* (The Metropolitan Museum of Art, 2000–October 2002).
146   *Among his possessions*: *Britten's Old Clocks and Watches and Their Makers,* 96.
146   *in 1714 the British government*: William Andrewes, "Pursuing Precision," *Patek Philippe,* no. 12 (Autumn/Winter 2002): 14–16.
147   *In 1764 the device known as the H4*: Ibid. A more in-depth account of the Longitude Prize can be found in Dava Sobel, *Longitude: The True Story of a Lone Genius Who Solved the Greatest Scientific Problem of His Time* (Penguin, 2005).
147   *men like the French clock maker Pierre Le Roy*: Frederick James Britten, *Former Clock and Watchmakers and Their Work: Including an Account of the Development of Horological Instruments from the Earliest Mechanism, with Portraits of Masters of the Art* (E. & F. N. Spon, 1894), 124, 229, 262, 271.

147 *In the middle of the sixteenth century*: For the early history of watchmaking in Geneva, I drew upon Landes, *Revolution in Time*; Federation of the Swiss Watch Industry history, " From the Roots Until Today's Achievements (digital version); Philip Vollmer, James Isaac Good, and William Henry Roberts, *John Calvin: Theologian, Preacher, Educator, Statesman* (Heidelberg Press, 1909).

148 *"crosses, chalices or other instruments"*: "Triumph over Tragedy," *Patek Philippe*, no. 12 (Autumn/Winter 2002): 12.

149 *While the French had introduced miniature painting*: Hans Boeckh, "Fired with Passion," *Patek Philippe*, no. 12 (Autumn/Winter 2002): 42–46.

150 *the Dutch scientist Huygens*: Lance Day and Ian McNeil, *Biographical Dictionary of the History of Technology* (Taylor & Francis, 1998), 634.

150 *Thomas Tompion*: For material on Tompion, Graham, and Mudge, I drew upon material from the British Museum; The Sir Harry and Lady Djanogly Gallery; Frederick James Britten, *Former Clock and Watchmakers and Their Work: Including an Account of the Development of Horological Instruments from the Earliest Mechanism, with Portraits of Masters of the Art* (E. & F. N. Spon, 1894), 83–97, 124–31.

151 *the French watchmaker François Pigeon*: Thomas Lips, "Wheels within Wheels," *Patek Philippe*, no. 12 (Autumn/Winter 2002): 54.

151 *One of the earliest iterations*: This account of Madame de Pompadour's watch is drawn from Martin Huber and Alan Banbery, *Patek Philipp Genève*, 2nd ed. (Andreas Huber, Munich, 1993), 40–41.

151 *Chief among them stood Abraham-Louis Breguet*: Among the wealth of material on Breguet, I drew upon Sir David Salomons, *Breguet* (Printed for the author, London, 1921); George Daniels, *The Art of Breguet* (Sotheby Parke-Bernet, 1975).

152 *"Custom ordains that watches"*: C. Montgomery M'Govern, *The American Watch Industry*, edited by Andrew Carnegie (Hall and Locke Co., 1911), 336–46.

152 *In 1876 the American watchmaking industry*: For history and background on the American watch-making industry, I drew upon "A Short History of American Watch Making," *Elgin*, October 4, 2002; Landes, *Revolution in Time*, 308–37.

153 *A horrific crash*: "1891 Train Wreck Led to Railroad Watches," *News and Courier* Charleston, S.C., January 25, 1958.

153 *a global marketplace*: John Reardon, *Patek Philippe in America* (Cefari, 2008), 12–14.

153 *a stunning pocket watch with eleven complications*: For this account of the Nostitz watch, I primarily used "Leroy No. 1," 16–18, Les Ateliers L. Leroy SAS, undated, Musée du Temps, Besançon, France.

154 La Merveilleuse: Gil Baillod, "Locomotives of Industrialization," *Watch Around* 6 (Autumn 2008–Winter 2009): 34.

154 *introduced the first pocket minute repeaters*: Patek Philippe, *Voyage to the End of Time*, 21.

154 *"A watch was only worth anything"*: Cited in ibid., 55.

154 *"smallest watch ever constructed"*: Ibid., 21.

155 *a blue enamel and rose-cut diamond pendant watch*: Reardon, *Patek Philippe in America*, 11.

156 *hoping to impress Marie Antoinette*: Alix Kirta, "Marie Antoinette: The Queen, Her Watch and the Master Burglar," *Daily Telegraph*, London, April 24, 2009.

157 *until the Leroy No. 1 was*: Patek Philippe, *Voyage to the End of Time*, 18–19; "Leroy No. 1," 16–18, undated, Les Ateliers L. Leroy SAS.

157 *"Create a watch that brings together"*: "Leroy 01: One of the most complicated watches in the world," *Horozima*, May 18, 2010.

CHAPTER ELEVEN: A GENTLEMAN'S WAR

The descriptions of James Ward Packard's watches in this chapter come from the Patek Philippe Archives; Patek Philippe Museum, Geneva; Martin Huber and Alan Banbery, *Patek Philippe Genève*, 2nd ed. (Andreas Huber, 1993); Henry B. Fried, *The Museum of the American Watchmakers Institute* (AWI Press, 1993); Henry B. Fried, *The James W. Packard Collection of Unusual and Complicated Watches* (Horological Institute of America, 1959); private collections.

The descriptions of Henry Graves, Jr.'s watches come from the Patek Philippe Archives; Patek Philippe Museum, Geneva; Huber and Banbery, *Patek Philippe Genève*; Sotheby's; Christie's; Antiquorum; Tiffany & Co. Archives Collection.

159 *The building, completed in 1890*: Huber and Banbery, *Patek Philippe Genève*, 24–26.
159 *In 1901 the company reorganized*: Patek Philippe, *Voyage to the End of Time* (Patek Philippe SA, Geneva, 1989), 67.
160 *"the opportunity to apply the skills"*: Ibid., 22.
161 *"First one, then the other"*: *Masterpieces of The Time Museum*, Sotheby's catalogue monograph, Alan Banbery, consultant and curator of the Patek Philippe Museum, Geneva, September 1999.
162 *the pair requested exclusive serial numbers*: Michael Korda, *Marking Time* (Barnes & Noble Books, 2004), 143.
164 *In a dictionary on*: Abraham Rees, *Universal Dictionary of Arts, Sciences and Literature* (Longman Hurst, 1820), "Clocks Watches, and Chronometers 1819–1820."
164 *Another observer visiting the Patek workshops*: Cited in John Reardon, *Patek Philippe in America* (Cefari, 2008), 22.
165 *Commodore Vanderbilt had gifted his son*: Korda, *Marking Time*, 19.
165 *Vacheron Constantin delighted King Fuad I of Egypt*: Timmy Tan, "Innovative Profits," *TimeWerke*, 338–41, online, http://www.thehourlounge.com/upload/article/file/Press%20Corner/338–347.pdf
165 *In 1963 the city of Berlin*: Kathleen McLaughlin, "Watches Form Timeless Display," *New York Times*, September 29, 1967.
165 *Adolf Hitler had rewarded*: Korda, *Marking Time*, 19.
166 *"To carry a fine Breguet watch"*: Sir David Lionel Salomons, *Breguet* (Printed for the Author, London, 1921), 5.
166 *"Evening after evening"*: "The Centenary of Breguet," *Jeweler's Circular*, November, 15, 1922.
166 *For the story of the Marie-Antoinette, I drew upon the accounts of Danny Rubenstein*, "Hickory Dickory Dock," *Haaretz*, November 11, 2007; Alix Kirsta, "Marie Antoinette: The Queen, Her Watch, and the Master Burglar," *Telegraph*, London, April 24, 2009.
167 *In 1910 J. P. Morgan purchased*: "Famous Watches Bought by J. P. Morgan," *New York Times*, April 24, 1910.
167 *The coal-mining baron*: "Watch Collection to Go at Auction," *New York Times*, December 18, 1927.
167 *Wheeler collected precisely*: "Historic Watches Put on Exhibition," *New York Times*, June 26, 1916.
167 *It might take him years to persuade*: "Millions in Watches Left by Suicide," *New York Times*, February 2, 1923.

CHAPTER TWELVE: TIME STOP

168   *"Your teeth need"*: Colgate advertisement circa 1923.

168   *Ford slashed automobile prices*: Philip Brown, Hugh Lauder, *Capitalism and Social Progress: The Future of Society in a Global Economy* (Palgrave Macmillan, 2001), 29.

168   *Before World War I*: U.S. Census Bureau and U.S. Agriculture report, February 8, 1988.

169   *Brooks Brothers*: Brooks Brothers Motor Section catalogue, 1907.

169   *"We now have music by machines"*: Willa Cather and L. Brent Bohlke, "Willa Cather in Person: Interviews, Speeches and Letters," in *Lincoln Evening State Journal* (University of Nebraska Press, 1990), 148.

170   *"To gain and hold the esteem of men"*: Thorstein Veblen, "Conspicuous Leisure," in *The Theory of the Leisure Class*, 1899, online.

170   *Seventy-five percent of all radios*: James Quinn, "The Roaring Twenties," *Economics Depression within a Depression, The Casey Report*, October 22, 2010.

170   *In 1923 when four million new cars*: Gary D. Best, *The Dollar Decade: Mammon and the Machine in 1920s America* (ABC-CLIO, 2004), 74.

170   *"When the making of millionaires"*: *Pittsburgh Gazette Times*, cited in Larry Samuel, *Rich* (Amacom, 2009), 41.

170   *bank deposits in New York totaled $1 billion*: Samuel, *Rich*, 42.

170   *sixty-three banks and trust companies held deposits*: Ibid.

171   *60 percent of the population still lived below the poverty line*: There are a number of sources discussing the economics of the Great Depression, notably John Kenneth Galbraith, *The Great Crash 1929* (Houghton Mifflin Harcourt, 2009); Robert Samuelson, "Great Depression," in *The Concise Encyclopedia of Economics, 2002*, online.

171   *Farm prices*: L. J. Norton, "The Land Market and Farm Mortgage Debts, 1917–1921," *Journal of Farm Economics* 24, no. 1 (1942): 168–77; Melissa Walker, *Country Women Cope with Hard Times: A Collection of Oral Histories* (University of South Carolina Press, 2004).

171   *the "family" bank*: "$30,000,000 Merger of N.Y. Banks Near," *New York Times*, July 14, 1920.

171   *"one of the most powerful"*: "Big Banking Merger at Informal Stage," *New York Times*, July 15, 1920.

171   *exceeded its resources*: "$30,000,000 Merger of N.Y. Banks Near," *New York Times*, July 14, 1920.

172   *On the evening of February 28, 1922*: The story of the robbery and resulting deaths comes from several accounts, including "$20,000 Jewels Is Laughing Burglar's Loot," *New York Tribune*, March 2, 1922; "Ardsley Forms Vigilantes," *New York Times*, March 3, 1922; "Burglar Gets Gems While Family Dines," *New York Times*, March 2, 1922; "2 Westchester Vigilantes Die in Auto Wreck," *New York Tribune*, March 22, 1922; "2 Ardsley Bankers Die in Motor Smash," *New York Times*, March 22, 1922.

174   *"He represented the highest type"*: *The Historical Register: A Biographical Record of the Men of Our Time Who Have Contributed to the Making of America* (Historical Register Biographical Record, 1922), 111.

CHAPTER THIRTEEN: THE FINAL WINDUP

The description and information on James Ward Packard's watches are drawn from the Patek Philippe Archives; Patek Philippe Museum, Geneva; Martin Huber and Alan Banbery, *Patek Philippe Genève*, 2nd ed. (Andreas Huber, 1993); Henry B. Fried, *The Museum of the American Watchmakers Institute* (AWI Press, 1993); Henry B. Fried, *The James W. Packard Collection of Unusual and Complicated Watches* (Horological Institute of America, 1959); private collections.

The description and information on Henry Graves, Jr.'s watches are drawn from the Patek Philippe Archives; Patek Philippe Museum, Geneva; Huber and Banbery, *Patek Philippe Genève*; Sotheby's; Christie's; Antiquorum; Tiffany & Co. Archives Collection.

176 *bequeathing his most complicated*: Last Will and Testament of James Ward Packard, November 1925, clause 14.
177 *"provide for the proper housing"*: Ibid.
178 *On June 5, 1926*: "Miss Graves Wed to R. H. Fullerton," *New York Times*, June 6, 1926.
178 *reporting every detail*: "Gwendolen Graves Engaged to Wed," *New York Times*, February 13, 1926; "Old New York Brought Back to Town Hall," *New York Times*, March 7, 1926; "Miss Graves's Bridal," *New York Times*, May 16, 1926; "Luncheon Given for Miss Graves," *New York Times*, May 20, 1926.
179 *Henry gifted his daughter*: "Real Estate Notes, Henry Graves Jr. Buys Co-Apartment in 1,030 Fifth Avenue," *New York Times*, April 20, 1927.
179 *Two years earlier Hamersley*: Christopher Gray, "The Construction of the 1899 Mansion That Was Once at 1030 Fifth Avenue," *New York Times*, November 2, 2003.
179 *organizing a benefit*: "Immigrants Will Give Concerts," *New York Times*, April 2, 1925.
179 *Henry sold Shadowbrook*: "Buys Graves Estate," *New York Times*, May 6, 1925; deed, June 1, 1925, filed Westchester County, New York.
180 *Engraved with the Fullerton coat of arms*: "Watches from the Collection of the Late Reginald H. Fullerton, Jr. and His Grandfather Henry Graves, Jr.," Sotheby's, June 14, 2014.
180 *Three years earlier, on June 26, 1923*: "Duncan Graves Weds Helen Johnson," *New York Times*, June 27, 1923.
180 *"New York's Most Aristocratic Hotel"*: Ambassador hotel pamphlet, New-York Historical Society.
180 *"some of the highest [rents]"*: "$6,000 a Room Will Be Asked in City's New Hotel," *New York Tribune*, February 1, 1920.
183 *"strictest secrecy"*: "The Contest," in *Masterpieces of The Time Museum*, Sotheby's catalogue monograph, Alan Banbery, consultant and curator of the Patek Philippe Museum, Geneva, September 1999.
183 *"the most complicated watch"*: Ibid.; Patek Philippe, *Voyage to the End of Time* (Patek Philippe SA, Geneva, 1989), 26; Lucien F. Trueb, "The Battle of the Super-Collectors," *Watch Time, Special Patek Philippe*, undated, 60.
183 *"had not been heard in the watchmaking industry"*: Patek Philippe, *Voyage to the End of Time*, 26.
183 *the design took shape*: Sotheby's, "Masterpieces of the Time Museum," December 2, 1999.
184 *Patek Philippe assembled the finest watchmakers*: Ibid.

CHAPTER FOURTEEN: GAME OVER

186 *Ward's cancer returned*: For the description of Packard's hospitalization, I drew upon the diaries of Elizabeth Gillmer Packard, 1926 to 1928.

186 *Dr. George Washington Crile*: For the background and history on Dr. Washington and the Cleveland Clinic, I drew upon John D. Clough, *To Act As a Unit: The Story of the Cleveland Clinic* (Cleveland Clinic Press, 2006).

187 *developed the first dosimeter*: Ibid., 48.

187 *developed the first condenser dosimeter*: Ibid.

188 *"Who will build it?"*: "Proceedings of the Dedication of the James Ward Packard Laboratory of Electrical and Mechanical Engineering," Lehigh University publication, vol. 5, no. 2 (February 1931), Special Collections, Lehigh University Libraries, Bethlehem, Pennsylvania.

188 *a "partial payment"*: Letter of November 20, 1926, cited in ibid., 8.

188 *"Will you please give me"*: Ibid.

188 *Ward agreed to donate $1 million*: Ibid.

188 *"Million to Lehigh"*: *New York World*, January 15, 1927.

188 *"Lehigh University's Windfall"*: *Philadelphia Record*, January 16, 1927.

188 *Charles Lindbergh had bounced down*: Edwin James, "Crowd Roars Thunderous Welcome," *New York Times*, May 22, 1927.

189 *"speedy decline of radio"*: Gerald Leinwand, *1927: High Tide of the 1920's, Part 4*, (Basic Books, 2002), 255.

189 *Dow Jones Industrial Average hit 200*: Dow Jones Industrial Average History.

189 *"I cannot help but raise a dissenting voice"*: Cited in Maureen Burton and Bruce Brown, *The Financial System and the Economy: Principles of Money and Banking* (M. E. Sharpe, 2009), 322.

189 *Ninety-five percent of the world's cars*: *Literary Digest*, November 5, 1927.

189 *After raw cotton and oil*: Ibid.

189 *More than four million people*: "Business & Finance: U.S. Motors Abroad," *Time*, July 22, 1929.

189 *When Ford retired the Model T*: Rudi Volti, *Cars and Culture: The Life Story of Technology* (Johns Hopkins University Press, 2006), 56.

190 *the number of major auto manufacturers*: Robert R. Ebert, "Consumers, Competition, and Consolidation: The Auto Industry Matures in the 1920s," PhD diss., Baldwin-Wallace College.

190 *Ford's emergence as an auto giant*: Charles K. Hyde, *Storied Independent Automakers* (Wayne State University Press, 2009), 86.

190 *three companies now produced*: "New Union Creates 'Big Three' in Autos," New York Times, May 31, 1928.

190 *"cool, self-possessed"*: "Business and Finance: Motor Week," *Time*, July 2, 1928.

190 *King Alexander I of Yugoslavia*: *Automotive News Special Supplement* (Packard 40th Anniversary Edition), February 5, 1940, 65–68.

190 *His Highness the Maharajah of Gwalior*: Ibid.

190 *When a mining engineer in Colombia*: Ibid.

191 *The company erected*: Packard Proving Grounds, Packard Motor Car Company Foundation.

191 *"so beautifully banked"*: "Quality Is Confirmed!," *Life*, October 20, 1952.

191 *the company would ship 32,122 cars*: "Motor Production Promises Records," *New York Times*, September 5, 1928.

191 *"Prospects never looked better"*: "Packard Raises Dividend," *New York Times*, November 10, 1927.

191 *skeleton pocket watch*: Henry B. Fried, *The Museum of the American Watchmakers Institute* (AWI Press, 1993), 92; Vacheron Constantin archives.

192 *eighteen-carat gold with richly carved rims*: Fried, *The Museum of the American Watchmakers Institute*, 74.

193 *The dial*: Ibid., 76–78.

195 *"to elevate and dignify the art"*: Paul Moore, "Noteworthy Watches from the Packard Collection," *Jeweler's Circular*, March 7, 1929.

195 *"While the factory system"*: Paul Moore, "Mysteries of Time," circa 1928. Roderick Blood Collection.

195 *"Time is as important as music"*: Moore, "Noteworthy Watches from the Packard Collection."

195 *"for educational work"*: Ibid.

195 *Ward and Elizabeth apparently agreed*: Henry B. Fried, *The James W. Packard Collection of Unusual and Complicated Watches,* (Horological Institute of America, 1959), 1; Moore, "Noteworthy Watches from the Packard Collection."

196 *"quite sure that Mr. Packard would not wish"*: "Proceedings of the Dedication of the James Ward Packard Laboratory of Electrical and Mechanical Engineering," 9, Special Collections of Lehigh University.

196 *"Mr. Packard realizing the importance of the undertaking"*: Ibid.

196 *"Seems so trivial at this time"*: Elizabeth Gillmer Packard diary, March 18, 1928.

197 *"peacefully and without realizing"*: Elizabeth Gillmer Packard diary, March 18, 19, 20, 1928.

197 *an estate valued at $7 million*: "Business Motors," *Time*, April 9, 1928.

197 *net annual profits*: Ibid.

197 *"With his passing"*: "James W. Packard, Auto Pioneer, Dies," *New York Times*, March 21, 1928.

198 *"if you detect any error of fact"*: Henry Luce cable to Packard News Service, March 24, 1928, Roderick Blood Collection, Larz Anderson Auto Museum, Brookline, Massachusetts.

198 *"Yet few of the men"*: "Business & Finance: Death of Packard," *Time, April 2, 1928.*

198 *On June 8 Ward's nephew*: Elizabeth Gillmer Packard diary, June 8, 1928; Cornerstone ceremony, Special Collections, Lehigh University.

CHAPTER FIFTEEN: ACROSS THE SEA

199 *In 1928*: RMS *Olympic* manifest, letters and itinerary of Henry Graves, Jr., June 1928–October 1928, Vacheron Constantin archives, period descriptions.

202 *"Mr. Packard's hobby"*: "James W. Packard Auto Pioneer Dies," *New York Times*, March 21, 1928.

202 *Patek Philippe had delivered*: Patek Philippe Archives; Patek Philippe Museum, Geneva; Tiffany & Co. Archives Collection.

202 *its own factory at the Place Cornavin*: John Reardon, *Patek Philippe in America: Marketing the World's Foremost Watch* (Cefari, 2008), 24.

203 *By 1927 Patek had begun producing*: Patek Philippe, "History of the Company," online.

203 *some of the first of what would be*: Information on Graves's wristwatches come from several sources, including Patek Philippe Archives; Patek Philippe Museum, Geneva; Sotheby's, "Watches from the Collection of the Late Reginald H. Fullerton, Jr. and His Grandfather Henry Graves, Jr.," New York, catalogue, June 14, 2012, 33, 34.

203 *a coin-form watch*: Sotheby's, "Watches from the Collection of the Late Reginald H. Fullerton, Jr. and His Grandfather Henry Graves, Jr.," 30.

204 *learned of a tourbillon chronometer*: Henry Graves, Jr., letter to Vacheron Constantin, June 20, 1928. Vacheron Constantin archives,

205 *"I will purchase the chronometer"*: Henry Graves, Jr., letter to Vacheron Constantin, June 24, 1928, Vacheron Constantin archives,

205 *"Please have the watch cleaned"*: Ibid.

205 *"I would be assured"*: Henry Graves, Jr., letter to Vacheron Constantin, June 28, 1928, Vacheron Constantin archives,

206 *"Do better if possible"*: Vacheron Constantin Company History.

206 *a unique skeletal watch*: Henry Graves, Jr., letter to Vacheron Constantin, July 18, 1928, Vacheron Constantin archives,

206 *"Also do not forget"*: Ibid.

207 *"[They] do not interest me at all"*: Henry Graves, Jr., letter to Vacheron Constantin, July 29, 1928, Vacheron Constantin archives,

CHAPTER SIXTEEN: THE JAMES W. PACKARD COLLECTION OF UNUSUAL AND COMPLICATED WATCHES

208 *an advertising executive*: "Packard Killed Taking Air Pilot Lesson," *Detroit Free Press*, August 27, 1929.

208 *"Feel my loss"*: Elizabeth Gillmer Packard diary, January 20, 1930.

208 *Lee catalogued nine pieces*: "Collection of Watches, Appraised at Cleveland Museum of Art, with Assistance of Geo. E. Lee, Expert of Cleveland, O."

209 *"They do not enter into the same realm"*: Henry B. Fried, *The James W. Packard Collection of Unusual and Complicated Watches* (Horological Institute of America, 1959), 2.

209 *valued at just $15,240*: "Collection of Watches, Appraised at Cleveland Museum of Art, with Assistance of Geo. E. Lee, Expert of Cleveland, O."

209 *declared the collection's value at $80,000*: Paul Moore, "Noteworthy Watches from the Packard Collection," *Jeweler's Circular*, March 7, 1929.

209 *Elizabeth took several of them*: Elizabeth Gillmer Packard diaries, November 16, 1929, December 26, 1929, and September 10, 1930.

209 *invited Warren Packard*: Elizabeth Gillmer Packard diary, July 26, 1928.

210 *"Much pleased with the display"*: Elizabeth Gillmer Packard diary, December 18, 1928.

210 *"the gem of the collection"*: "The Packard Watch Collection," catalogue at the fourth annual Jewelry Show, August 5–9, 1929, Chicago.

210 *"No man who has any appreciation of fine work"*: Ibid.

211 *he took to the skies*: "Packard Killed Taking Air Pilot Lesson," *Detroit Free Press*, August 27, 1929.

CHAPTER SEVENTEEN: A SUPERCOMPLICATION

212 *"In the domestic field"*: President Calvin Coolidge, State of the Union Address, December 4, 1928.

212 *In 1927 brokers borrowed $4 billion*: Adam Shell, "More Investors Take Risk of Buying Stocks with Loans," *USA Today*, May 7, 2007; Karen Blumenthal, *Six Days in October: The Crash of 1929* (Simon & Schuster, 2002).

212 *"There is no reason"*: Bernard Baruch's interview in *The American Magazine,* June 1929, cited in "A Storm Unseen, Always about to Pass," *New York Times,* October 1, 2008.

213 *the Dow peaked at 381.17*: William M. LeFevre, "The '29 Crash; Seeking Clues to the Market's Future," *New York Times,* December 27, 1987.

213 *Anthony Campagna secured*: "$275,00 Apartment Bought by Banker," *New York Times,* November 14, 1930.

213 *Henry and Florence were one of the first ten*: "Cooperative Opening at 834 Fifth Avenue," *New York Times,* September 20, 1931.

213 *Rupert Murdoch purchased the triplex penthouse*: William Neuman, "Murdoch Set to Pay Record $44 Million for 5th Ave. Triplex," *New York Times,* December 17, 2004.

214 *"Stock prices have reached"*: Irving Fisher, cited in John Kenneth Galbraith, *The Great Crash 1929* (Houghton Mifflin Harcourt, 2009), 70.

214 *"Black Thursday"*: Claire Suddath, "The Crash of 1929," *Time,* October 29, 2008.

214 *investors had lost*: "October 29, 1929: 'Black Tuesday,'" CNN, March 10, 2003.

214 *William Durant*: "Timeline: A Selected Wall Street Chronology," *The American Experience,* PBS, http://www.pbs.org/wgbh/americanexperience/features/timeline/crash/

214 *Durant found himself bankrupt*: Paul Arculus, *Durant's Right-Hand Man* (Friesen Press, 2011), 278.

214 *its developers purchased the mansion*: "Haggin Home Sold for New Building," *New York Times,* April 25, 1930.

214 *"the most pedigreed building"*: Max Abelson, "You're The Top!" *New York Observer,* May 1, 2007.

215 *paid out eight percent dividends*: "Atlas Cement Co. Passes Its Dividend," *New York Times,* July 30, 1910.

215 *Atlas Portland operated*: "U.S. Steel to Become Chief Cement Maker," *New York Times,* December 17, 1929.

215 *U.S. Steel Corporation acquired*: "U.S. Steel to Acquire Atlas Portland Cement for Common Stock Valued at $31,320,000," *New York Times,* December 16, 1929.

216 *the family's fortunes*: "Setting the Standard," *Patek Philippe,* no. 8 (Autumn/Winter 2000): 78.

216 *Edward Hale had died*: "Edward Hale Graves," *New York Times,* June 3, 1930.

216 *Duncan Graves's firstborn, Henry*: John S. Wurts, *Magna Charta,* part 5, *A Collection of Baronial Pedigrees* (Brookfield, 1946), 1234.

216 *George Coe, suddenly*: "G. C. Graves Dies on Ship in Pacific," *New York Times,* December 15, 1932.

216 *he left one-quarter of his estate*: File 4, George Coe Graves last will, May 9, 1932, George Coe Graves Correspondence, Office of the Secretary Records, The Metropolitan Museum of Art Archives.

216 *he had gone to court*: Atlantic Reporter, "Fidelity Union trust Co. v. Graves et al. 147/282, Court of Chancery of New Jersey," 1947, 750–58.

216 *An intrepid explorer*: Explorers Club of New York Archives.

218 *"indebted"*: Field Museum News 2, no. 2 (1931).

218 *George traveled to Khabarovsk*: "Brings Back Specimens," *New York Times,* July 18, 1930.

218 *South Asiatic Hall*: "Museum Dedicates South Asiatic Hall," *New York Times,* November 18, 1930.

219 *Patek Philippe had sent the watch*: "Complication for Mr. Graves' Supercomplicated Watch #198'385."

219 *severe financial realities*: Lucien Trueb, "An Independent Streak," *Watch Time,* April 2005, 89–90.

220 *Charles Stern sent his son Henri*: Ibid.

220 *he sold his shares to the brothers*: John Reardon, *Patek Philippe in America: Marketing the World's Foremost Watch* (Cefari, 2008), 42.

220 *only at the end of World War II*: Trueb, "An Independent Streak," 89.

220 *From a very young age*: Ibid.

221 *died of a sudden heart attack*: "Dexter Wright Hewitt," *New York Times*, September 4, 1933.

221 *a Pacaraima-Venezuela Expedition*: "Exploration Is Put Off," *New York Times*, July 24, 1931.

222 *to become a partner*: "Financial Notes," *New York Times*, July 5, 1933.

222 *George had been killed*: "George C. Graves Dies after Auto Accident," *New York Times*, November 5, 1934.

222 *valued at nearly $1 million*: "C. G. Graves Assets Listed," *New York Times*, April 28, 1936.

223 *"Henry Graves, Jr., has one of the most expensive hobbies"*: C. B. Driscoll, *"New York Day by Day," Billings (Montana) Gazette*, June 21, 1938.

223 *the kidnapping*: For an account of the kidnapping, see A. Scott Berg, *Lindbergh* (Berkeley Trade, 1999).

224 *"the biggest story since the Resurrection"*: Cited in ibid., 308.

225 *The owner-shareholders of 834 Fifth Avenue*: For the financial history of 834 Fifth Avenue, I drew upon several sources, including "Two Large Flats Go to Mortgages," *New York Times*, July 24, 1936; "All Tenants Participate in 5th Ave Co-op Plan," *New York Times*, July 10. 1952; "L. S. Rockefeller Buys on Fifth Avenue, *New York Times*, November 5, 1946; "Estate Appraisal of Henry Graves, Jr. 1953," P.J. Curry Co., *Schmidlapp v. Commissioner of Internal Revenue*, 1931; William Neuman, "Murdoch Set to Pay Record $44 Million for 5th Ave. Triplex," *New York Times*, December 17, 2004; Christopher Gray, "Mr. Murdoch Buys His Dream House," *New York Times*, December 30, 2007.

225 *turned into a plot of tract homes*: "Eighteen Colonial Residences to Be Built on Graves Estate in Essex County," *New York Times*, April 29, 1934.

225 *"No other collection"*: "Masterpieces of Engraving and Etching, Collection of Henry Graves, Jr.," catalogue, April 3, 1936, America Art Association Anderson Galleries Inc.

226 *Adam and Eve*: "Durer Engraving Sold for $10,000," *New York Times*, April 4, 1936.

226 *The entire sale realized $79,635*: Ibid.

226 *unemployment hovered at nearly 17 percent*: U.S. Department of Labor, Bureau of Labor Statistics.

226 *FDR raised the income tax*: Bruce Bartlett, "Taxes, Bailouts, and Class Warfare," *Forbes*, March 20, 2009.

226 *"Adirondack Lodge will make its strongest appeal"*: Adirondack Lodge, Eagle Island, Upper Saranac Lake, New York, 1937 real estate ad.

226 *the couple offered to sell Eagle Island*: "Big Estate Is Given to the Girl Scouts," *New York Times*, August 30, 1937.

227 *in the pages of* Life *magazine*: "Watches: These Are the Best in the World," *Life*, December 23, 1940.

CHAPTER EIGHTEEN: AGE OF QUARTZ

228 *"finest type of mechanicians"*: Paul Moore, "Noteworthy Watches from the Packard Collection," *Jewelers' Circular*, May 7, 1929.

229   *took the prize at the Neuchâtel Observatory*: Sotheby's, "Important Watches," cata-
        logue, June 15, 1999.
229   *"always listen carefully"*: Cited in John Reardon, *Patek Philippe in America: Marketing
        the World's Foremost Watch* (Cefari, 2008), 78.
229   *Patek Philippe is known to have produced only*: Patek Philippe archives; Patek Philippe
        Museum, Geneva; Sotheby's, "Watches from the Collection of the Late Reginald H.
        Fullerton, Jr. and His Grandfather Henry Graves, Jr.," New York, catalogue, June 14,
        2012, 32.
229   *an eighteen-carat gold perpetual calendar chronograph wristwatch*: Patek Philippe
        archives; Patek Philippe Museum; Sotheby's, "Important Watches," catalogue, June
        15, 1999.
229   *the first perpetual calendar chronograph*: "Understanding the Patek Philippe Reference
        1518: The World's First Perpetual Calendar Chronograph," *Hodinkee*, May 24, 2010.
229   *a pink gold rectangular wristwatch*: Patek Philippe archives; Patek Philippe Museum;
        Sotheby's, "Important Watches," catalogue, June 15, 1999.
230   *The platinum wristwatch*: Sotheby's "Watches from the Collection of the Late Regi-
        nald H. Fullerton, Jr. and His Grandfather Henry Graves, Jr." 38.
231   *eighteen-carat gold split-second chronograph wristwatch*: Ibid., 46.
232   *Graves's nearly $8 million estate*: Surrogate's Court, County of New York, *In the Matter of
        the Appraisal under the Estate Tax Law, the Estate of Henry Graves, Jr.* August 19, 1954.
232   *Laurance Rockefeller had reorganized*: "All Tenants Participate in 5th Ave Co-op
        Plan," *New York Times*, July 10, 1952; "L. S. Rockefeller Buys on Fifth Avenue, *New
        York Times*, November 5, 1946.
232   *Henry owned 1,300 shares*: Appraiser's Report, P. J. Curry Co. Surrogate's Court,
        County of New York, *In the Matter of the Appraisal under the Estate Tax Law, the
        Estate of Henry Graves, Jr.*, August 19, 1954.
232   *The collection with the rare Double Eagles*: Ibid.
233   *"one of the outstanding private collections"*: *Notable American Prints: The Collection of
        Henry Graves, Jr.*, catalogue, Kennedy Galleries, May 9, 1959.
233   *"Many of the buildings"*: Dore Ashton, "Old New York Prints," *New York Times*, May
        9, 1959.
233   *only twenty-eight timepieces were accounted for*: Appraiser's Report, P. J. Curry Co.
234   *failed it during the postwar boom*: For a history of the end of the Packard Motor Car
        company after World War II, I drew upon James A. Ward, *The Fall of The Packard
        Motor Car Company* (Stanford University Press, 1995).
235   *in stark contrast*: "Examining Chrysler's 1979 Rescue," NPR, November 12, 2008.
235   *when it propped up Chrysler*: Sharon Terlep, "GM Squeezes after Bailout," *Wall Street
        Journal*, February 17, 2012.
235   *"In the years following"*: David Halberstam, *The Fifties* (Ballantine Books, 1993), x.
237   *the Seiko Crystal Chronometer*: Joe Thompson, "QC-951 1969: Seiko's Breakout Year,"
        Watch Time, December 20, 2009.
237   *Between 1970 and 1984*: Federation of the Swiss Watch Industry, history, "From the
        Roots Until Today's Achievements" (digital version).
237   *led by a Norwegian refugee*: David S. Landes, *Revolution in Time: Clocks and the Mak-
        ing of the Modern World* (Harvard University Press, 2000), 339–41.
237   *By 1962 one out of every three*: Ibid.
238   *At the time, Gérard Bauer*: Lucien Trueb, "An Independent Streak," *Watch Time*, April
        2005, 90.
238   *introduced its first quartz clock*: Ibid., 90, 94.
238   *"I took the decision"*: Philippe Stern, interviewed by Melvyn Teillol-Foo in *iW* maga-
        zine, December 21, 2007.

238 *At a retrospective*: "Watches Form Timeless Display," *New York Times*, September 29, 1967.

238 *in 1949*: "Unusual Watches Placed on Display," *New York Times*, April 15, 1949.

239 *"most complicated watch"*: Ibid.

CHAPTER NINETEEN: COLLECTING TIME

240 *Winthrop Kellogg Edey*: Adrian Dannatt, "Winthrop Edey," *Independent*, London, March 12, 1999; Wendy Moonan, "Antiques: Life of Clocks Measured in Passion. Kelly Edey, in Life and Memoriam," *New York Times*, December 21, 2001; Frick Art Reference Library, 1937–99.

241 *George Daniels was another seminal keeper*: For Daniels's career and biography I drew upon several accounts including: Kate Youde, "Father Time: Why George Daniels is the World's Best Horologist," *Independent*, London, September 5, 2010; "George Daniels," *The Telegraph*, London, October 24, 2011.

242 *The son of a prominent*: "Seth Glanville Atwood," *Chicago Tribune*, February 25, 2010.

242 *"Why is there a universe"*: Seth Atwood, "Masterpieces of the Time Museum," Sotheby's, catalogue, December 2, 1999.

242 *"I decided to try to collect a few items"*: Ibid.

243 *"You can't just walk into a gallery"*: Wendy Moonan, "Timeless Value of Timepieces," *New York Times*, June 14, 2002.

244 *the price, £4,400*: "Masterpieces from the Time Museum Make $11.8 million," *Independent*, Ireland, July 17, 2002.

244 *"I saw things I wanted"*: Wendy Moonan, "Timeless Value of Timepieces," *New York Times*, June 14, 2002.

244 *He acquired one of the earliest*: Seth Atwood and William Andrewes, *The Time Museum* (Johnson Press, 1983).

244 *was sold to a Swiss dentist*: Jeanne Schinto, "Time Is Money at Sotheby's: Masterpieces from The Time Museum, October 13–15, 2004," *Maine Antique Digest*, January 2005.

245 *all four surviving examples*: Owen Gingerich, "Seth Atwood's Vision," in "Masterpieces of the Time Museum," Sotheby's, catalogue, December 2, 1999, 14.

245 *a British man named Colonel Quill*: Schinto, "Time Is Money at Sotheby's."

245 *it mysteriously turned up*: Ibid.

245 *Atwood then commissioned*: Ibid.

245 *"tremendously enriched"*: Atwood, "Masterpieces of the Time Museum."

246 *incorporated a secular perpetual calendar*: Patek Philippe Genève, *Voyage to the End of Time* (Patek Philippe SA, 1989), 30.

246 *fetching $1,042,338*: Christie's, "Important Pocket Watches and Wristwatches," November 12, 2007, Geneva Lot 222.

247 *120 antique pieces*: Philippe Stern, "The World's Most Beautiful Museum of Horology," *Patek Philippe Museum* Tribune Des Arts, Le Tribune de Genève, undated.

247 *"My grandfather and then my father"*: Ibid.

248 *he had once sat for hours*: Gregory Katz, "The Time of Their Lives," *American Way*, July 1, 2006.

248 *In 1973, Stern began adding*: Laurie Kahle, "Keeping Time" *Robb Report*, February 1, 2002.

248 *"It was rather easy"*: Ibid.

249 *"The first was the case"*: James D. Malcolmson, "Patek Philippe," *Robb Report*, September 18, 2009.

249 *Duke Ellington bought*: John Reardon, *Patek Philippe in America: Marketing the World's Foremost Watch* (Cefari, 2008), 124.

249 *"representing the pinnacle"*: Patek Philippe, *Voyage to the End of Time,* 11.

250 *As a boy, he often watched*: Sotheby's, "Watches from the Collection of the Late Reginald H. Fullerton, Jr. and His Grandfather Henry Graves, Jr.," catalogue, June 14, 2012, 8.

250 *Fullerton married*: "Kathleen Knudson Is Married to Reginald Fullerton Jr.," *New York Times,* January 28, 1968.

251 *five months after Henry died*: Sotheby's, "Watches from the Collection of the Late Reginald H. Fullerton, Jr. and His Grandfather Henry Graves, Jr.," 8, 18, 19, 20.

251 *Fullerton later wrote about conducting séances*: Sotheby's, "Watches from the Collection of the Late Reginald H. Fullerton, Jr. and His Grandfather Henry Graves, Jr.,"19.

251 *$200,000*: The amount paid for the Supercomplication is reported by Patek Philippe in *Star Caliber 2000* (Editions Scriptar SA, 2000), 23.

CHAPTER TWENTY: BACK IN TIME

253 *"People want to see the art of the mechanics"*: David Stout, "Henry B. Fried," *New York Times,* March 12, 1996.

253 *he was known to break into an explanation*: Ibid.

253 *"They are watches with a soul"*: Nancy Wolfson, "Pocket Watches," *Cigar Aficionado,* Winter 1994–95.

253 *"that the skilled horologist"*: Henry B. Fried, *The James W. Packard Collection of Unusual and Complicated Watches* (Horological Institute of America, 1959).

254 *"In ten to twenty years"*: "The High Cost of Going Back in Time," *New York Times,* September 21, 1986.

254 *"Wealth is back in style"*: *U.S. News & World Report,* cited in Larry Samuel, *Rich: The Rise and Fall of American Wealth Culture* (Amacom, 2009), 182.

254 *Old Money had dwindled*: Samuel, *Rich,* 185.

254 Some of the figures concerning wealth culture are cited in Samuel, *Rich,* 183–85.

255 *a 1955 Patek Philippe*: "The High Cost of Going Back in Time," *New York Times,* September 21, 1986.

255 *once owned by the late King Talal*: Ibid.

255 *it sold six wristwatches*: Ibid.

258 *Philippe Stern for one*: Stacy Meichtry, "How Top Watchmakers Intervene in Auctions," *Wall Street Journal,* October 8, 2007.

258 *In 2007 Antiquorum's board*: Matthew Malone, "Time Bomb," *Portfolio,* November 18, 2007; *Wall Street Journal,* October 8, 2007.

258 *In 1998 Christie's sold a trio*: Christie's results, June 23, 1998.

258 *At "The Art of Patek Philippe" sale*: Antiquorum sale results, April 9, 1989; Christie's sale notes on historical information regarding its sale in 2008.

258 *realized $19,800*: Sotheby's 1990 sale result notes, Patek Philippe Archives.

259 *"It's my obsession"*: Philippe Stern, interviewed by Melvyn Teillol-Foo in *iW* magazine, December 21, 2007.

259 *The meticulous process*: Nancy Wolfson, "A Tradition of Precision," *Cigar Aficionado,* November/December 1999.

259 *"Today, exceptional watchmaking"*: Ibid.

259 *"know-how and not mathematical rule"*: Melvyn Teillol-Foo, "Philippe Stern: Keeper of the Patek Philippe Spirit," *International Watch,* December 21, 2007.

259 *manufactured fewer than ten thousand*: Lucien F. Trueb, "An Independent Streak," *Watch Time,* April 2005, 94.

259 *In June 1979 Stern met*: For the story of the construction of the Calibre 89, I primar-

ily relied on Patek Philippe Genève, *Voyage to the End of Time* (Patek Philippe SA, 1989), 35–37.

260 *"most complicated portable timepiece"*: As billed by Patek Philippe and noted in Bruce Weber, "Works in Progress: High Tick," *New York Times* magazine, January 15, 1989.

260 *the constellation of the eagle*: Ibid., 47.

261 *Over dessert*: Douglas Martin, "A Timepiece for the Tasteful," *New York Times*, March 15, 1989.

261 *"Occasionally an object"*: Ibid.

261 *won the bid for the Calibre*: Rita Reif, "Auctions," *New York Times*, February 1, 1991.

261 *"Watches are part of human cultural development"*: Laurie Kahle, "Keeping Time," *Robb Report*, February 1, 2002.

261 *Sotheby's shattered records*: Rita Reif, "Auctions," *New York Times*, February 1, 1991.

262 *jumped in price*: Norma Buchanan, "Present at the Creation," *Watch Time*, April 2008, 112.

262 *Swiss watch exports*: "Swiss Watch Exports Set Record in 1990," AP, February 1, 1991; Swiss Watchmaker's Association, statistics.

262 *"modern quartz"*: James Sterba, "By 1990 Modern Quartz Will Have Nearly Silenced the Tick," *New York Times*, February 9, 1982.

263 *Nicolas Hayek*: The account of Hayek comes primarily from Michael Clerizo, "Nicolas Hayek: Time Bandit," *Wall Street Journal*, June 10, 2010; Margarlit Fox, "Nicolas Hayek Dies at 82," *New York Times*, June 29, 2010; "Message and Muscle: An Interview with Swatch Titan Nicolas Hayek," *Harvard Business Review*, March 1993.

264 *"not unlike the first gatherings of the average"*: Henry B. Fried, *The James W. Packard Collection of Unusual and Complication Watches*, monograph (Horological Institute of America, 1959), 44.

CHAPTER TWENTY-ONE: THE COMEBACK

266 *In 1994 Blancpain*: Peter Passel, "Watches That Time Hasn't Forgotten," *New York Times*, November 24, 1995.

267 *It was Graves's prizewinning*: Antiquorum sale results, "Important Watches, Wristwatches and Clocks," Geneva, October 31, 1998.

267 *The astonishing*: Nancy Wolfson, "A Tradition of Precision," *Cigar Aficionado*, November/December 1999.

267 *Banbery paid $950,000*: Ibid.

268 *fetched $96,000*: Sotheby's sale results, February 9, 1999; Patek Philippe Archives; Patek Philippe Museum.

268 *The entire sale realized $5,082,386*: Sotheby's sale results, February 9, 1999.

268 *When Atwood struck upon the shocking decision*: For the account of Atwood's decision to sell off his Time Museum, among the sources that I drew upon are Lewis Lazare, "Time Is Money," *Chicago Reader*, April 20, 2000; Douglas Martin, "Justice Shepro, Connoisseur of Old Clocks, Dies at 74," *New York Times*, April 17, 2000.

268 *"There has never been another collection"*: Douglas Martin, "Justice Shepro, Connoisseur of Old Clocks, Dies at 74," *New York Times*, April 17, 2000.

269 *Of all the magnificent timepieces*: Rachel Ramsay, "Sale of a Master's Collection," *New York Times*, March 8, 2012.

271 *the first of the two gold pocket watches*: Sotheby's sale results.

271 *on May 18, 2004*: Christie's sale results.

272 *the Packard dual-time pocket watch*: Ibid.

272 *the watchmaker scored*: Antiquorum sale results.

272 *On November 14, 2005*: Christie's sale results.
272 *on May 14*: Ibid.
273 *After sitting undisturbed*: Ibid.
273 *"Hence at the beginning of the 20th century"*: Patek Philippe Museum, Geneva.
274 *eighteen-carat gold Wittnauer*: eBay results.
274 *came in at $47,100*: Ibid.
274 *sold for $630,000*: Antiquorum sale results.
275 *First manufactured in 1913*: "Important Watches, New York," catalogue Christie's, December 14, 2010, Patek Philippe Archives abstract.
275 *sold for $242,500*: Christie's sale results, "Important Watches" sale, New York, December 14, 2010.
275 *In 2006 a sheaf of yellowing letters*: Correspondence of Henry Graves, Jr., and Vacheron Constantin, June 1928–October 1928. Vacheron Constantin archives.
276 *The* grande complication *earned first prize*: Vacheron Constantin Heritage Department, Geneva, Switzerland.

CHAPTER TWENTY-TWO: THE FINAL GAVEL

278 *the obituary*: Warren Packard III, *Chattanoogan,* October 25, 2010.
279 *dropping 778 points*: Alexandra Twin, "Stocks Crushed," CNN/*Money,* September 29, 2008.
280 *brought in $7.2 million*: "Patek Philippe Sales Break Records at Antiquorum," *National Jeweler,* March 30, 2007.
280 *During a Sotheby's sale of pocket watches*: Stacy Perman, "The Day That Changed Watchmaking," *Businessweek,* December 1, 2009.
280 *In November 2009*: Ibid.
280 *realizing $91.2 million*: Christie's sales results, January 27, 2011.
280 *reaching $115.6 million*: Christie's sales results, February 1, 2012.
283 *On May 10, 2010*: Nazanin Lankarani, "Collectors' Items Bring Spectacular Returns," *New York Times,* March 24, 2011.
283 *reported $257.2 million*: Simon de Burton, "Auctions: No Time Like the Present to Sell Masterpiece Watches," *Financial Times,* June 8, 2012.
283 *jumped 49 percent in China*: Xu Junqian, "Luxury Watches Mainly Fashion Statement: Study," *China Daily,* April 11, 2012.
283 *fetched a stunning $1,273,265*: Christie's results, "Important Watches," Geneva, November 14, 2011.
285 *The entire sale brought $8,339,813*: Sotheby's sale results, June 14, 2012.
287 *came to power in 1970*: Judith Miller, "Creating Modern Oman," *Foreign Affairs,* May/June 1997.
287 *"Oman's Renaissance man."*: Brian Whitaker, "Oman's Sultan Qaboos," *The Guardian* (UK), March 4, 2011.
287 *Named as head of Qatar's*: Louise Baring, "Scandal of the Sheikh and his £1 bn Shopping Spree," *The Telegraph* (UK), April 30, 2005.
287 *This endeavor had in fact turned the sheikh*: Georgina Adam and Charlotte Burns, "Qatar Revealed As The World's Biggest Contemporary Art Buyer," *The Art Newspaper,* Issue 226, July-August 2011.

EPILOGUE

289   *Sheikh Saud Bin Mohammed bin Ali Al-Thani had pledged the Graves Supercomplica-tion*: Miles Weiss and Katya Kazakina, "Qatari Sheik Pledges Most Expensive Watch to Sotheby's, *Bloomberg*, November 9, 2012.

289   *some $42 million*: Ibid.

289   *"moved through markets like a whirlwind"*: Georgina Adam, "Meet Sheikh Saud Al Thani of Qatar," *The Art Newspaper*, May 1, 2004.

289   *accusing him of skipping out $19.8 million*: Martin Robinson, "Qatari Sheikh Who Spent More Than £1 billion on his Art Collection is Accused of Leaving 'Extraordi-nary Trail' of Unpaid Debt Around World's Top Auction Houses, *Daily Mail*, Novem-ber 1, 2012.

289   *a London High Court judge ruled to freeze $15 million*: Riah Pryor, "Asset freeze on bid-defaulting Sheikh extended by London judge," *The Art Newspaper* online, November 9, 2012.

289   *"There are no doubt timing issues"*: "Qatari Sheikh Who Spent More Than £1 billion on his Art Collection is Accused of Leaving 'Extraordinary Trail' of Unpaid Debt Around World's Top Auction Houses, *Daily Mail*, November 1, 2012.

289   *his house arrest in 2005*: "World's Biggest Art Collector Arrested in Qatar, *The Art Newspaper*, March 13, 2005.

290   *set a world record when he purchased Gustave LeGray's*: Louis Baring, "Scandal of the Sheikh And His £1bn Shopping Spree," *Telegraph* (UK), April 30, 2005.

290   *A year later he parted with $15 million*: Ibid.

290   *$8.8 million to possess a complete set*: Ibid.

290   *During an Islamic art sale*: Georgina Adam. "World's Biggest Art Collector Arrested In Qatar, *The Art Newspaper*, March 13, 2005.

290   *On a visit to Qatar in 2004*: Georgina Adam, "Meet Sheikh Al Thani of Qatar," *The Art Newspaper*, May 1 2004.

SELECT BIBLIOGRAPHY

PUBLISHED WORKS

Ball, Norman R., and John N. Vardalas. *Ferranti-Packard: Pioneers in Canadian Electrical Manufacturing.* McGill-Queens, 1994.

Banbery, Alan, and Martin Huber. *Patek Philippe Genève.* Patek Philippe SA, Geneva, 1993.

Bancroft, Hubert Howe. *The Book of the Fair; an Historical and Descriptive Presentation of the World's Science, Art, and Industry, As Viewed through the Columbian Exposition At Chicago in 1893,* vol. 1. Bancroft, 1895.

Best, Gary D. *The Dollar Decade: Mammon and the Machine in 1920s America.* ABC-CLIO, 2004.

Britten's Old Clocks and Watches and Their Makers, Being an Historical and Descriptive Account of the Different Styles of Clocks and Watches of the Past in England and Abroad to Which Is Added a List of Eight Thousand Makers. B. T. Batsford, 1899.

Butler, Joseph Green. *History of Youngstown and the Mahoning Valley, Ohio,* vol. 2. American Historical Society, 1921.

Carosso, Allistorck Vincent P. *Investment Banking in America.* Harvard University Press, 1970.

Cashman, Sean Dennis. *America in the Gilded Age: From the Death of Lincoln to the Rise of Theodore Roosevelt.* New York University Press, 1993.

"Catalogue of the Collection of Watches, the Property of J. Pierpont Morgan." Chiswick Press, London, 1912.

Clerizo, Michael. *Masters of Contemporary Watchmaking.* Thames & Hudson, 2009.

Clough, John D. *To Act As a Unit: The Story of the Cleveland Clinic.* Cleveland Clinic Press, 2006.

A Complete History of the New York and Brooklyn Bridge from Its Conception in 1866 to Its Completion in 1883. S. W. Gree's Son, 1883.

Davies, Pete. *American Road: The Story of an Epic Transcontinental Journey at the Dawn of the Motor Age.* Macmillan, 2003.

Deluxe Illustrated Catalogue of the Notable Art Treasures Collected by the Late Henry Graves, Orange, New Jersey. J. J. Little & Ives, 1909.

De Sormo, Maitland C. *Summers on the Saranacs.* Adirondack Yesteryears, 1980.

Einstein, Arthur, and Steven Rossi. *Ask the Man Who Owns One: An Illustrated History of Packard Advertising.* McFarland, 2010.

Foulkes, Nick. *High Society: The History of America's Upper Classes.* Assouline, 2008.

Fried, Henry B. *The James W. Packard Collection of Unusual and Complicated Watches.* Monograph. Horological Institute of America, 1959.

Fried, Henry B. *The Museum of the American Watchmakers Institute.* AWI Press, 1993.

Galbraith, John Kenneth. *The Great Crash 1929*. Houghton Mifflin Harcourt, 2009.

Gilborn, Craig. *Adirondack Camps: Homes Away from Home, 1850–1950*. The Adirondack Museum/Syracuse University Press, 2000.

Hadley, Earl J. *The Magic Powder: History of the Universal Atlas Cement Company and the Cement Industry*. G. P. Putnam & Sons, 1945.

Heydt, George Frederic. *Charles L. Tiffany and the House of Tiffany & Company*. Tiffany & Co., 1893.

History of Trumbull and Mahoning Counties, vol. 1. H. Z. William & Bros., 1882.

Hyde, Charles K. *Storied Independent Automakers*. Wayne State University Press, 2009.

Ide, Evan P. *Packard Motor Car Company*. Arcadia Publishing, 2003.

Jonnes, Jill. *Empires of Light: Edison, Tesla, Westinghouse, and the Race to Electrify the World*. Random House, 2003.

Josephson, Matthew. *The Robber Barons*. Harvest Book, 1934.

Kimes, Beverly. *Packard: A History of the Motorcar and the Company*. Automobile Quarterly, 2005.

Kirschenbaum, Howard. *The Story of Sagamore*. Sagamore First Raquette Lake, 1996.

Landes, David S. *Revolution in Time: Clocks and the Making of the Modern World*. Belknap Press of Harvard University Press, 1983.

Madsen, Axel. *John Jacob Astor: America's First Multimillionaire*. Wiley, 2001.

Martin, Terry. *Packard: The Warren Years*. Historical Publishing Network and The National Packard Museum, 2009.

Matuz, Roger. *Albert Kahn: Builder of Detroit*. Wayne State University Press, 2002.

Norton, Thomas E., *100 Years of Collecting in America: The Story of Sotheby Parke Bernet*. Harry N Abram, 1987.

Ostrander, Stephen M. *A History of the City of Brooklyn and Kings County*. Published by subscription, Brooklyn, 1894.

Patek Philippe Genève. *Star Caliber 2000*. Editions Scriptar SA, Lausanne, Switzerland, 2000.

Patek Philippe Genève. *Voyage to the End of Time*. Philippe SA, Geneva, 1989.

Reardon, John. *Patek Philippe in America: Marketing the World's Foremost Watch*. Cefari Publishing, 2008.

Renehan, Edward J. *Dark Genius of Wall Street: The Misunderstood Life of Jay Gould, King of the Robber Barons*. Basic Books, 2005.

Samuel, Larry. *Rich: The Rise and Fall of American Wealth Culture*. Amacom, 2009.

Scott, Michael, G. H. *Packard: The Complete Story*. Tab Books, 1985.

Secrest, Meryle. *Duveen: A Life in Art*. Knopf, 2004.

Ulmann, Albert. *Maiden Lane: The Story of a Single Street*. Maiden Lane Historical Society, 1931.

Ward, James. *The Fall of the Packard Motor Company*. Stanford University Press, 1995.

Warren, George B., and Thomas Benedict Clarke. *Catalogue of Antique Chinese Porcelains*. Merrymount Press, 1902.

Watts, Steven. *The People's Tycoon: Henry Ford and the American Century*. Knopf, 2005.

White, Richard. *Railroaded: The Transcontinentals and the Making of Modern America*. Norton, 2011.

ARCHIVES AND COLLECTIONS

Adirondack Collection, Saranac Lake Free Library, Saranac, New York

American Museum of Natural History Museum Archives, New York

Roderick Blood Collection, Larz Anderson Museum, Brookline, Massachusetts

George Coe Graves Correspondence and Graves family records, Office of the Secretary
     Records, The Metropolitan Museum of Art Archives, New York
New York Explorers Club Archives, New York
Patek Philippe Archives, Geneva, Switzerland
Special Collections, Lehigh University Libraries, Bethlehem, Pennsylvania
Special Collections of the National Packard Museum, Warren, Ohio
Tiffany & Co. Archives Collection, New York
Vacheron Constantin, Geneva, Swtizerland

INDEX

The letter *n* after a page number refers to "footnote" text.

About the Author

Stacy Perman is an award-winning journalist and the author of the *New York Times* bestseller *In-N-Out Burger*. The recipient of a MacDowell fellowship, she is a former writer with *BusinessWeek* and *Time*. Her work has appeared in the *Los Angeles Times*, the *Wall Street Journal*, and many other publications. She lives in New York City.